MECHANICAL ENGINEERING

A Series of Textbooks and Reference Books

Editor

L. L. Faulkner

*Columbus Division, Battelle Memorial Institute
and Department of Mechanical Engineering
The Ohio State University
Columbus, Ohio*

1. *Spring Designer's Handbook,* Harold Carlson
2. *Computer-Aided Graphics and Design,* Daniel L. Ryan
3. *Lubrication Fundamentals,* J. George Wills
4. *Solar Engineering for Domestic Buildings,* William A. Himmelman
5. *Applied Engineering Mechanics: Statics and Dynamics,* G. Boothroyd and C. Poli
6. *Centrifugal Pump Clinic,* Igor J. Karassik
7. *Computer-Aided Kinetics for Machine Design,* Daniel L. Ryan
8. *Plastics Products Design Handbook, Part A: Materials and Components; Part B: Processes and Design for Processes,* edited by Edward Miller
9. *Turbomachinery: Basic Theory and Applications,* Earl Logan, Jr.
10. *Vibrations of Shells and Plates,* Werner Soedel
11. *Flat and Corrugated Diaphragm Design Handbook,* Mario Di Giovanni
12. *Practical Stress Analysis in Engineering Design,* Alexander Blake
13. *An Introduction to the Design and Behavior of Bolted Joints,* John H. Bickford
14. *Optimal Engineering Design: Principles and Applications,* James N. Siddall
15. *Spring Manufacturing Handbook,* Harold Carlson
16. *Industrial Noise Control: Fundamentals and Applications,* edited by Lewis H. Bell
17. *Gears and Their Vibration: A Basic Approach to Understanding Gear Noise,* J. Derek Smith
18. *Chains for Power Transmission and Material Handling: Design and Applications Handbook,* American Chain Association
19. *Corrosion and Corrosion Protection Handbook,* edited by Philip A. Schweitzer
20. *Gear Drive Systems: Design and Application,* Peter Lynwander

MECHANISM ANALYSIS

Additional Volumes in Preparation

Mechanical Engineering Software

MECHANISM ANALYSIS

SIMPLIFIED GRAPHICAL AND ANALYTICAL TECHNIQUES

SECOND EDITION, REVISED AND EXPANDED

LYNDON O. BARTON

Senior Engineer
E.I. du Pont de Nemours & Co.
Wilmington, Delaware

Marcel Dekker, Inc.　　　　New York • Basel • Hong Kong

Library of Congress Cataloging-in-Publication Data

Barton, Lyndon O.,
   Mechanism analysis : simplified graphical and analytical
techniques / Lyndon O. Barton. -- 2nd ed., rev. and expanded.
      p.   cm. -- (Mechanical engineering ; 81)
   Includes bibliographical references and index.
   ISBN 0-8247-8794-3
   1. Machinery, Kinematics of. I. Title. II. Series: Mechanical
engineering (Marcel Dekker, Inc.) ; 81.
TJ175.B29   1993
621.8′11--dc20   92-37563
                                                                         CIP

MARCEL DEKKER, INC.

270 Madison Avenue, New York, New York 10016

Current printing (last digit):
10   9   8   7   6   5   4   3   2   1

PRINTED IN THE UNITED STATES OF AMERICA

3

To my wife Olive, my children Rhonda, Loren, Carol, and
Leon, and my mother Clarice.

Preface

The first edition of this book, published in 1984, was well received. Written primarily for mechanical design engineers and mechanical design engineering students concerned with the design of machines and mechanisms, the book presented a wide assortment of graphical and analytical techniques for analyzing basic mechanisms, with emphasis being placed on linkages such as the four-bar and slider-crank linkages and their variations.

An important feature of that edition was the presentation of new and innovative simplification techniques such as:

The rectilinear and angular motion diagrams (Chapter 2)
The link extension concept for velocity analysis of mechanisms by the instant center method (Chapter 5)
The generalized procedure for constructing the acceleration polygon (Chapter 9)
The parallelogram method for slider-crank acceleration analysis (Chapter 12)
The simplified vector method (Chapters 14–18) for analyzing various classes of mechanisms by employing basic algebra and trigonometry instead of calculus or other sophisticated forms of mathematics. This simplified mathematical approach also made it possible to include several closed form analytical solutions rarely found, except in more advanced kinematics textbooks.

Another important feature of the first edition was the complete listings of computer as well as hand-held calculator programs for analyzing basic mechanisms.

This second edition retains all the essential elements of form and content of the first edition, plus there are three notable additions:

1. Three chapters have been added: Gear Fundamentals (Chapter 20), Gear Train Fundamentals (Chapter 21), and Cam Fundamentals (Chapter 22). With the addition of these chapters, the book may now be used as a single-source reference volume or independent teaching text in mechanisms, inasmuch as it covers the three major classes of mechanisms (linkages, gears, and cams) ordinarily taught in a first course of this subject. Basic in scope, the material presented in these chapters is intended primarily to: (1) serve as an introduction or brief review of the fundamentals, (2) illustrate the application of motion principles, from the study of linkages in the first edition to the analysis and design of simple gears, gear trains, and cams, and (3) lay the foundation for more advanced study in any of these areas, if so desired by the student.

   An important feature of this presentation is that alternative techniques in both graphical and analytical approaches, where applicable to problem solutions or theory development, are presented to provide the student or designer choices of techniques or approaches.

2. Several chapters have been expanded by adding new topics to provide, in each case, a more comprehensive treatment of that subject area. As in the first edition, several of these topics contain new and innovative approaches, such as:

Application of the link extension concept to velocity analysis of rolling elements by the instant center method (Chapter 5)

Acceleration analysis of sliding contact mechanisms by the effective component method (Chapter 8)

Development of algebraic expressions for acceleration analysis of rolling contact elements, using the relative acceleration method (Chapter 9)

Analysis of four-bar acceleration by the parallelogram method (Chapter 12)

Determination of center of curvature by the mean proportional method (Chapter 12)

Development of cam equations for modified uniform velocity motion, using circular arcs (Chapter 22).

A new computer program, written in Basic language and interactive, to calculate design data for a gear and pinion assembly has been added to the computer program listings (Appendix A).

The book is written for easy reading and comprehension without reliance on any other source. Needed background materials on such topics as uniform accelerated motion (Chapter 2), vectors (Chapter 3), complex algebra (Chapter 13), and trigonometry (Appendix B) is provided for review. Also, concepts are presented as concisely as possible, employing numerous illustrative examples, graphical aids, and step-by-step procedures for virtually all graphical constructions. In addition, the topics are arranged in a logical sequence corresponding to that ordinarily followed in teaching a course in kinematics. Also, chapters are organized to permit flexible use of the text as part of a mechanical or machine design course, or part of a course in dynamics. Finally, numerous examples and practice problems are included to crystalize and reinforce the concepts presented.

Some of the material in this book is based on several technical papers that the author has previously published. Other material has been drawn from class notes that the author has developed and used over several years of teaching kinematics of mechanisms as part of an engineering technology college curriculum.

Lyndon O. Barton

Acknowledgments

The author gratefully acknowledges his indebtedness to the E. I. du Pont de Nemours and Company Engineering Department and the Delaware Technical Community College Mechanical Engineering Department for providing the practical engineering and teaching opportunities, respectively, that have enabled him to pursue and accumulate the knowledge and experience that form the basis of this book.

Grateful acknowledgments and appreciation are also extended to

Penton Publishing Company, publishers of *Machine Design* magazine, for permission to reprint portions of previously published articles (including illustrations);

American Society for Engineering Education, publishers of *Engineering Design Graphics* journal for permission to reprint portions of previously published articles (including illustrations);

Mr. Albert A. Stewart, for the valuable assistance he has rendered in the calculator program development;

Gordon F. Walls, for his assistance with the artwork;

Teachers, relatives, and friends who have been a source of inspiration and encouragement in the author's career; and

A devoted family, for their love, understanding, and support, always.

Lyndon O. Barton

Contents

MECHANISM ANALYSIS

1

INTRODUCTORY CONCEPTS

Mechanism analysis (or kinematics of machines) is inherently a vital part in the design of a new machine or in studying the design of an existing machine. For this reason, the subject has always been of considerable importance to the mechanical engineer. Moreover, if we consider the tremendous advances that have been made within recent years in the design of high-speed machines, computers, complex instruments, automatic controls, and mechanical robots, it is not surprising that the study of mechanisms has continued to attract greater attention and emphasis than ever before.

Mechanism analysis may be defined as a systematic analysis of a mechanism based on principles of kinematics, or the study of motion of machine components without regard to the forces that cause the motion. To better appreciate the role of mechanism analysis in the overall design process, consider the following. Typically, the design of a new machine begins when there is a need for a mechanical device to perform a specific function. To fulfill this need, a conceptual or inventive phase of the design process is required to establish the general form of the device. Having arrived at a concept, the designer usually prepares a preliminary geometric layout of the machine or mechanism for a complete kinematic analysis. Here the designer is concerned not only that all components of the machine are properly proportioned so that the desired motions can be achieved (synthesis phase), but also with the analysis of the components themselves to determine such characteristics as displacements, velocities, and accelerations (analysis

1

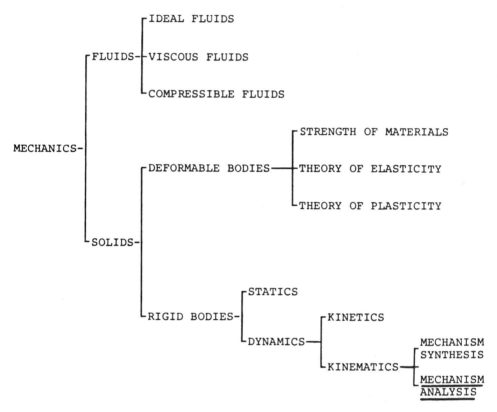

Figure I.1   Mechanism analysis and other branches of mechanics.

phase). At the completion of this analysis, the designer is ready to proceed to the next logical step in the design process: **kinetic analysis**, in which individual machine members are analyzed further to determine the forces resulting from the motion.

Mechanism analysis therefore serves as a necessary prerequisite for the proper sizing of machine members, so that they can withstand the loads and stresses to which they will be subjected. Figure I.1 shows the relationship of mechanism analysis to other branches of mechanics.

1
Kinematic Terminology

1.1 MECHANICAL CONCEPTS

A **mechanism** is a combination of rigid bodies so connected that the motion of one will produce a definite and predictable motion of others, according to a physical law. Alternatively, a mechanism is considered to be a kinematic chain in which one of the rigid bodies is fixed. An example of a mechanism is the slider crank shown in Figure 1.1(a). Instruments, watches, and governors provide other examples of mechanisms.

A **machine** is a mechanism or group of mechanisms used to perform useful work. A machine transmits forces. An example of a machine is the internal combustion engine unit shown in Figure 1.1(b).

The term "machine" should not be confused with "mechanism" even though in actuality, both may refer to the same device. The difference in terminology is related primarily to function. Whereas the function of a machine is to transmit energy, that of a mechanism is to transmit motion. Stated in other words, the term "mechanism" applies to the geometric arrangement that imparts definite motions to parts of a machine.

A **pair** is a joint between two bodies that permits relative motion.

1. A **lower pair** has surface contact, such as a hinge or pivot. Surface contact is a characteristic of sleeve bearings, piston rings, screwed joints, and ball-and-socket joints. These joints are called **revolute** or **turning pairs**.

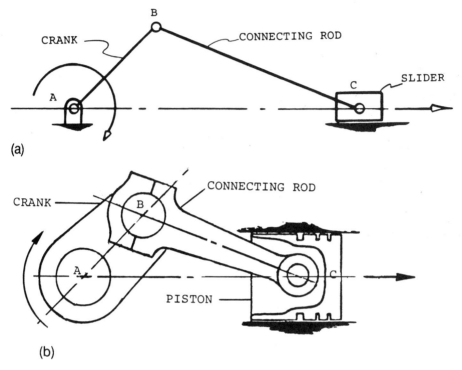

Figure 1.1a   Slider-crank mechanism. 1.1b   Internal combustion engine unit.

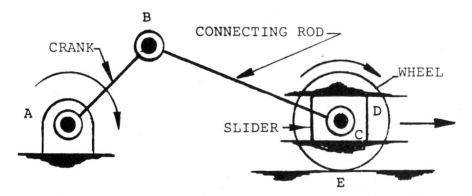

Figure 1.2   Slider-crank with turning, sliding, and rolling pairs.

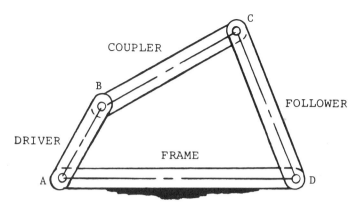

Figure 1.3   Four-bar linkage mechanism.

2. A **higher pair** has line or point contact between the surface elements. Point contact is usually found in ball bearings. Line contact is characteristic of cams, roller bearings, and most gears. Thus, a higher pair may also be defined as a **rolling contact pair**.

Stated another way, the common types of pairs in planar motion are as follows:

1. **Turning pairs**, such as A, B, and C in Figure 1.2, which permit only rotation of one body relative to another.
2. **Sliding contact pairs**, such as D in Figure 1.2, which permit only linear displacement of one body relative to another.
3. **Rolling contact pairs**, such as E in Figure 1.2, which permit only rolling motion of one body relative to another. In a rolling contact pair, the two surfaces must be held in contact by some external force or constraint, and there must be sufficient friction to prevent relative sliding between the surfaces.

A **link** is a rigid body that serves to transmit force from one body to another or to cause or control motion, such as the connecting rod or crank arm in Figure 1.1. Alternatively, a link is defined as a rigid body having two or more pairing elements.

A **linkage** is a mechanism in which all connections are lower pairs. The four-bar and slider-crank linkages shown in Figures 1.3 and 1.1a are two examples.

A **kinematic chain** is a group of links connected by means of pairs to transmit motion or force. There are three types of chains: locked chain, constrained chain, and unconstrained chain.

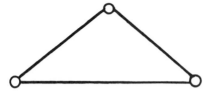

Figure 1.4   Locked chain.

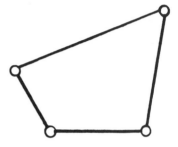

Figure 1.5   Constrained chain.

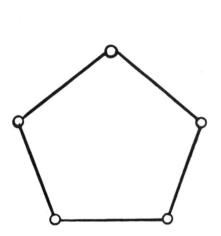

Figure 1.6   Unconstrained chain.

1. A **locked chain** has no relative motion between the links. An ex-
    ample of this is the three-link chain shown in Figure 1.4.
2. A **constrained chain** is one in which there is definite relative mo-
    tion between the links. For example, if one link is fixed and another
    link is put in motion, the points on all the other links will move
    in definite paths and always in the same paths regardless of the
    number of lines the motion is repeated. An example is the four-bar
    mechanism shown in Figure 1.5.
3. An **unconstrained chain** is one in which, with one link fixed and
    another link put in motion, the points on the remaining links will
    not follow definite paths. An example of an unconstrained chain
    is the five-link mechanism shown in Figure 1.6.

Gruebler's Equation and the Kinematic Chain

An easy way to determine what type of chain a particular linkage is is to apply Gruebler's equation, which states

$$f = 3(n - 1) - 2j \qquad (1.1)$$

where

f = degrees of freedom

n = number of links

j = number of joints

When $f = 0$, the chain is locked. For example, in Figure 1.4,

$$f = 3(3 - 1) - 2(3) = 0$$

When $f = 1$, the chain is constrained. For example, in Figure 1.5,

$$f = 3(4 - 1) - 2(4) = 1$$

When $f > 1$, the chain is unconstrained. For example, in Figure 1.6,

$$f = 3(5 - 1) - 2(5) = 2$$

When applying Eq. (1.1), it is important to note the following assumptions:

1. The linkage must be a closed pin-jointed chain.
2. A simple joint involving two links is equivalent to one joint ($j = 1$), whereas a double joint involving three links is equivalent to two simple joints ($j = 2$).

A *simple mechanism* is one that consists of three or four links, whereas a *compound mechanism* consists of a combination of simple mechanisms and is usually made up of more than four links.

A *structure* is a combination of rigid bodies capable of transmitting forces or carrying loads, but having no relative motion in them. Alternatively, a structure may be thought of as a locked kinematic chain. A *frame* is a stationary structure that supports a machine or mechanism. Normally, it is the fixed link of a mechanism (e.g., link 1 in Figure 1.3).

A *driver* is that part of a mechanism which causes motion, such as the crank in Figure 1.3. A *follower* is that part of a mechanism whose motion is affected by the motion of the driver, such as the slider in Figure 1.3.

Modes of Transmission

Motion can be transmitted from driver to follower by:

1. Direct contact
   a. Sliding
   b. Rolling
2. Intermediate connectors
   a. Rigid: links
   b. Flexible: belts, fluids
3. Nonmaterial: magnetic forces

A *gear* is a machine member, generally circular, whose active surface is provided with teeth to engage a similar member to impart rotation from one shaft to another. An example is the spur gear shown in Figure 1.7.

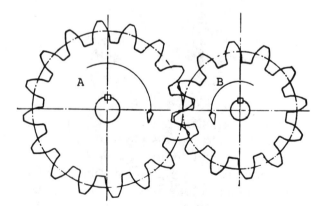

Figure 1.7   Spur gears.

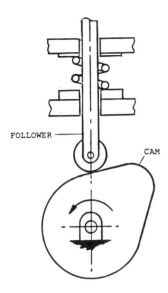

FOLLOWER

CAM

Figure 1.8   Cam-follower mechanism.

A *cam* is a rotating or sliding machine member whose function is to impart a predetermined motion to another part called the *follower*, that rolls or slides along the surface of the member. An example is shown in Figure 1.8.

Inversion

Inversion is the process of fixing different links in a chain to create different mechanisms. Many useful mechanisms may be obtained by the inversion of various kinematic chains. An example of such inversion can be seen in the slider crank chain shown in Figure 1.9:

1. By making link 1 of the chain fixed, we obtain a steam engine mechanism [Figure 1.9(a)].
2. By fixing the crank, link 2, we obtain the Whitworth quick-return mechanism, used in various types of metal shapers [Figure 1.9(b)].
3. By fixing the connecting rod, link 3, we obtain the oscillating cylinder engine, once used as a type of marine engine [Figure 1.9(c)].
4. Finally, by fixing the slider, link 4, we obtain the mechanism shown in Figure 1.9(d). This mechanism has found very little practical application. However, by rotating the figure 90° clockwise, the mechanism can be recognized as part of a garden pump.

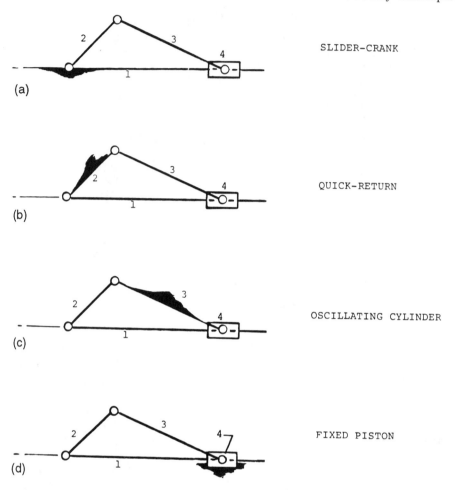

Figure 1.9   Slider-crank chain: (a) Slider crank; (b) quick-return mechanism; (c) oscillating cylinder; (d) fixed piston.

It is important to keep in mind that the inversion of a mechanism does not change the relative motion between the links, but does alter their absolute motions.

1.2   MOTION CLASSIFICATION

Definitions

Motion is the act of changing position. The change of position can be with respect to some other body that is either at rest or moving.

Rest is a state in which the body has no motion.

Absolute motion is the change of position of a body with respect to another body at absolute rest.

Relative motion is the change of position of a body with respect to another body that is also moving with respect to a fixed frame of reference.

Types of Motion

Plane Motion

In plane motion all points on a body in motion move in the same plane or parallel planes. All points of the body or system of bodies remain at a constant distance from a reference plane throughout the motion. Typical examples are the connecting rod on a slider-crank mechanism (Figure 1.1) and the side rod on a locomotive (Figure 1.10). There are three classes of plane motion: (1) rotation, (2) translation, and (3) combined translation and rotation.

1. ***Rotation***: When one point in a body remains stationary while the body turns about that point, the body is said to be in rotation. That is, all points in the body describe circular paths about a stationary axis that is perpendicular to the plane of rotation. Crank AB in Figure 1.1 has rotary motion.
2. ***Translation***: When a body moves without turning, that body is said to be in translation, or the distances between particles of the body remain unchanged. There are two types of translation:
   a. ***Rectilinear translation***, where all points in the body move along parallel straight paths, such as the slider C in Figure 1.1.

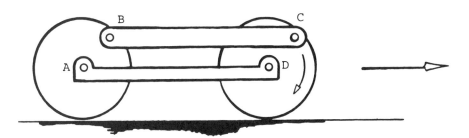

Figure 1.10   Locomotive side rod drive. Link BC undergoes curvilinear motion; link AD undergoes rectilinear translation.

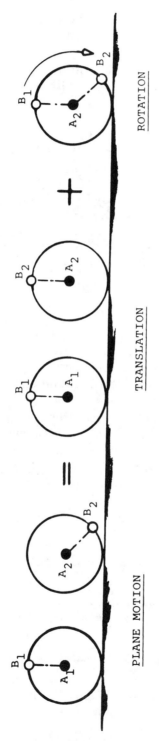

Figure 1.11   Rolling wheel.

b. *Curvilinear translation*, where all points in the body move along similar (or parallel) curved paths. Curvilinear translation is not to be confused with rotation, where all paths on the body are in the form of concentric circles. A good example is the locomotive side rod in Figure 1.10. Note that the paths of any two points on the rod, say, B and C, have the same curvature.

3. *Combined motion*: When a body undergoes simultaneous translation and rotation, that body is said to be in combined motion. That is, all points in the body change position and all lines turn as the body moves.

A common example of a body that undergoes combined motion is the connecting rod BC in Figure 1.1, where one end B rotates about the crank axle A, whereas the other end C translates along a straight path as defined by the slider motion. Hence, every other point on the member experiences part rotation and part translation.

Another common example is that of the rolling wheel (Figure 1.11), where it can be seen that the resultant motion of point B is the summation of its translation and rotation motions.

Three-Dimensional Motion

There are two types of three-dimensional motion:

1. *Helical motion*: When a body has rotation combined with translation along the same axis of rotation, that body is said to be in helical motion. The most common example is the turning of a nut on a screw (see Figure 1.12). The nut rotates and at the same time translates along the axis of rotation.

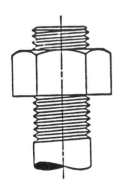

Figure 1.12   Screw and nut.

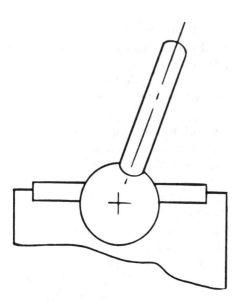

Figure 1.13   Ball joint.

2. **Spherical motion**: When a body moves in three-dimensional space such that each part of that body remains at a constant distance from a fixed point, the body is said to be in spherical motion. A common example is the ball-and-socket joint, shown in Figure 1.13. A point on the ball or handle, which is rigidly attached to the ball, moves in space without changing its distance from the center of the sphere.

Other Types of Motion

Additional types of motion include:

1. **Continuous motion**: When a point continues a move indefinitely along a given path in the same sense, such a motion is said to be continuous. An example of this is a rotating wheel, in which the path of a point away from the axis returns on itself.
2. **Reciprocating motion**: When a point traverses the same path and reverses its motion at the end of such path, the motion is said to be reciprocating. An example is the slider in a typical slider-crank mechanism (Figure 1.1). **Oscillation** is reciprocating circular motion, as in a pendulum.
3. **Intermittent motion**: When the motion of a point is interrupted by periods of rest, such motion is said to be intermittent, as in a ratchet.

1.3 MOTION CHARACTERISTICS

The **path** is the locus of a point as it changes from one position to another, **distance** is a measure of the path through which a point moves, and **displacement** is a measure of the net change in position of a point. There are two types of displacement:

1. **Linear displacement** (Δs) is the change position of a point as it moves along a straight line. Linear displacement is typically expressed in terms of inches, feet, or miles in a specified direction. Hence, it is a vector quantity.
2. **Angular displacement** (Δθ) is the angle between two positions of a rotating line or body. It has both magnitude and sense, either clockwise or counterclockwise. Angular displacement is typically expressed in terms of degrees, radians, or resolutions in a specified rotational sense.

   Velocity is the rate of change of position of a point with respect to time, or displacement per unit time.

1. **Linear velocity** (v) is displacement per unit of time of a point moving along a straight line. The average linear velocity is given by the expression

$$v = \frac{\Delta s}{\Delta t}$$

   where Δs is the change in linear displacement and Δt the time interval. Linear velocity is typically expressed in terms of inches per second, feet per second, or miles per hour in a specified direction. Hence, it is a vector quantity.
2. **Angular velocity** (ω) is the angular displacement per unit time of a line or body in rotation and has both magnitude and sense, either clockwise or counterclockwise. The average angular velocity is given by the expression

$$\omega = \frac{\Delta \theta}{\Delta t}$$

   where Δθ is the average angular velocity and Δt the time interval. Angular velocity is typically expressed in terms of radians per second or revolutions per minute in a specified rotational sense.

Acceleration is the rate of change of velocity with respect to time, or the rate of speedup.

1. **Linear acceleration** (a) is the change in linear velocity per unit of time. Average linear acceleration is given by the expression

$$a = \frac{\Delta v}{\Delta t}$$

   where Δv is the change in linear velocity and Δt the time interval. Linear acceleration is typically expressed in terms of inches per second per second or feet per second per second in a specified direction. Hence, it is a vector quantity.

2. **Angular acceleration** (α) is the change in angular velocity per unit of time. Average angular acceleration is given by the expression

$$\alpha = \frac{\Delta \omega}{\Delta t}$$

   where $\Delta \omega$ is the change in angular velocity and Δt the time interval. Angular acceleration is typically expressed in terms of radians per second per second or revolutions per minute per minute.

Deceleration or **retardation** is negative acceleration or rate of slowing down.

Speed is the rate of motion in any direction, or the total distance covered in one unit of time. Speed is not to be confused with velocity, which is a vector quantity. Speed is typically expressed in magnitude terms such as feet per second, inches per second, or miles per hour, without regard to direction. Hence, it is a scalar quantity.

Angular speed is the rate at which a body turns about an axis. Typical units are revolutions per minute and radians per second, without any reference to rotational sense.

Phase describes the relative positions of links in a mechanism or a machine at any instant. This is usually defined by the angle of one of the links of the mechanism, for example, the angle of crank AB in Figure 1.1.

When a mechanism moves through all its possible phases and returns to its starting position, it has completed a *cycle*, or a motion cycle. Virtually all mechanisms have a cyclic pattern in which the cycle repeats itself over and over again. The time required for a motion to complete one motion cycle is called a *period*.

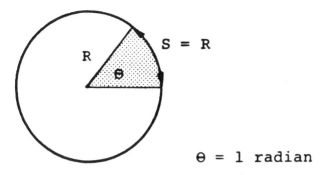

Figure 1.14 Geometric representation of a radian.

A body that rotates a counterclockwise direction is assumed to have a *positive sense*. This is because angular displacement is normally measured counterclockwise. Thus in Figure 1.1, the crank AB has a positive sense, and its angular velocity is also positive. Conversely, a body that rotates in a clockwise direction is assumed to have a *negative sense*, in which case the angular velocity of crank AB in Figure 1.1 is negative.

Finally, *radian* is the angle subtended by an arc that is equal in length to the radius of the circle (see Figure 1.14). There are 2π radians in a circle.

2
Uniformly Accelerated Motion

2.1 RECTILINEAR MOTION

Displacement, Velocity, Acceleration, and Time Relationships

The analysis of rectilinear or straight-line motion with constant or uniform acceleration is relatively common in the study of kinematics. To establish some basic relationships, let us consider a body moving with uniform acceleration where the velocity changes from an initial value to a final value after some time, while completing some distance. Let

s = distance completed, ft

v_1 = initial velocity, ft/sec

v_2 = final velocity, ft/sec

a = acceleration, ft/sec²

t = time, sec

By definition,

Distance = average velocity × time

$$s = \frac{v_1 + v_2}{2} t \qquad (2.1)$$

18

After 1 sec, initial velocity is increased to $v_1 + a$.

After 2 sec, initial velocity is increased to $v_1 + 2a$.

After 3 sec, initial velocity is increased to $v_1 + 3a$.

After t sec, initial velocity is increased to $v_1 + at$.

This means that

$$v_2 = v_1 + at \qquad (2.2)$$

Now, substitute for v_2 in Eq. (2.1) and obtain

$$s = \frac{v_1 + (v_1 + at)}{2} t$$

$$s = \frac{2v_1 + at}{2} t$$

$$s = v_1 t + \frac{1}{2} at^2 \qquad (2.3)$$

From Eq. (2.2),

$$v_2 - v_1 = at \qquad (2.4)$$

and from Eq. (2.1),

$$v_1 + v_2 = \frac{2}{t} s$$

or

$$v_2 + v_1 = \frac{2}{t} s \qquad (2.5)$$

Now, multiply Eqs. (2.4) and (2.5) and obtain

$$(v_2 - v_1)(v_2 + v_1) = at \times \frac{2}{t} s$$

which gives

$$v_2^2 - v_1^2 = 2as$$

or

$$v_2^2 = v_1^2 + 2as$$

In summary, for uniform linear acceleration

$$s = \frac{v_1 + v_2}{2} t$$

$$s = v_1 t + \frac{1}{2} at^2$$

$$v_2 = v_1 + at$$

$$v_2^2 = v_1^2 + 2as$$

For uniform linear velocity $(a = 0)$

$$s = v_1 t$$

Also, if the body is accelerating, a is positive; if the body is decelerating, a is negative.

EXAMPLE 2.1

A car passes a certain point A with a velocity of 30 ft/sec and another point B 1 mile away with a velocity of 60 ft/sec. If the acceleration is uniform, determine

a.   The average velocity of the car
b.   The time taken to travel from A to B
c.   The acceleration of the car

Given

$$v_1 = 30 \text{ ft/sec}$$

$$v_2 = 60 \text{ ft/sec}$$

$$s = 5280 \text{ ft}$$

Required

$$\bar{v} = ? \quad t = ? \quad a = ?$$

SOLUTION

Average velocity

$$\overline{v} = \frac{v_1 + v_2}{2}$$

$$= \frac{30 + 60}{2}$$

$$= 45 \text{ ft/sec}$$

Time taken

$$t = \frac{\text{distance}}{\text{average velocity}}$$

$$= \frac{5280}{45}$$

$$= 117.3 \text{ sec or } 1 \text{ min, } 57.3 \text{ sec}$$

Acceleration

$$v_2^2 = v_1^2 + 2as$$

$$60^2 = 30^2 + 2a(5280)$$

$$a = \frac{60^2 - 30^2}{2(5280)}$$

$$= 0.256 \text{ ft/sec}^2$$

EXAMPLE 2.2

In coming to a stop, a train passes one signal with a speed of 60 mph and a second signal 30 sec later. During this period the brakes are applied to give a uniform acceleration. If the signals are 2400 ft apart, find

a. The velocity of the train passing the second sign
b. The magnitude of the deceleration

Given

$$v_1 = 60 \text{ mph} = 88 \text{ ft/sec}$$

$$t = 30 \text{ sec}$$

$$s = 2400 \text{ ft}$$

Required

$$v_2 = ? \qquad a = ?$$

SOLUTION

Final velocity v_2 is obtained from

$$s = \frac{v_1 + v_2}{2} t$$

$$2400 = \frac{88 + v_2}{2} 30$$

$$v_2 = \frac{2400(2)}{30} - 88$$

$$= 72 \text{ ft/sec}$$

Deceleration is obtained from

$$s = v_1 t + \frac{1}{2} a t^2$$

$$2400 = 88(30) + \frac{1}{2} a(30^2)$$

$$a = \frac{(2400 - 2640)(2)}{900}$$

$$= -0.53 \text{ ft/sec}^2$$

Rectilinear Motion Relationship Diagram

The importance to the study of kinematics of the rectilinear relationships just derived cannot be overemphasized. However, for many students, and even designers, quick recall of these expressions can be

difficult. Furthermore, if the appropriate references are not readily available, precious time can be lost in trying to derive the desired expressions. To aid in such situations, the simple diagram presented in Figure 2.1 can be useful. Here

v_1 (side of the smaller square) represents the initial linear velocity
v_2 (side of the larger square) represents the final linear velocity
at (a times t) represents the difference bewteen v_1 and v_2, where
t is time
as (a times s) represents the area of each trapezoidal section, where
s is the displacement

Considering the larger square, we can write the expression for any one of its sides as follows:

$$v_2 = v_1 + at$$

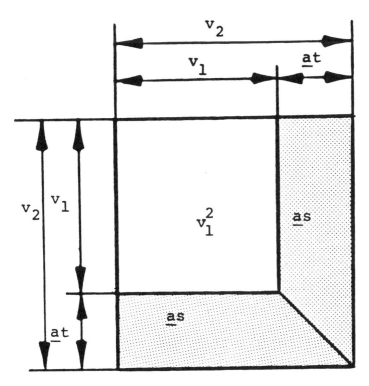

Figure 2.1   Rectilinear motion.

An expression for the midpoint of the side (or average linear velocity $\bar{v}$) is given by

$$\bar{v} = \frac{v_1 + (v_1 + at)}{2}$$

or

$$\bar{v} = \frac{v_1 + v_2}{2}$$

Also, considering the area of the larger square, we can write an expression for this area in terms of its constituent parts

$$v_2^2 = v_1^2 + 2as$$

Finally, considering one of the trapezoids, we can determine its area from the relationship

$$\text{Area} = \frac{1}{2} \text{ (sum of parallel sides)} \times \text{width}$$

Therefore,

$$as = \frac{1}{2} (v_1 + v_2)at$$

or

$$s = \frac{1}{2} (v_1 + v_2)t$$

Alternatively, the area of a trapezoid may also be expressed as the sum of its rectangular and triangular portions, as follows:

$$as = v_1at + \frac{1}{2} (at)^2$$

$$s = v_1t + \frac{1}{2} at^2$$

2.2   ANGULAR MOTION

Displacement, Velocity, Acceleration, and Time Relationships

As with rectilinear motion, to establish the basic angular relationships, let us consider a rotating body a uniform angular acceleration where the angular velocity changes from an initial value to a final value after some time t, while turning through some angle. Let

θ = angle turned through, rad

ω_1 = initial angular velocity, rad/sec

ω_2 = final angular velocity, rad/sec, after t sec

t = time, sec

By definition,

Displacement = average velocity × time

$$\theta = \frac{\omega_1 + \omega_2}{2} t \qquad (2.6)$$

After 1 sec, initial velocity is increased to $\omega_1 + \alpha$.

After 2 sec, initial velocity is increased to $\omega_1 + 2\alpha$.

After 3 sec, initial velocity is increased to $\omega_1 + 3\alpha$.

After t sec, initial velocity is increased to $\omega_1 + t\alpha$.

This means that

$$\omega_2 = \omega_1 + \alpha t \qquad (2.7)$$

Now substitute for ω_2 in Eq. (2.6) and obtain

$$\theta = \frac{\omega_1 + (\omega_1 + \alpha t)}{2} t$$

$$\theta = \frac{2\omega_1 + \alpha t}{2} t$$

$$\theta = \omega_1 + \frac{1}{2} \alpha t^2 \qquad (2.8)$$

Again from Eq. (2.7),

$$\omega_2 - \omega_1 = \alpha t \tag{2.9}$$

and from Eq. (2.6),

$$\omega_1 + \omega_2 = \frac{2\theta}{t}$$

or

$$\omega_2 + \omega_1 = \frac{2\theta}{t} \tag{2.10}$$

Now, multiply Eqs. (2.9) and (2.10) and obtain

$$(\omega_2 - \omega_1)(\omega_2 + \omega_1) = \alpha t \times \frac{2\theta}{t}$$

which gives

$$\omega_2^2 - \omega_1^2 = 2\alpha\theta$$

or

$$\omega_2^2 = \omega_1^2 + 2\alpha\theta \tag{2.11}$$

In summary, for uniform angular acceleration

$$\theta = \frac{\omega_1 + \omega_2}{2} t$$

$$\omega_2 = \omega_1 + \alpha t$$

$$\theta = \omega_1 t + \frac{1}{2} \alpha t^2$$

$$\omega_2^2 = \omega_1^2 + 2\alpha\theta$$

For a uniform angular velocity ($\alpha = 0$)

$$\theta = \omega_1 t$$

Also, if the body is accelerating, α is positive; if the body is decelerating, α is negative.

EXAMPLE 2.3

A motor starting from rest develops a speed of 3000 rpm in 15 sec. If the acceleration is uniform, determine

a.   The rate of acceleration
b.   The number of revolutions made in coming up to speed

Given

$$\omega_1 = 0 \quad (\text{rest})$$

$$\omega_2 = 3000 \text{ rpm} = 3000 \left(\frac{2\pi}{60}\right) = 314.2 \text{ rad/sec}$$

$$t = 15 \text{ sec}$$

Required

$$\theta = ? \qquad \alpha = ?$$

SOLUTION

Acceleration

$$\omega_2 = \omega_1 + \alpha t$$

$$314.2 = 0 + \alpha(15)$$

$$\alpha = \frac{314.2}{15}$$

$$= 20.9 \text{ rad/sec}^2$$

Number of revolutions

$$\theta = \frac{\omega_1 + \omega_2}{2} t$$

$$= \frac{0 + 314.2}{2}(15)$$

$$= 2356.5 \text{ rad}$$

$$= \frac{2356.5}{2\pi}$$

$$= 375 \text{ rev}$$

EXAMPLE 2.4

A crank shaft rotating at 50 rpm has an angular deceleration of 1 rad/min/sec. Calculate its angular velocity after 20 sec and the number of revolutions it makes

a.   In 40 sec
b.   In coming to rest

Given

$$\omega_1 = 50 \text{ rpm} = 50 \left(\frac{2\pi}{60} \right) = 5.24 \text{ rad/sec}$$

$$\alpha = -1 \text{ rad/min/sec} = \frac{-1}{60} \text{ rad/sec/sec}$$

$$t = 20 \text{ sec}$$

Required

$$\omega_2 = ? \qquad \theta = ?$$

SOLUTION

Angular velocity after 20 sec

$$\omega_2 = \omega_1 + \alpha t$$

$$= 5.24 + \frac{-1}{60} (40) = 5.24 - 0.33$$

$$= 4.9 \text{ rad/sec}$$

Number of revolutions in 40 sec

$$\theta = \omega_1 t + \frac{1}{2} 2\alpha t^2$$

$$= 5.24(40) + \frac{1}{2} \left(\frac{-1}{60} \right) (40)^2$$

$$= 196.26 \text{ rad}$$

$$= \frac{196.26}{2\pi}$$

$$= 31.23 \text{ rev}$$

Number of revolutions in coming to rest

Data:

$\omega_1 = 5.24$ rad/sec

$\omega_2 = 0$ (rest)

$\alpha = \dfrac{-1}{60}$ rad/sec²

Required:

$\theta = ?$ (angular displacement)

Equations:

$$\omega_2^2 = \omega_1^2 + 2\alpha\theta$$

$$0 = 5.24^2 + 2\left(\frac{-1}{60}\right)\theta$$

$$\theta = 5.24^2\left(\frac{60}{2}\right)$$

$$= 823.7 \text{ rad}$$

$$= \frac{823.7}{2\pi}$$

$$= 131.1 \text{ rev}$$

Angular Motion Relationship Diagram

As in the rectilinear case, a similar diagram (Figure 2.2) can be used for the angular relationships. Here

ω_1 (side of the smaller square) represents the initial angular velocity
ω_2 (side of the larger square) represents the final angular velocity
αt (α times t) represents the difference between ω_1 and ω_2, where
α is the uniform angular acceleration, and
t is time
$\alpha\theta$ (α times θ) represents the area of each trapezoidal section, where
θ is the angular displacement

Considering the larger square, we can write an expression for any one of its sides as follows:

$$\omega_2 = \omega_1 + \alpha t$$

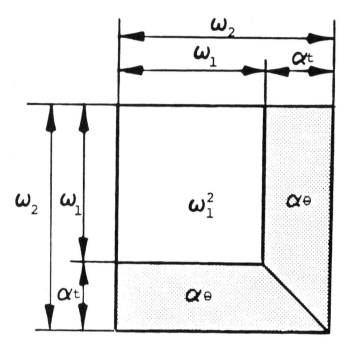

Figure 2.2   Angular motion.

and an expression for the midpoint of the side (or average angular
velocity $\bar{\omega}$) is given by

$$\bar{\omega} = \frac{\omega_1 + (\omega_1 + \alpha t)}{2}$$

or

$$\bar{\omega} = \frac{\omega_1 + \omega_2}{2}$$

Also, considering the area of the larger square, we can write an expression for this area in terms of its constituent parts

$$\omega_2^2 = \omega_1^2 + 2\alpha\theta$$

Finally, considering the trapezoidal area, we can write

$$\alpha\theta = \frac{1}{2}(\omega_1 + \omega_2)\alpha t$$

from which

$$\theta = \frac{1}{2}(\omega_1 + \omega_2)t$$

Alternatively, the area of one trapezoid can be expressed as the sum of its rectangular and triangular sections. Thus,

$$\alpha\theta = \omega_1\alpha t + \frac{1}{2}(\alpha t)^2$$

or

$$\theta = \omega_1 t + \frac{1}{2}\alpha t^2$$

2.3 CONVERSION BETWEEN ANGULAR AND RECTILINEAR MOTION

There are many situations where it is necessary to convert from angular to rectilinear motion, and vice versa. To establish the basic relationships, consider the pulley-and-belt arrangement in Figure 2.3. If the pulley turns through an angle θ, if we assume that the belt moves without slipping, the corresponding displacement of the belt is equivalent to the length of arc $r\theta$, or

$$s = r\theta \tag{2.12}$$

If the pulley is rotating at uniform velocity ω where the angular displacement $\theta = \omega t$, the total belt displacement is

$$s = r(\omega t) \tag{2.13}$$

Also, the belt moves at uniform speed; its velocity v is related to the displacement by

$$s = vt \tag{2.14}$$

Therefore, Eq. (2.13) can be written as

$$vt = r\omega t \tag{2.15}$$

or

$$v = r\omega \tag{2.16}$$

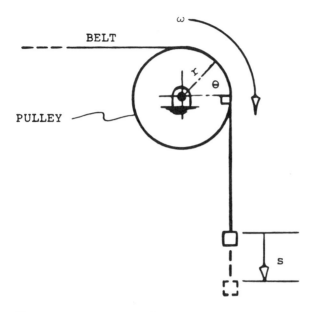

Figure 2.3   Conversion from angular to rectilinear motion.

This expression is probably the most useful in velocity analysis. In effect, it states that *if a point has uniform motion in a circular path, the linear speed at any instant is equal to the distance of that point from the center of rotation multiplied by the angular speed.* Also, since velocity is a directed speed, it is clear from the illustration that the only direction the velocity of the point can have at that instant is tangential to the circular path in the same sense as the angular motion.

Another way of looking at the velocity direction is to consider any object, such as a ball, being rotated about one end of a string while the other end is held fixed. Suppose that the string were to be suddenly released; the ball will then "take off" in a direction tangential to the circular path that it obtained before the release.

Similarly, it can be shown that for uniform acceleration of the pulley where the angular acceleration is α, the linear acceleration a, of the belt is given by

$$a = r\alpha \qquad\qquad\qquad (2.17)$$

This acceleration is known as the **tangential acceleration** since the linear velocity acts in a direction tangential to the path of rotation. Hence, to convert from angular motion to linear motion, we multiply the respective values by radius r.

EXAMPLE 2.5

In Figure 2.4 the pulley D is belt-driven by pulley B, which is fastened
to pulley C. Starting from rest, the body A falls 60 ft in 4 sec. For each
pulley, determine

a.   The number of revolutions
b.   The angular velocity
c.   The angular acceleration

SOLUTION

Calculate linear values for the pulleys, then convert to angular values.
Given

r_B = 8 in.

r_C = 6 in.

r_D = 10 in.

S = 60 ft

v_1 = 0 ft/sec

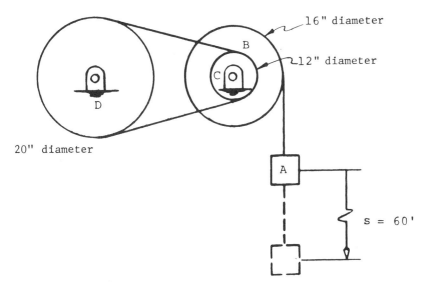

16" diameter

12" diameter

20" diameter

s = 60'

Figure 2.4   Example problem.

Required

$$\theta_B = ? \qquad \theta_C = ? \qquad \theta_D = ?$$

$$\alpha_B = ? \qquad \alpha_C = ? \qquad \alpha_D = ?$$

Revolutions

Pulley B:

$$s = r_B \theta_B$$

$$60 = \left(\frac{8}{12}\right) \theta_B$$

$$\theta_B = 60 \left(\frac{12}{8}\right)$$

$$= 90 \text{ rad}$$

$$= \frac{90}{2\pi}$$

$$= 14.3 \text{ rev}$$

Pulley C:

$$\theta_C = 14.3 \text{ rev} \quad (\text{same as } \theta_B)$$

Pulley D:

$$\frac{\theta_D}{\theta_C} = \frac{\text{radius of pulley } C}{\text{radius of pulley } D}$$

$$\frac{\theta_D}{14.3} = \frac{6}{10}$$

$$\theta_D = 14.3 \left(\frac{6}{10}\right)$$

$$= 8.58 \text{ rev}$$

Angular velocities
Pulley B: The angular velocity of pulley B is obtained from the following relationship:

$$v_B = r_B \omega_B$$

where

v_B = linear velocity after 4 sec (unknown)

ω_B = angular velocity after 4 sec (required)

Therefore, to find v_B, we first need to determine the linear acceleration a_B. Since the motion is not uniform or free-falling, we use the relationship

$$s = v_1 t + \frac{1}{2} a t^2$$

where

$s = 60$ ft

$v_1 = 0$

$t = 4$ sec

$a = a_B$

Therefore,

$$60 = 0 + \frac{1}{2} a_B (4)^2$$

and

$$a_B = \frac{60(2)}{16}$$

$$= 7.5 \ \text{ft/sec}^2$$

Linear velocity v_B is given by the relationship

$$v_2 = v_1 + at$$

where

$v_2 = v_B$

$v_1 = 0$

$a = a_B = 7.5 \ \text{ft/sec}^2$

$t = 4$ sec

Therefore,

$$v_B = 0 + 7.5(4)$$

$$= 30 \text{ ft/sec}$$

Required angular velocity ω_B can now be determined by substituting the value found for v_B.

$$30 = \left(\frac{8}{12}\right) \omega_B$$

$$\omega_B = 30 \left(\frac{12}{8}\right)$$

$$= 45 \text{ rad/sec}$$

Pulley C:

$\omega_C = 45$ rad/sec   (same as ω_B, since pulleys C and B are attached)

Pulley D:

$$\frac{\omega_D}{\omega_C} = \frac{\text{radius of pulley } C}{\text{radius of pulley } D}$$

$$\frac{\omega_D}{45} = \frac{6}{10}$$

$$\omega_D = 27 \text{ rad/sec}$$

Angular acceleration

Pulley B:

$$a_B = r_B \alpha_B$$

where

$$a_B = 7.5 \text{ ft/sec}^2 \quad \text{(found)}$$

$$r_B = \frac{8}{12}$$

Therefore,

$$7.5 = \frac{8}{12} \alpha_B$$

and

$$\alpha_B = 7.5 \left(\frac{12}{8}\right)$$

$$= 11.25 \text{ rad/sec}^2$$

Pulley C:

$\alpha_C = 11.25$ rad/sec^2 (same as α_B since pulleys B and C are attached)

Pulley D:

$$\frac{\alpha_D}{\alpha_C} = \frac{\text{radius of pulley } C}{\text{radius of pulley } D}$$

$$\frac{\alpha_D}{11.25} = \frac{6}{10}$$

$$\alpha_D = 11.25 \left(\frac{6}{10}\right)$$

$$= 6.75 \text{ rad/sec}^2$$

2.4   VELOCITY-TIME GRAPH SOLUTIONS

In many cases its has been found more convenient and simpler to solve some motion problems graphically, using the velocity-time diagram concept. The velocity diagram is a graph in which the velocity of a point is plotted against a time base. Figure 2.5 shows the three possible conditions that can exist

1. Uniform velocity [Figure 2.5(a)]
2. Uniform acceleration [Figure 2.5(b)]
3. Variable velocity [Figure 2.5(c)]

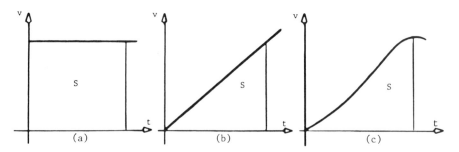

Figure 2.5   Velocity-time curves: (a) Uniform velocity; (b) uniform acceleration; (c) variable velocity.

Note that in case 1 the velocity-time curve has no slope, and therefore
the point has no acceleration. This motion is normally refered to as
uniform motion.

If the motion is not uniform, but the acceleration is constant, the
point is said to have **uniformly accelerated motion**, as in case 2. Otherwise, the **motion is variable**, as in case 3, where the acceleration
changes from one instant to another.

Also, note that acceleration may be depicted as positive or negative
depending on whether the slope of the velocity curve is positive (upward to the right or downward to the left) or negative (upward to the
left or downward to the right). In each case, the distance covered by
the point is given by the area under the curve corresponding to the
time period during which the motion took place.

EXAMPLE 2.6

The maximum acceleration of a Ferris wheel at a park is 1 rad/min/
sec and the maximum deceleraton is 2 rad/min/sec. Determine the
minimum time it will take the wheel to complete 15 revolutions going
from rest to rest.

SOLUTION

Let

$$T = \text{total time}$$

$$t = \text{time to accelerate}$$

$$T - t = \text{time to decelerate}$$

Given

$$\alpha_A = 1 \text{ rad/min/sec} = \frac{1}{60} \text{ rad/sec}^2$$

$$\alpha_D = -2 \text{ rad/min/sec} = \frac{-1}{30} \text{ rad/sec}^2$$

$$\theta = 15 \text{ rev} = 15(2\pi) = 94.24 \text{ rad}$$

Required

$$T = ?$$

Let triangle imf in Figure 2.6 represent the starting velocity v_i through maximum velocity v_m to final velocity v_f. This triangle then represents the total angular displacement θ of the wheel in time T. That is,

$$\theta = \text{area of triangle imf}$$

Now for the acceleration portion of the curve, consider triangular segment ima and find ω_m in terms of t' using the relationship

$$\omega_2 = \omega_1 + \alpha t$$

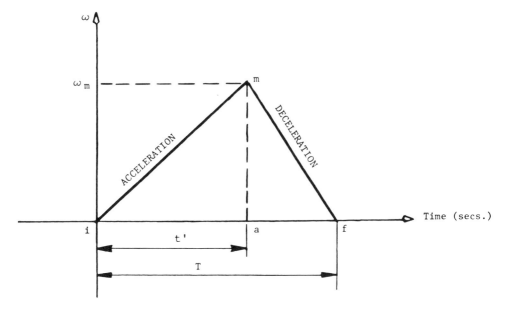

Figure 2.6   Velocity-time graph.

where

$$\omega_2 = \omega_m \text{ rad/sec}$$

$$\omega_1 = \omega_i = 0 \text{ rad/sec}$$

$$\alpha = \alpha_A = \frac{1}{60} \text{ rad/sec}^2$$

$$t = t' \text{ sec} \quad \text{(acceleration time)}$$

Hence,

$$\omega_m = 0 + \left(\frac{1}{60}\right) t'$$

$$= \frac{t'}{60} \text{ rad/sec} \tag{2.18}$$

Similarly, for the deceleration portion of the curve, consider triangular segment amf and find t' in terms of T, using the relationship

$$\omega_2 = \omega_1 + \alpha t$$

where

$$\omega_2 = \omega_f = 0 \text{ rad/sec}$$

$$\omega_1 = \omega_m = \frac{t'}{60} \text{ rad/sec}$$

$$\alpha = \alpha_D = \frac{-1}{30} \text{ rad/sec}^2$$

$$t = (T - t') \text{ sec} \quad \text{(deceleration time)}$$

Hence,

$$0 = \frac{t'}{60} + \frac{-1}{30} (T - t')$$

$$= \frac{t'}{60} - \frac{1}{30} (T - t')$$

from which

$$T - t' = \frac{t'}{2}$$

$$T = \frac{t'}{2} + t'$$

$$= \frac{3}{2} t'$$

$$t' = \frac{2}{3} T \qquad\qquad\qquad (2.19)$$

Using the area relationship for a triangle to define angular displacement, we note that

$$\theta = \frac{1}{2} \omega_m T$$

or

$$94.24 = \frac{1}{2} \omega_m T$$

After substituting for ω_m and T from Eqs. (2.18) and (2.19), we obtain

$$94.24 = \frac{1}{2} \left(\frac{1}{60} \right) \left(\frac{2}{3} T \right) T$$

from which

$$T^2 = 94.24 \left(\frac{2}{1} \right) \left(\frac{60}{1} \right) \left(\frac{3}{2} \right)$$

$$= 16{,}963.2 \ \text{sec}^2$$

$$T = 130.2 \ \text{sec or 2 min, 10.2 sec}$$

EXAMPLE 2.7

A car, traveling between two stoplights 4 miles apart, does the distance in 10 min. During the first minute, the car moves at a constant acceleration, and during the last 40 sec, it comes to rest with uniform deceleration. For the remainder of the journey the car moves at a uniform speed. Find

a. The uniform speed
b. The acceleration
c. The deceleration
d. The distances covered during uniform velocity, acceleration, and deceleration

SOLUTION

Construct a velocity-time curve (Figure 2.7) to describe the motion of the car. Given

$$v_a = 0 \quad \text{(at rest)}$$

$$T = 600 \text{ sec}$$

$$t_1 = 60 \text{ sec}$$

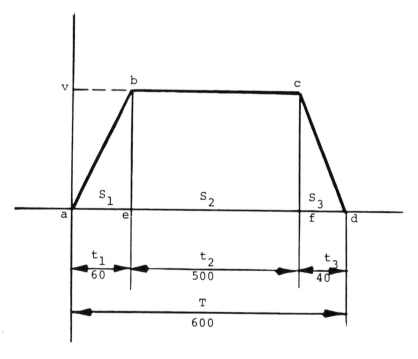

Figure 2.7   Example problem.

$$t_2 = 600 - 60 - 40 = 500 \text{ sec}$$

$$S_T = 4(5280) \text{ ft} \quad \text{(total area under curve)}$$

Let the uniform velocity be v. Then the total distance covered S_T may be computed from

$$S_T = \text{area of trapezium abcd}$$

$$= \frac{1}{2} (bc + ad)v$$

$$4(5280) = \frac{1}{2} (500 + 600)v$$

$$v = \frac{4(5280)(2)}{1100}$$

$$= 38.4 \text{ ft/sec}$$

Acceleration

$$v_2 = v_1 + at$$

where

$$v_1 = v_a = 0$$

$$v_2 = 38.4 \text{ ft/sec}$$

$$t = t_1 = 60 \text{ sec}$$

$$38.4 = 0 + a(60)$$

and

$$a = \frac{38.4}{60}$$

$$= 0.64 \text{ ft/sec}^2$$

Deceleration

$$v_2 = v_1 + at$$

where

$$v_2 = 0$$

$$v_1 = v_f = 38.4 \text{ ft/sec}$$

$$t = t_3 = 40 \text{ sec}$$

$$0 = 38.4 + a(40)$$

and

$$a = -\frac{38.4}{40}$$

$$= -0.96 \text{ ft/sec}^2$$

Note: Negative sign denotes deceleration.

Distance covered during acceleration

$$s_1 = \text{area of triangle abe}$$

$$= \frac{1}{2} vt_1$$

$$= \frac{1}{2} (38.4)(60)$$

$$= 1152 \text{ ft}$$

Distance covered during uniform velocity

$$S_2 = \text{area of rectangle ebcf}$$

$$= vt_2$$

$$= 38.4(500)$$

$$= 19,200 \text{ ft}$$

Distance covered during deceleration

$$S_3 = \text{area of triangle fcd}$$

$$= \frac{1}{2} vt_3$$

$$= \frac{1}{2} (38.4)(40)$$

$$= 768 \text{ ft}$$

2.5   SUMMARY OF MOTION FORMULAS

Linear and Angular Relationships

	Linear		Angular	
	Symbol	Units	Symbol	Units
Displacement	s	ft	θ	rad
Initial velocity	v_1	ft/sec	ω_1	rad/sec
Final velocity	v_2	ft/sec	ω_2	rad/sec
Average velocity	$\overline{v}$	ft/sec	$\overline{\omega}$	rad/sec
Acceleration	a	ft/sec^2	α	rad/sec^2
Time	t	sec	t	sec

$$v_2 = v_1 + at \qquad\qquad \omega_2 = \omega_1 + \alpha t$$

$$v_2^2 = v_1^2 + 2as \qquad\qquad \omega_2^2 = \omega_1^2 + 2\alpha\theta$$

$$\overline{v} = \frac{v_1 + v_2}{2} \qquad\qquad \overline{\omega} = \frac{\omega_1 + \omega_2}{2}$$

$$s = \frac{v_1 + v_2}{2}\, t \qquad\qquad \theta = \frac{\omega_1 + \omega_2}{2}\, t$$

$$s = v_1 t + \frac{1}{2}\, at^2 \qquad\qquad \theta = \omega_1 t + \frac{1}{2}\, \alpha t^2$$

Conversion from Angular to Linear

Displacement: $s = r \times \theta$ or $(r\theta)$

Velocity: $v = r \times \omega$ or $(r\omega)$

Acceleration: $a = r \times \alpha$ or $(r\alpha)$

3
Vectors

3.1 PROPERTIES OF VECTORS

In mechanics, quantities are classified as either vectors or scalars. A *vector* has been defined as a quantity that has magnitude and direction. Examples of vector quantities are displacement, velocity, acceleration, and force.

A quantity that has magnitude but no direction is called a *scalar*. Examples of scalar quantities are time, volume, area, speed, and distance.

Graphically, a vector is represented by an arrow, as in Figure 3.1, where the length, usually drawn to scale, represents the magnitude and the arrowhead indicates the direction. The arrowhead is commonly called the *head or terminus* of the vector, whereas the opposite end is called the *tail or origin*. The direction is usually specified as the angle in degrees that the arrow makes usually with some known reference line. For example, in Figure 3.1(a) the vector $\overline{V}$ represents a velocity having a magnitude of five units in the direction of θ and in Figure 3.1(b) the vector $-\overline{V}$ represents a velocity of the same magnitude but in opposite direction. Thus, to convert a vector from positive to negative, we reverse its direction.

In this text, vector quantities are normally denoted by capital letters with bars above (e.g., $\overline{V}$ and $\overline{A}$) to distinguish them from their scalar counterparts (e.g., V and A), denoted by capital letters without bars. With this notation, the normal addition and subtraction signs, $+$

(a)

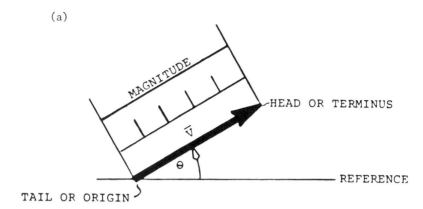

(b)

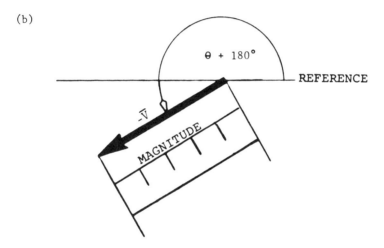

Figure 3.1 Properties of a vector: (a) Vector $\overline{V}$; (b) vector $-\overline{V}$.

and $-$, can be used without risk of confusion between vector and scalar operations. Lowercase letters (e.g., v and a) are also used in some instances to denote scalar quantities, particularly when vectors are not involved.

3.2 VECTOR ADDITION

There are two methods for adding two vectors.

1. The **triangular method**, in which the vectors are connected head to tail and the **resultant** is determined by a third vector, which extends from the tail to the head of the connected vectors.

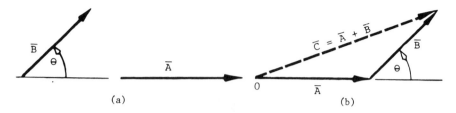

Figure 3.2   Triangular method.

For example, consider the vectors $\overline{A}$ and $\overline{B}$ in Figure 3.2(a). To determine the sum or resultant of these vectors, start at some point O, called the origin, and connect vector $\overline{A}$ to vector $\overline{B}$ as shown in Figure 3.2(b). Then, from the same origin or the tail of vector $\overline{A}$, draw a third vector $\overline{C}$ to close the triangle at the arrowhead of vector $\overline{B}$. Vector $\overline{C}$ is therefore the required sum or resultant of vectors $\overline{A}$ and $\overline{B}$.

Note that the resultant vector always tends to oppose the general sense of the summed vectors. We could think of it as a "counterbalancing vector," since it appears to have a counterbalancing effect on the loop, which in this case is a triangle.

Note also that the resultant is the same for $\overline{B} + \overline{A}$ as for $\overline{A} + \overline{B}$. This means that vector addition is **commutative**.

2. The **parallelogram method**, in which the vectors are connected tail to tail so that they form two adjacent sides of a parallelogram. The **resultant** is found by drawing a third vector extending from the connected point to form a diagonal of the parallelogram.

For example, consider the same vectors $\overline{A}$ and $\overline{B}$ in Figure 3.2(a). To determine the sum or resultant of these vectors, connect the vectors $\overline{A}$ and $\overline{B}$ tail to tail as shown in Figure 3.3. Define this point of connection as O (origin). Using these vectors as adjacent sides, complete

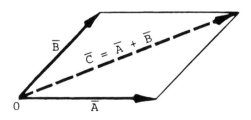

Figure 3.3   Parallelogram method.

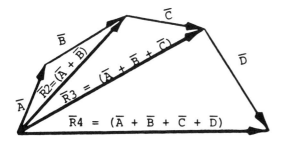

Figure 3.4   Triangular method.

the parallelogram as shown. Then from the point of origin O, draw a third vector $\overline{C}$, also originating at O, to form a diagonal of the parallelogram. This vector $\overline{C}$ is the required resultant of the summed vectors.

Both the parallelogram and triangular methods are also applicable to vector addition involving more than two vector quantities. For example, consider the summation and vectors $\overline{A}$, $\overline{B}$, $\overline{C}$, and $\overline{D}$ shown in Figure 3.4. From the triangular method, we note that

$$\overline{R}_2 = \overline{A} + \overline{B}$$

$$\overline{R}_3 = (\overline{A} + \overline{B}) + \overline{C}$$

$$\overline{R}_4 = (\overline{A} + \overline{B} + \overline{C}) + \overline{D}$$

$$= \overline{A} + \overline{B} + \overline{C} + \overline{D}$$

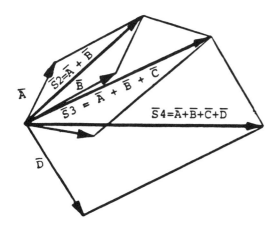

Figure 3.5   Parallelogram method.

Also, in Figure 3.5, we note, from the parallelogram method, that

$$S_2 = \overline{A} + \overline{B}$$

$$S_3 = (\overline{A} + \overline{B}) + \overline{C}$$

$$S_4 = (\overline{A} + \overline{B} + \overline{C}) + \overline{D}$$

$$= \overline{A} + \overline{B} + \overline{C} + \overline{D}$$

3.3 VECTOR SUBTRACTION

Triangular Method

To subtract one vector from another, we simply reverse the direction of that vector and sum both vectors normally. For example, consider vectors $\overline{A}$ and $\overline{B}$ in Figure 3.6(a). To determine the resultant of $\overline{A}$ − $\overline{B}$, we reverse the direction of vector $\overline{B}$, which in effect changes the vector from $+\overline{B}$ and $-\overline{B}$. With this change made, we now add both vectors by placing the tail of $-\overline{B}$ at the head of $\overline{A}$ and connecting the tail of $\overline{A}$ and head of $\overline{B}$ to obtain vector $\overline{A}$ − $\overline{B}$, the resultant [see Figure 3.6(b)]. In equation form, vector subtraction can be expressed as

$$\overline{A} - \overline{B} = \overline{A} + (-\overline{B})$$

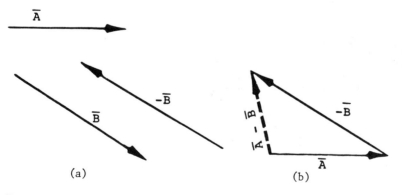

Figure 3.6   Vector subtraction: triangular method.

Note that vector subtraction is just a specialized case of vector addition, in which the subtracted vector is reversed in direction and treated as a positive vector.

Parallelogram Method

An alternative method of vector subtraction is to join both vectors tail to tail, then draw a third vector to connect the termini of the given vectors. This third vector, **when properly directed**, represents the difference of the two vectors. The direction of the third vector is obtained by directing the arrow toward the vector from which the subtraction is made.

As an example, consider again the vectors $\overline{A}$ and $\overline{B}$ in Figure 3.7(a). To find $\overline{A} - \overline{B}$, we join the tail of $\overline{A}$ to that of $\overline{B}$, and the magnitude of vector $\overline{A} - \overline{B}$ is given by a line connecting the terminus of $\overline{A}$ to that of $\overline{B}$ [see Figure 3.7(b)]. The direction is given by directing the vector $\overline{A} - \overline{B}$ from the terminus of $\overline{B}$ to the terminus of $\overline{A}$. In other words, the vector $\overline{A} - \overline{B}$ is directed from $\overline{B}$ to $\overline{A}$.

Note that this method, in effect, is a variation of the parallelogram method used in vector addition. In that case, the difference is obtained by completing the "other" diagonal that joins the termini of the given vectors.

EXAMPLE 3.1

Given vectors $\overline{A}$, $\overline{B}$, and $\overline{C}$ in Figure 3.8, in which magnitudes and directions are as shown, determine the following:

a. $\overline{A} + \overline{B} - \overline{C}$
b. $\overline{C} - \overline{B} + \overline{A}$

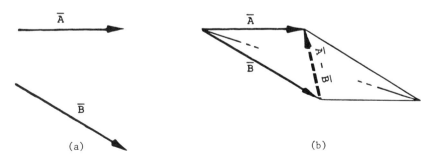

Figure 3.7   Vector subtraction: parallelogram method.

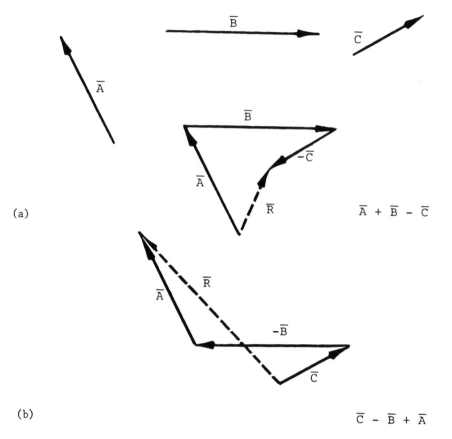

Figure 3.8   Determining resultants of vector systems.

SOLUTION

See construction given in Figures 3.8(a) and (b).

3.4   THE VECTOR POLYGON

The **vector polygon** is the configuration that results from addition or subtraction of more than two vectors graphically. The polygon can be considered to be a closed loop consisting of the vectors that are added or subtracted and the resultant vectors. Each vector polygon can therefore be represented by an algebraic expression in terms of the vector

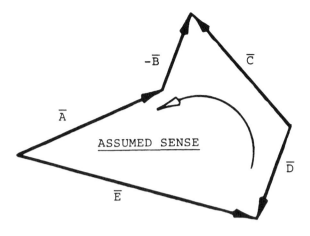

$$\overline{E} = \overline{D} - \overline{C} + (-\overline{B}) + \overline{A}$$

Figure 3.9   Vector polygon.

components and their resultant. For example, consider the vector polygon shown in Figure 3.9. Let it be required to write an algebraic expression for vector $\overline{E}$.

For convenience, it may be assumed that all vectors having one sense (clockwise or counterclockwise) with respect to the closing of the loop are positive, and vectors having the opposite sense are negative. Then it is easy to see that the sense of vector $\overline{E}$ opposes that of vectors $\overline{D}$, $\overline{B}$, and $\overline{A}$, whereas it is the same as that of vector $\overline{C}$. Therefore, from the rules of vector addition and subtraction discussed, we can immediately write the equation

$$\overline{E} - \overline{D} + \overline{C} - (-\overline{B}) - \overline{A} = 0 \qquad (3.1)$$

from which

$$\overline{E} - \overline{D} + \overline{C} + \overline{B} - \overline{A} = 0$$

or

$$\overline{E} = \overline{D} - \overline{C} - \overline{B} + \overline{A} \qquad (3.2)$$

Similarly, the expressions for vectors $\overline{A}$, $\overline{B}$, $\overline{C}$, and $\overline{D}$ can be derived from first principles, as follows:

$$-\overline{A} - (-\overline{B}) + \overline{C} + (-\overline{D}) + \overline{E} = 0 \tag{3.3}$$

$$\overline{A} = B + \overline{C} - \overline{D} + \overline{E} \tag{3.4}$$

$$-(-\overline{B}) + \overline{C} - \overline{D} + \overline{E} - \overline{A} = 0 \tag{3.5}$$

$$\overline{B} = \overline{D} - \overline{E} + \overline{A} - \overline{C} \tag{3.6}$$

$$\overline{C} - \overline{D} + \overline{E} - \overline{A} - (-\overline{B}) = 0 \tag{3.7}$$

$$\overline{C} = \overline{D} - \overline{E} + \overline{A} - \overline{B} \tag{3.8}$$

$$-\overline{D} + \overline{C} - (-\overline{B}) - \overline{A} + \overline{E} = 0 \tag{3.9}$$

$$\overline{D} = \overline{C} + \overline{B} - \overline{A} + \overline{E} \tag{3.10}$$

Alternatively, the expressions for $\overline{A}, \overline{B}, \overline{C}$, and $\overline{D}$ can be obtained directly from Eq. (3.2).

3.5   VECTOR RESOLUTION

Recalling the rule on summation of vector quantities, we saw that a number of vectors added together was equivalent to a single vector called a **resultant**. Or, stated another way, we can say that the resultant vector is a summation of a number of component vectors. When a vector is represented as a summation of other vectors, that vector is said to be resolved, and the vectors being summed are the **components** of the resolved vector. The vectors $\overline{A}_1$, $\overline{A}_2$, and $\overline{A}_3$ in Figure 3.10, for example, are components of resolved vector $\overline{A}$.

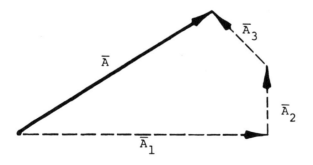

Figure 3.10   Vector components.

Although the components of a vector can be limitless, it is normally more useful to resolve a vector into just two components along specific axes. In such a case, it is useful to recall the two methods of vector addition—the triangular method and the parallelogram method—and note that the process of vector resolution is simply a reversal of the addition process.

Suppose that we wish to resolve vector $\overline{A}$ in Figure 3.11(a) so that its two components, $\overline{C}$ and $\overline{B}$, have the orientation shown by the dashed lines b-b and c-c. Either of the following methods can be employed:

1. ***Triangular method*** [Figure 3.11(b)]: Through the origin and terminus of vector $\overline{A}$, draw lines parallel to b-b and c-c to intersect at some point. This point, according to the triangular method, defines the terminus of vector $\overline{C}$, which extends from the tail of vector $\overline{A}$ to the tail of vector $\overline{B}$, which in turn extends to the terminus of vector $\overline{A}$.
2. ***Parallelogram method*** [Figure 3.11(c)]: An alternative resolution approach is to draw through the tail of vector $\overline{A}$ two axes, each parallel to lines b-b and c-c, then through the terminus or head of $\overline{A}$ draw a line parallel to axis c-c intersect b-b and similarly, another line parallel to axis b-b through the terminus to intersect axis c-c. These points of intersection on axes b-b and c-c define, respectively, the termini of vector components $\overline{B}$ and $\overline{C}$.

3.6 ORTHOGONAL COMPONENTS

A specialized, but commonly encountered case is the resolution of a vector into orthogonal or rectangular components along mutually perpendicular axes. For example, given orthogonal axes x-x and y-y and vector $\overline{V}$ in Figure 3.12, suppose that we wish to determine the components of $\overline{V}$ along x-x and y-y.

Using the parallelogram method discussed above, we can readily determine these components to be $\overline{V}_x$ along axis x-x, and $\overline{V}_y$ along axis y-y, by dropping perpendicular lines from the terminus of vector $\overline{V}$ to axes x-x and y-y, respectively. Here it is useful to note, from trigonometry, that the magnitudes of component vectors $\overline{V}_x$ and $\overline{V}_y$ are

$$V_x = V \cos \theta$$

and

$$V_y = V \sin \theta$$

where θ is the angle that vector $\overline{V}$ makes with the x axis.

(a)

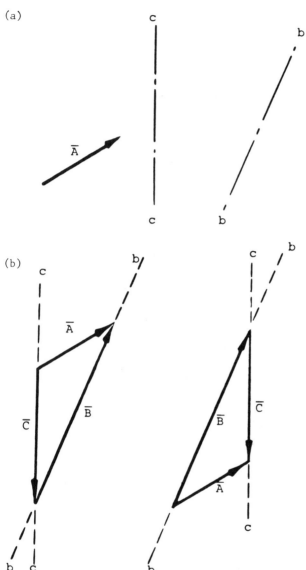

(b)

Figure 3.11   (a) and (b) Vector resolution: Triangular method.

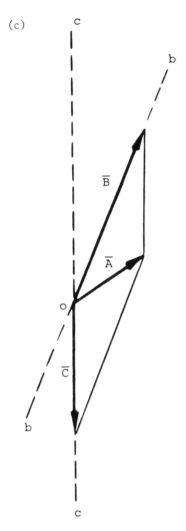

Figure 3.11   (c) Vector resolution: Parallelogram method.

3.7 TRANSLATIONAL AND ROTATIONAL COMPONENTS

Orthogonal components, when applied to link motion, may be described in terms of translational and **rotational components**, whereby

The **translational component** is defined as that component of the vector which tends to cause translation of the link along its own axis.

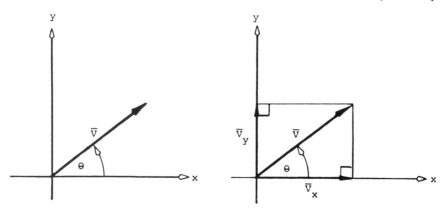

Figure 3.12   Orthogonal components of a vector.

The **rotational component** is defined as that component of the vector
   which tends to cause rotation of the link about some center located
   on the link axis.

Consider the vector $\overline{V}$ depicted in Figure 3.13(a) and (b). Observe that
the translational component $(\overline{V}^t)$ in Figure 3.13(b) is equivalent to the
x component $(\overline{V}_x)$ in Figure 3.13(a), or

$$\overline{V}_x = \overline{V}^t \quad \text{translational component}$$

and the rotational component $(\overline{V}^r)$ [Figure 3.13(b)] is equivalent to the
y component $(\overline{V}_y)$ in Figure 3.13(a), or

$$\overline{V}_y = \overline{V}^r \quad \text{rotational component}$$

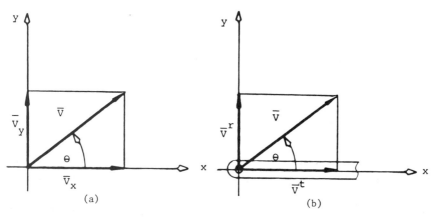

Figure 3.13   Translational and rotational components of a vector.

3.8 EFFECTIVE COMPONENTS

The ***effective component*** of a vector may be defined as the projection of that vector along the axis where the effect is to be measured. To determine the effective component of any vector, we simply drop a perpendicular line from the terminus of the vector to a line drawn through the origin of the vector along which the effect is to be measured. The point at which these two lines meet defines the terminus of the required effective component. For example, if we consider again the vector $\overline{V}$ shown in Figure 3.13, $\overline{V}_x$ is the effective component of this vector along the x axis. In other words, the effective component of vector $\overline{V}$ along axis x-x is the value $\overline{V}_x$. Similarly, if we consider the same vector with respect to another axis, y-y, $\overline{V}_y$ is the effective component of $\overline{V}$ along that axis.

   In general, if we consider vector $\overline{V}_P$ oriented with respect to axes a-a and b-b as shown in Figure 3.14, then by dropping perpendiculars from the terminus of this vector to these axes, we obtain, respectively,

$\overline{V}_P^{aa}$ effective component along a-a
$\overline{V}_P^{bb}$ effective component along b-b

where the superscripts aa and bb are used to indicate the axes along which the effective components are considered.

   From this example, the following observations can be made:

1. A perpendicular drawn from the terminus of an effective component of a vector to intersect the line of action of that vector will define the magnitude of that vector. Therefore, if one effective component and direction of a vector are known, the vector can readily be determined.

2. The line drawn from the terminus of the absolute vector must be at a right angle to the effective component, not to the absolute vector. This is because the latter construction will yield an effective component that is "greater" in magnitude than the absolute vector, which is impossible. ***No component can be greater than its whole***.

3. A vector can have an effective component in any conceivable direction, except that which is perpendicular to the vector itself. In this context, it might be well to note that the effective component in the direction of an absolute vector is the vector itself.

4. The orthogonal components of the vector are merely a specialized set of effective components of that vector along mutually perpendicular axes. Compare Figures 3.12 and 3.13.

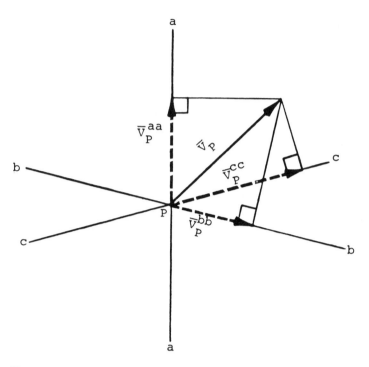

Figure 3.14   Effective component of a vector.

5.   If two effective components of a vector are known, that vector can
     be completely determined by constructing perpendiculars from the
     terminus of each effective component vector such that they intersect
     each other. This point of intersection defines the terminus or mag-
     nitude of the required vector. For example, in Figure 3.14, the
     perpendiculars drawn from the termini of $\overline{V}_P^{aa}$ **and** $\overline{V}_P^{bb}$ **define the**
     terminus $\overline{V}_P$. Note that this is precisely the reverse of the procedure
     that is followed in finding the two effective components $\overline{V}_P^{bb}$ and
     $\overline{V}_P^{aa}$ when the absolute vector $\overline{V}_P$ is known.

II

GRAPHICAL TECHNIQUES

Graphical techniques offer a convenient way to solve velocity and acceleration problems relating to mechanism. Compared to analytic techniques, they are simpler, faster, and more easily understood. Accuracy, however, depends on precision of **line work**, **measurements**, and a wise **choice of scales** for the vectors.

In this section a variety of graphical methods are presented for analyzing mechanisms to determine the velocities and accelerations of their members. These include such methods as instant centers, effective component of velocity and acceleration, relative velocity and acceleration, graphical differentiation and integration, and some special constructions.

As one would expect, each method has advantages and disadvantages. No single method is suitable for solving all problems, even though some are more versatile than others. Generally, the most suitable method is one that yields the required information in the shortest time with the least amount of effort. The objective here is to illustrate the variety of graphical methods used to analyze a mechanism that are available to mechanical design engineers and students—enabling them, it is hoped, to make an appropriate selection when confronted with a practical problem.

II.A

GRAPHICAL TECHNIQUES: VELOCITY ANALYSIS

Velocity is inherently an important factor in dynamic analysis. Since force is proportional to acceleration, which is the rate of change of velocity, velocity analysis becomes a necessary prerequisite to the acceleration and force analysis of a machine member. In high-speed machines, forces generated during impact and sudden changes of velocity can limit the operating speed of the machine. Also, as the operating speed increases, it requires greater and greater forces to make various links move through their intended cycles. Drive torques must be increased correspondingly as speeds are increased. As with impact, this can result in increased deformation and vibration within the machine.

Also, as speeds increase, lubrication and wear become more critical. For example, in the crankshaft bearing of an automobile, wear depends on the speed of the crankshaft and the pressure between the crank pins and bearings. Similarly, the cutting speeds of machine tools and the flow rates of fluids in engines and pumps are all functions of the velocities of the output members. For these reasons, methods of determining relative and absolute velocities are of great importance in making a complete analysis of the motions of parts of a machine.

4
Effective Component of Velocity Method

4.1 INTRODUCTION

The effective component method of velocity is based on two principles. The first is the use of the effective components of a vector, and the second is the rigid-body principle. As discussed earlier, the **effective component** of a vector in any direction is the projection of that vector along a line drawn through the vector origin in the direction of interest. This direction is usually defined for the convenience of applying the rigid-body principle.

4.2 THE RIGID-BODY PRINCIPLE

The **rigid-body principle** may be stated as follows: **In a rigid body, the distance between two points remains constant and the velocity components along a line joining these points must be the same at both points**. This principle is easily explained, in that if the velocity components were different at the two points, the link would change in length and therefore not remain rigid. Thus, if we know the velocity of one point of a rigid body, we can find the velocity of any other point on that body by resolving the known velocity into components along and perpendicular to the line joining the two points and making the velocity component of the unknown velocity equal to that of the known component along the joining line.

4.3   VELOCITIES OF END POINTS ON A LINK

Consider link BC in Figure 4.1, where the velocity of point B is completely known (in magnitude as well as direction). The line of action for the velocity of point C is also known. We want to find $\overline{V}_C$.

   To obtain the velocity at point C, we need the effective component of $\overline{V}_C$ along BC, which has the same magnitude as the effective component of $\overline{V}_B$ along BC. We must first obtain $\overline{V}_B^{BC}$, then $\overline{V}_C^{BC}$, and finally $\overline{V}_C$.

PROCEDURE

1. We determine the effective component of $\overline{V}_B$ along BC, that is, $\overline{V}_B^{BC}$. This is determined by dropping a perpendicular from the terminus of $\overline{V}_B$ to intersect a line joining B and C (line BC).
2. Since the link is a rigid body and therefore all points along BC *must* experience the same velocity as $\overline{V}_B^{BC}$, we can immediately lay out the effective component of $\overline{V}_C$ along BC, that is, $\overline{V}_C^{BC}$.

$$\overline{V}_B^{BC} = \overline{V}_C^{BC}$$

3. With the effective component $\overline{V}_C^{BC}$ completely defined and the direction of $\overline{V}_C$ also known, we simply construct a perpendicular from the terminus of $\overline{V}_C^{BC}$ to intersect the line of action $\overline{V}_C$. This point of intersection defines the magnitude of $\overline{V}_C$.

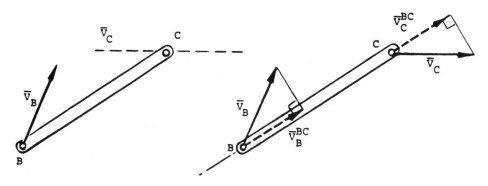

Figure 4.1   Velocities of end points on a link.

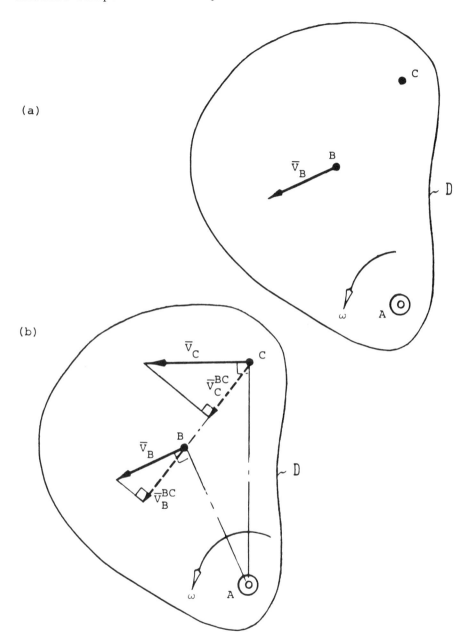

Figure 4.2 Velocities of points on a body with pure rotation.

4.4 VELOCITIES OF POINTS ON A ROTATING BODY

Now consider body D, which rotates about point O (Figure 4.2). The velocity of point B or $\overline{V}_B$ is given as shown, and the velocity of point C ($\overline{V}_C$) is required.

A close look at this problem will suggest that the procedure should be identical to that used in the preceding problem, except that now the direction of $\overline{V}_C$ is not given explicitly. However, as we noted earlier, since the body is rotating, the linear velocity of point C must be tangential to a circular path described by C. Therefore, the direction of $\overline{V}_C$ is perpendicular to the radial line AC.

PROCEDURE

1. Determine the effective component of $\overline{V}_B$ along BC, that is, $\overline{V}_B^{BC}$. Here it should be noted that the effective component is obtained by dropping the perpendicular from the terminus of $\overline{V}_B$ to extension of a line joining B and C (line BC).
2. Lay out the effective component of $\overline{V}_C$ along BC, that is, $\overline{V}_C^{BC}$.

$$\overline{V}_C^{BC} = \overline{V}_B^{BC}$$

(From the rigid-body principle, all points along BC must have the same velocity.)
3. Determine the required velocity $\overline{V}_C$. From the terminus of $\overline{V}_C^{BC}$, project a perpendicular line to intersect the line of action of $\overline{V}_C$. This point of intersection defines the magnitude of $\overline{V}_C$.

4.5 VELOCITY OF ANY POINT ON A LINK

Sometimes it is necessary to determine the velocity of a point on a link other than the two end points. This can easily be obtained as follows.

Point on Centerline of Two Known Velocities

Consider link BC in Figure 4.3, where the velocities of points B and C are known. Find the velocity of point D located on the link.

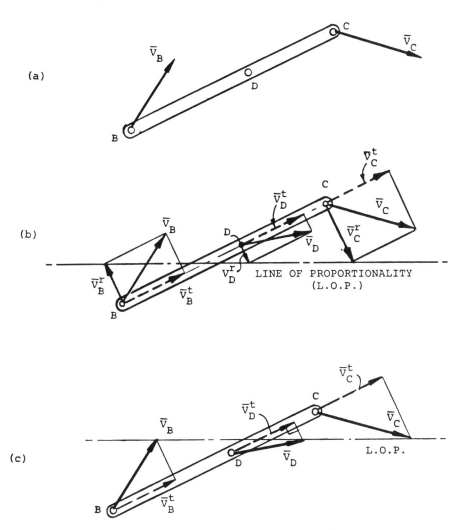

Figure 4.3   Proportionality of velocities.

The velocity of point D $(\overline{V}_D)$ can be obtained from the summation of two components: (1) a translation component $\overline{V}_D^t$, which must be the same for all points along BC (according to the rigid-body principle) and (2) a rotational component $\overline{V}_D^r$, which must be proportional to $\overline{V}_B^r$ or $\overline{V}_C^r$ based on its distance from a center of rotation (according to the rotation principle). In summary,

$$\overline{V}_D = \overline{V}_D^t + \overline{V}_D^r$$

where

$$\overline{V}_D^t = \overline{V}_D^{BC}$$

$$\overline{V}_D^r = \frac{DO}{BO}\,\overline{V}_B^r$$

$$= \frac{DO}{CO}\,\overline{V}_C^r$$

PROCEDURE

1. Resolve the given velocity vectors $\overline{V}_B$ and $\overline{V}_C$ into their orthogonal
    (rotational and translational) components with respect to link BC.
    The rotational components are obtained by dropping perpendicular
    lines from the termini of $\overline{V}_B$ and $\overline{V}_C$ to lines drawn normal to BC
    through points B and C. These components are designated $\overline{V}_B^r$ and
    $\overline{V}_C^r$ [see Figure 4.3(b)]. The translational components are obtained
    by dropping perpendiculars from the termini of $\overline{V}_B$ and $\overline{V}_C$ to line
    BC or BC extended. Note that these components, designated $\overline{V}_B^t$ and
    $\overline{V}_C^t$, are the same as the effective components of $\overline{V}_B$ and $\overline{V}_C$. That
    is,

    $$\overline{V}_B^t = \overline{V}_B^{BC} \quad\text{and}\quad \overline{V}_C^t = \overline{V}_C^{BC}$$

2. Lay out the translation component of $\overline{V}_D$, that is, $\overline{V}_D^t$ or $\overline{V}_D^{BC}$, along
    BC. From the rigid-body principle, all points experience the same
    velocity along a straight line. Therefore,

    $$\overline{V}_D^t \quad\text{or}\quad \overline{V}_D^{BC} = \overline{V}_B^t \quad\text{or}\quad \overline{V}_B^{BC} = \overline{V}_C^t \quad\text{or}\quad \overline{V}_C^{BC}$$

3. Determine the rotational component of $\overline{V}_D$ $(\overline{V}_D^r)$. Because the point
    D lies on the same straight line as B and C, its rotational component
    must be proportional to that of point B as well as point C. Therefore,
    draw a straight line to connect the **terminus** of the rotational com-
    ponent of $\overline{V}_B$ to the **terminus** of $\overline{V}_C$. This line is the **line of pro-
    portionality** for rotational velocity components of all points on BC.
    Therefore, the required rotational component $\overline{V}_D^r$ is obtained by
    constructing a perpendicular from point D to meet the proportion-
    ality line [see Figure 4.3(c)].

4. Determine the velocity $\overline{V}_D$. Having determined both the rotational and translational components of $\overline{V}_D$, we can now obtain the resultant vector by graphically summing both the rotational and translational components of $\overline{V}_D$.

$$\overline{V}_D = \overline{V}_D^t + \overline{V}_D^r$$

Alternate Method

Figure 4.3(c) shows an alternate method for obtaining $\overline{V}_D$. Here, the proportionality line is established first connecting the termini of vectors $\overline{V}_D$ and $\overline{V}_C$. Then by dropping a perpendicular from the terminus of the effective component vector $\overline{V}_D^t$ to meet the proportionality line, the terminus of $\overline{V}_D$ is thus defined.

Point Offset from Centerline of Two Known Velocities

In the preceding section, the point of unknown velocity was located on the centerline connecting two points of known velocity. Let us now consider the expanded link BCD in Figure 4.4 where the velocities of points B and C are given as shown, and the velocity of point D, offset from the centerline BC, is required.

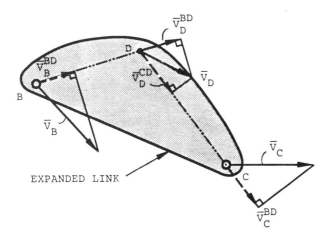

Figure 4.4 Velocity of any point on an expanded link.

To find this velocity, we again apply the rigid-body rule, first between the offset point D and one of the known velocity points, say, point B, to find the effective component of velocity of point D along the line joining B and D. Then, in like manner, find another effective component D by considering this point and the other known velocity point C. With two effective components of the same velocity ($\overline{V}_D$) thus defined, that velocity can then be determined.

PROCEDURE

1. Join points B and D with a straight line.
2. Lay out the effective component of $\overline{V}_B$ along BD (or $\overline{V}_B^{BD}$).
3. Lay out the effective component of $\overline{V}_D$ along BD (or $\overline{V}_D^{BD}$), applying the rigid-body principle that requires

$$\overline{V}_D^{BD} = \overline{V}_B^{BD}$$

Note that since the line of action of $\overline{V}_D$ is not known, this velocity cannot be determined from $\overline{V}_D^{BD}$ alone. To determine $\overline{V}_D$ completely, we need another effective component that can be obtained from $\overline{V}_C$.

4. Join points C and D with a straight line.
5. Lay out the effective components of $\overline{V}_C$ along CD (or $\overline{V}_C^{CD}$).
6. Lay out the effective component of $\overline{V}_D$ along CD, noting that $\overline{V}_D^{CD} = \overline{V}_C^{CD}$.
7. With two effective components of $\overline{V}_D$ known, this vector can now be defined completely by drawing perpendiculars from the termini of $\overline{V}_D^{BD}$ and $\overline{V}_D^{CD}$ until they intersect. The point of intersection will define the terminus of the required vector $\overline{V}_D$.
8. Draw vector $\overline{V}_D$.

4.6 VELOCITY ANALYSIS OF A SIMPLE MECHANISM

Crank AB of the slider-crank mechanism shown in Figure 4.5 rotates clockwise at 0.5 rad/sec. We want to determine the velocity of the slider.

In the preceding sections, the linear velocity of point B in link BC was given, and the direction of the velocity of point C was either known or determinable from the constraints of the link motion. Here, although

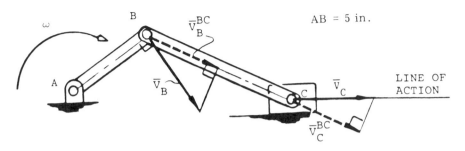

Figure 4.5 Slider-crank mechanism.

$\overline{V}_B$ is not given directly, it is determinable since we know that its magnitude is given by

$$V_B = AB\omega_{AB}$$

and that its direction must be perpendicular to AB (in the same sense as ω_{AB}). Further, the direction of $\overline{V}_C$, although not given explicitly, is obviously horizontal, if we consider that the line of action of the slider of which point C is a part must be along the slot, which is horizontal. Therefore, in principle, the procedure for determining the velocity of point C is basically the same as before.

PROCEDURE

1. Lay out the velocity of point B, that is, $\overline{V}_B$ (direction and magnitude), using a convenient scale. The magnitude of this vector is given as

$$V_B = \omega_{AB} \times AB$$

$$= 0.5(5) = 2.5 \text{ in./sec}$$

2. Determine the effective component of $\overline{V}_B$ along BC. Drop a perpendicular line from the terminus of $\overline{V}_B$ to link BC.

3. Lay out the effective component of V_C along BC.

$$\overline{V}_C^{BC} = \overline{V}_B^{BC}$$

4. Determine $\overline{V}_C$.
   a. Construct a perpendicular line from the terminus of $\overline{V}_C^{BC}$ to intersect the known line of action of $\overline{V}$.

b. Scale the magnitude of $\overline{V}_C$.

$$V_C = 2.3 \text{ in./sec}$$

Hence,

$$\overline{V}_C = 2.3 \text{ in./sec} \quad \text{(directed as shown)}$$

EXAMPLE 4.1

Consider the four-bar linkage in Figure 4.6, where crank AB rotates counterclockwise with an angular velocity of 1 rad/sec, as shown. Let it be required to find the angular velocity of the follower CD.

SOLUTION

To find the angular velocity of the follower, we must first determine the linear velocity of point C. The linear velocity of point C is obtained basically in the same manner as for the slider crank in Figure 4.5. There $\overline{V}_C$ was obtained using the effective component of $\overline{V}_C$ along link BC and the line of action of $\overline{V}_C$, which was known. Here the only difference to

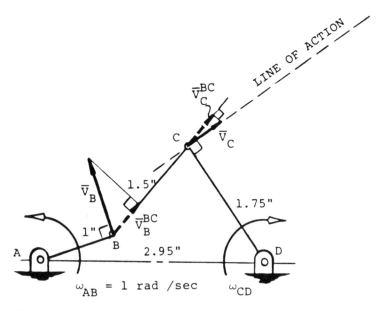

Figure 4.6   Velocity analysis of a four-bar mechanism.

be noted is in the line of action of the velocity of point C, which must be perpendicular to follower arm CD. This is because C can rotate only about pivot D. If we use this line of action of $\overline{V}_C$ and the effective component of $\overline{V}_C$ along BC, $\overline{V}_C$ is easily found as before. We therefore proceed as follows:

1. Lay out the velocity of point B, that is, $\overline{V}_B$ (magnitude and direction), using a convenient scale. The magnitude of this vector is given as

$$V_B = \omega_{AB} \times AB$$

$$= 1(1) = 1 \text{ in./sec}$$

2. Determine the effective component of $\overline{V}_B$ along link BC ($\overline{V}_B^{BC}$). This is obtained by dropping a perpendicular line from the terminus of $\overline{V}_B$ to meet BC extended (at right angles).
3. Locate the effective component of $\overline{V}_C$ along BC. Since from the rigid-body principle, the velocities of all points along BC must be the same, along BC lay out $\overline{V}_C^{BC}$ equal to $\overline{V}_B^{BC}$.
4. Determine the velocity of point C ($\overline{V}_C$). Since point C is constrained to move in a circular path, the line of action of $\overline{V}_C$ is known to be perpendicular to link CD. Therefore, $\overline{V}_C$ is found by dropping a perpendicular line from the terminus of $\overline{V}_C^{BC}$ to intersect the line of action of $\overline{V}_C$. This point of intersection defines the magnitude of $\overline{V}_C$.
5. Scale the magnitude of $\overline{V}_C$.

$$V_C = 0.45 \text{ in./sec}$$

6. Determine the angular velocity of CD (ω_{CD}). This is found from

$$\omega_{CD} = \frac{V_C}{CD}$$

$$= \frac{0.45}{1.75}$$

$$= 0.26 \text{ rad/sec}$$

4.7 VELOCITIES OF SLIDING CONTACT MECHANISMS

An important rule in the analysis of velocities in sliding contact states as follows: *If two bodies are in sliding contact, their velocities perpendicular to the sliding path are equal.* As an example, consider the Scotch yoke mechanism shown in Figure 4.7. Here slider S and yoke Y are members in sliding contact, and the sliding path is T–T. Also, P is a point on slider S as well as crank arm OP; hence, no relative motion exists between these two members at this point. This means that

$$\overline{V}_{P(C)} = \overline{V}_{P(S)}$$

Now, since Y can have only vertical motion, all points in contact with this member, including point P, must have velocities identical to that of Y in the same vertical direction. This means that $\overline{V}_Y$ must be the same as the vertical component of $\overline{V}_P$ on S or V_P on C, or

$$\overline{V}_Y = \overline{V}^{NN}_{P(C)} = \overline{V}^{NN}_{P(S)}$$

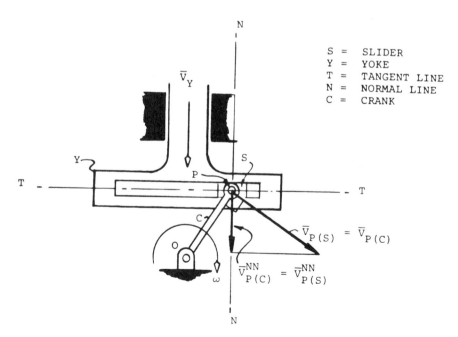

Figure 4.7   Scotch yoke mechanism.

Thus, in accordance with the sliding contact rule, both bodies, slider S and yoke Y, have equal velocities in direction N–N perpendicular to sliding path T–T.

EXAMPLE 4.2

In the quick-return mechanism shown in Figure 4.8, crank OP rotates clockwise at 7 rad/sec, slider S, to which it is attached, slides on whereas follower F. Determine the angular velocity of the follower.

SOLUTION

Like the Scotch yoke just discussed, point P on the slider is the same as point P on the crank arm. However, note that points P on S and P

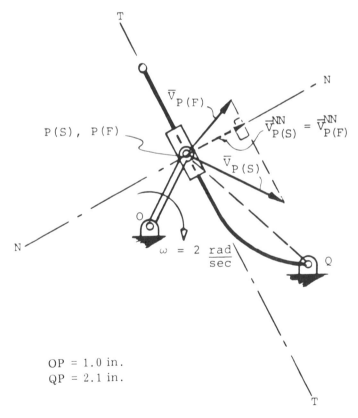

Figure 4.8   Velocity analysis of a quick-return mechanism.

on F, although coincidentally located, do not have identical velocities. In fact, it is for precisely this reason that sliding occurs between the members.

Nevertheless, according to the rule on sliding, these velocities do share a common component in the direction normal to the sliding path. Accordingly, if we were to determine the direction normal to the sliding path and the effective component of the known velocity in this direction, we could use this component to determine the unknown velocity. We therefore proceed as follows:

1. Determine and lay out the velocity of P on S [$\overline{V}_{P(S)}$]. The magnitude of this vector is given as

$$V_{P(S)} = OP\omega$$

$$= 7(1)$$

$$= 7 \text{ in./sec}$$

2. Determine the direction of velocity of P on F [$\overline{V}_{P(F)}$]. Connect P to Q with a straight line. This defines the radius arm with which point P(F) on the follower is rotating at the given instant. Therefore, through point P, draw a line perpendicularly to QP to represent the direction of $\overline{V}_{P(F)}$.
3. Determine the effective component of the velocity $\overline{V}_{P(S)}$ in the direction normal to sliding. The rule states that the velocities or velocity components of $\overline{V}_{P(F)}$ and $\overline{V}_{P(S)}$ in the direction perpendicular to sliding must be equal.
    a. Accordingly, through point P, construct the coordinate axis TT to indicate the path of sliding and another coordinate NN perpendicular to the path of sliding.
    b. Then drop a perpendicular line from $\overline{V}_{P(S)}$ to NN to define the effective component $\overline{V}_{P(S)}^{NN}$. Note that

$$\overline{V}_{P(S)}^{NN} = \overline{V}_{P(F)}^{NN}$$

4. Find the velocity magnitude of P on F [or $V_{P(F)}$]. From the terminus of $\overline{V}_{P(S)}^{NN}$, construct a perpendicular to intersect the line of action of $\overline{V}_{P(F)}$, determined in Step 2. This point of intersection defines the magnitude of the velocity $\overline{V}_{P(F)}$.

$$V_{P(F)} = 4.4 \text{ in./sec} \quad \text{(scaled)}$$

5. Find the angular velocity of the follower or ω_{QP}:

$$\omega_{QP} = \frac{V_{P(F)}}{QP} = \frac{4.4}{2.1}$$

$$\omega_{QP} = 2.1 \text{ rad/sec}$$

Note that the **velocity of sliding** is given by the vectorial difference between $\overline{V}_{P(S)}$ and $\overline{V}_{P(F)}$, or the scaled distance between the termini of these two vectors.

4.8 VELOCITY ANALYSIS OF A COMPOUND MECHANISM

Consider the mechanism in Figure 4.9, where wheel W turns clockwise at ω rad/sec. We want to determine the velocity of pin C.

Earlier, we saw that if the direction of a velocity and its effective component were known, the magnitude of that velocity could be readily determined. Alternatively, it has been shown (see Chapter 3) that if two effective components of a vector are known, that vector can be determined completely.

In this case, since the direction of $\overline{V}_C$ is not known nor is the path of C readily determined, it is necessary to determine $\overline{V}_C$ using two effective components of this vector: (1) the effective component along link BC ($\overline{V}_C^{BC}$) and (2) the effective component along link CD ($\overline{V}_C^{CD}$). Hence, we use the velocity $\overline{V}_B$ to obtain $\overline{V}_C^{BC}$ and the velocity $\overline{V}_E$ to obtain $\overline{V}_C^{CD}$ via point D.

PROCEDURE

1. Starting at point B, lay out the velocity $\overline{V}_B$, whose magnitude is obtained as

$$V_B = \omega AB$$

2. Find the effective component $\overline{V}_B^{BC}$ along BC.
3. Locate the effective component $\overline{V}_C^{BC}$ on BC.

$$\overline{V}_C^{BC} = \overline{V}_B^{BC}$$

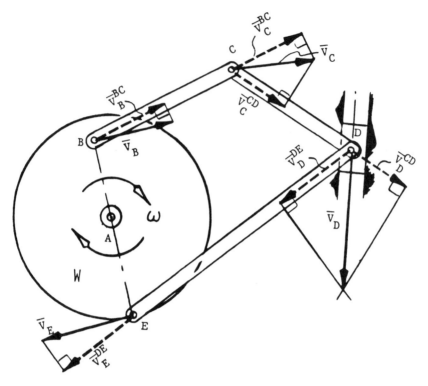

Figure 4.9   Velocities of points on a compound mechanism.

Since the motion of C is not known, $\overline{V}_C$ cannot be determined directly from $\overline{V}_C^{BC}$. However, by returning to the wheel and stepping off from point E in the opposite direction, we can obtain the additional information needed to define $\overline{V}_C$.

4. Lay out the velocity $\overline{V}_E$.
5. Find the effective component $\overline{V}_E^{DE}$ along DE.
6. Locate the effective component $\overline{V}_D^{DE}$ along DE.

$$\overline{V}_D^{DE} = \overline{V}_E^{DE}$$

7. Find the velocity of slider V_D. Since the motion of slider D is known, $\overline{V}_D$ is determined directly from $\overline{V}_D^{DE}$.
8. Using $\overline{V}_D$, find the effective component of this vector along link CD, that is $\overline{V}_D^{CD}$.
9. Locate the effective component $\overline{V}_C^{CD}$ along CD.

$$\overline{V}_C^{CD} = \overline{V}_D^{CD}$$

10. Now that we have determined two effective components for the velocity at pin C, $\overline{V}_C^{BC}$ and $\overline{V}_C^{CD}$, the absolute velocity $\overline{V}_C$ is determined by projecting perpendicular lines from the terminus of these vectors until they intersect. This point of intersection defines the magnitude of $\overline{V}_C$.

4.9 SUMMARY

The effective component method is particularly suitable for analyzing velocities of sliding members and is very useful when the instant centers of a mechanism are outside the limits of the drawing paper. However, a disadvantage of this method is the need for the analysis of velocities of points from link to link. For complex mechanisms, this could result in considerable drawing time and some loss of accuracy. Another disadvantage of the method is that it does not provide relative velocities of points on the mechanism, which are essential to the acceleration analysis.

5

Instant Center Method

5.1 INTRODUCTION

One of the most effective techniques for analyzing velocities of members or links in a mechanism is the method of instant centers. In simple terms, the **instant center** (or instantaneous center, as it is sometimes called) has been defined as that point about which a body may be considered to be rotating relative to another body at a given instant.

Applying this concept to a moving link of a mechanism makes it convenient to describe its motion, at any given instant, in terms of pure rotation about an instant center. In this context, it may be noted that even a link that undergoes translation may be considered as rotating about an instant center located at infinity. In other words, any straight line may be considered to be an arc of a circle of infinite radius.

The ability to describe any motion in terms of pure rotation greatly simplifies the analysis of a complex mechanism by making it more convenient to determine the velocity of any point on such a mechanism. Consequently, it is clear that the key to successful application of the method of instant centers must depend on one's ability first, to locate all possible instant centers of a mechanism, and second, to use these centers effectively to determine the required velocities.

Generally, for a simple mechanism that consists of four links, this analysis presents little or no problem. However, for a complex mechanism with more than four links, experience has shown that locating all of the instant centers from first principles can be a painstaking exercise. Moreover, once the instant centers have been found and documented, the resulting diagram is often so complex that it does not allow straightforward analysis of velocities.

In this chapter we show how the velocity analysis by instant centers can be simplified by employing graphical aids such as **circle diagrams** and **link extensions**. Circle diagrams are used to help locate instant centers that cannot be found easily by inspection, whereas link extensions aid in the visualization of relatinships between the links of a mechanism.

5.2 PURE ROTATION OF A RIGID BODY

To comprehend the concept of instant centers more fully, it is necessary first to consider some basic principles regarding pure rotation of a rigid body. We have already established the fact that if a body has rotary motion, that motion can be converted to rectilinear motion using the radius as a multiplying factor. Also, the linear velocity of any point on that body acts in a direction tangential to the path of rotation and has the same sense as the rotation. For example, consider point B on the rotating body in Figure 5.1(a). If A is fixed, the linear velocity of B is given by the relationship

$$\overline{V}_B = AB\omega \quad \text{(directed perpendicular to } AB) \tag{5.1}$$

where

$$AB = \text{radius of } B$$

$$\omega = \text{angular velocity}$$

Therefore, the angular velocity for the same body is given by

$$\omega = \frac{V_B}{AB} \tag{5.2}$$

Similarly,

$$\overline{V}_C = AC\omega \quad \text{(directed perpendicular to } AC) \tag{5.3}$$

and

$$\omega = \frac{V_C}{AC} \tag{5.4}$$

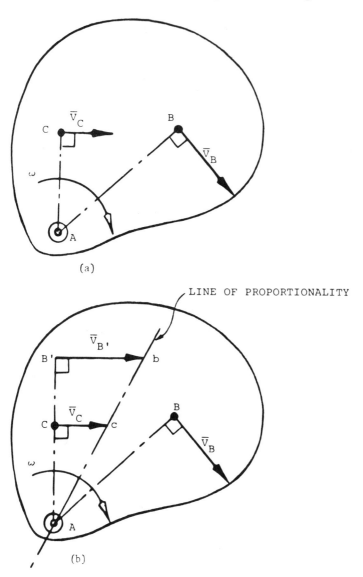

Figure 5.1   Velocities of points on a body in pure rotation.

From Eqs. (5.2) and (5.4),

$$\frac{V_B}{AB} = \frac{V_C}{AC} \tag{5.5}$$

or

$$\frac{V_B}{V_C} = \frac{AB}{AC} = \frac{r_B}{r_C} \tag{5.6}$$

which means that $\overline{V}_B$ is proportional to $\overline{V}_C$, as r_B is to r_C. The velocities $\overline{V}_B$ and $\overline{V}_C$ are represented vectorially in Figure 5.1(b).

The proportionality of the linear velocities $\overline{V}_B$ and $\overline{V}_C$ to their respective distances from the center of rotation is easily verified graphically [see Figure 5.1(b)] by rotating the velocity vector $\overline{V}_B$ about center A from point B to a point B' on the radial line AC and constructing a straight line through the center A to touch the termini of $\overline{V}_C$ and $\overline{V}_{B'}$ ($\overline{V}_B$ relocated). This straight line is normally called the **line of proportionality** between triangles $AB'b$ and ACc, where

$$\frac{B'b}{Cc} = \frac{AB'}{AC}$$

or

$$\frac{V_{B'}}{V_C} = \frac{AB'}{AC}$$

or

$$\frac{V_B}{V_C} = \frac{AB}{AC} \quad \text{(as before)}$$

EXAMPLE 5.1

Consider the 24-in.-diameter rotating disk shown in Figure 5.2, where A is the axis of rotation. B and C are points located on the radial lines AB and AC, as shown. If the linear velocity $\overline{V}_B$ is 2 ft/sec, as shown, determine the angular velocity of the disk and linear velocity of point C. Also, determine the distance of any point E on the disk where the linear velocity of $\overline{V}_E$ is 1.3 ft/sec. Locate velocity vectors $\overline{V}_B$, $\overline{V}_C$, and $\overline{V}_E$.

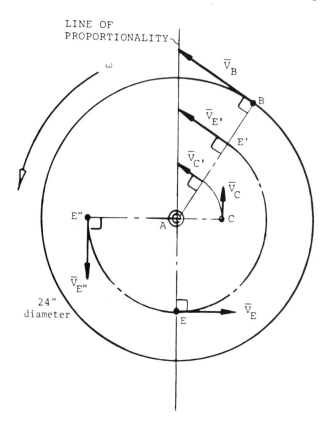

Figure 5.2   Rotating disk.

SOLUTION

The angular velocity is obtained from

$$V_B = AB\omega$$

$$\omega = \frac{V_B}{AB}$$

$$= \frac{2}{12/12}$$

$$= 2 \text{ rad/sec}$$

The linear velocity of point C is obtained from

$$V_C = AC\omega$$

$$= \frac{4}{12}(2) = 0.66 \text{ ft/sec}$$

The radius of E (or AE) is obtained from

$$V_E = AE\omega$$

$$AE = \frac{V_E}{\omega}$$

$$= \frac{1.3}{2} = 0.66 \text{ ft} = 8 \text{ in.}$$

Figure 5.2 shows the location of point E and linear velocities $\overline{V}_B$, $\overline{V}_C$, and $\overline{V}_E$. Note that all other points (e.g., E' and E'') on the same path described by point E have the same linear velocity.

In summary, then, it is useful to note the following principles:

1. The linear velocity is always tangential to the path of rotation of the point or perpendicular to the radial line joining the point to the center of rotation.
2. The direction of the velocity is always in the same general sense as that of rotation.
3. The magnitude of the velocity is always proportional to the distance of the point from the center of rotation.

Consequently, if the velocity of one point of a rotating body is known, the velocity of any other point on that body can be determined, provided that the center of rotation is known.

To illustrate these principles further, if the directions of velocities and the two points are given, the center of rotation lies at the intersection of lines drawn from the two points perpendicular to the velocity directions, as shown in Figure 5.3(a). However, if these directions are the same, the location of the center of rotation also depends on the relative magnitudes of the velocities. For example:

If the velocities are unequal, the center of rotation lies at the intersection of the common perpendicular drawn from the tails of the two vectors, with a line joining the termini of the same vectors. See the construction in Figure 5.3(b)

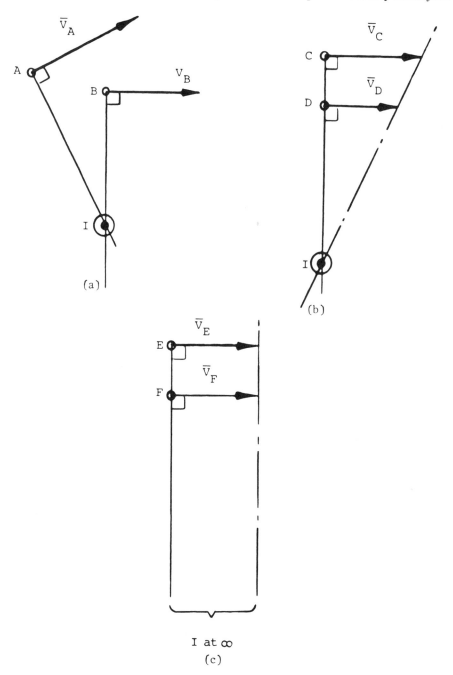

Figure 5.3   Locating the center of rotation.

If the velocities are equal, the center of the rotation lies at infinity, which means that the body has no rotation and is therefore translating. See the construction in Figure 5.3(c).

5.3 COMBINED MOTION OF A RIGID BODY

Although it is relatively simple to determine velocities on a body in pure rotation about a fixed axis, it is certainly not as straightforward to find the velocities on a body in combined motion. This is because a body in combined motion is simultaneously rotating and translating and therefore has no fixed axis of rotation. However, this problem can be simplified by considering the motion to conform, just for an instant, to that of pure rotation about some center of rotation, just for that instant, and thus the velocities in the body can be found using the same principles as those applicable to pure rotation. The center of rotation in this case is aptly called the *instant center of rotation*, since its position changes continuously from one instant to another.

The validity of this approach can be demonstrated by considering link *AB* in Figure 5.4(a), which moves with combined motion. Imagine that the link moves from position 1 to position 2 in a very small time interval. Then, if we use the construction shown in Figure 5.4(b), the same motion can be seen to conform to a circular path where point *I* is the center. This point is the instant center of link *AB*. Further, if we connect points *A* and *B* (or points *A'* and *B'*) to the center *I* by radial lines as shown in Figure 5.5, we see that the velocity directions of these points are indeed perpendicular to the four radial lines drawn, which is consistent with the principles of pure rotation.

Thus, if we know the velocity directions of two points on a floating link, the instant center of rotation must lie at the intersection of lines drawn from those points perpendicular to the velocity directions. Therefore, if we use the instant center, the combined motion of a link or any rigid body can be conveniently reduced to pure rotation, thereby simplifying the velocity analysis.

5.4 VELOCITY OF A BODY WITH ROLLING CONTACT

A common example of combined motion occurs with a rolling wheel. Consider the wheel rolling to the right in Figure 5.6. Here, as the wheel

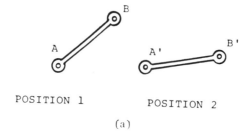

POSITION 1   POSITION 2

(a)

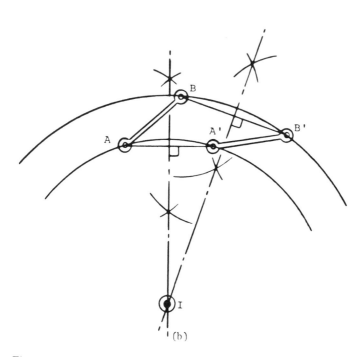

Figure 5.4   Instant center of link in plane motion: (a) Link in plane motion; (b) construction.

moves from one position to another, there is both rotational and translational motion.

To visualize this combined motion, it is convenient to think of each motion as though it occurred independently, then superpose the two effects to obtain the final result. For example, the combined motion can be expressed graphically as shown in Figure 5.6, where $\overline{V}^r$ and $\overline{V}^t$

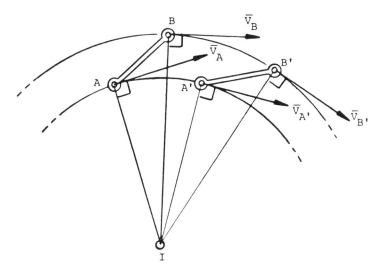

Figure 5.5   Velocity of a link in plane motion.

are the rotational and translational velocities for the respective points on the rim. From the summation of the rotational and translational components, it is noted that

1. Point A is a point of zero velocity.
2. The velocity of each point on the rim is perpendicular to a line joining that point to point A.
3. The magnitude of the velocity of each point is proportional to its distance from point A.

Thus, point A is considered the instant center of rotation of the rolling wheel.

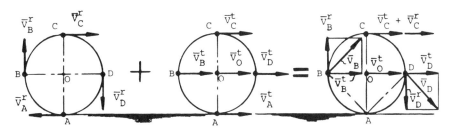

Figure 5.6   Velocities of points on a rolling wheel.

In summary, the instantaneous center of a wheel rolling without slipping lies at the point of contact. All points of the wheel have velocities perpendicular to their radii from the instant center and these velocities are proportional in magnitude to their respective distances from that center.

5.5   TYPES OF INSTANT CENTERS

Instant centers are of three types:

1. **Fixed** (type 1), that is, a stationary point in one body about which another body actually turns. This is normally a fixed axis of rotation on a mechanism.
2. **Permanent** (type 2), that is, a point common to two bodies having the same velocity in each body, such as a hinged joint connecting two moving links of mechanism. The term "permanent" implies that the relative position between the connected links is always the same, regardless of the change in position of the mechanism.
3. **Imaginary** (type 3), that is, a point within or outside the mechanism that can be visualized as having the same characteristics as either a fixed center (type 1) or permanent center (type 2) at any given instant. When this center behaves like a fixed center about which the body tends to turn, it is considered an **instant axis of rotation**.

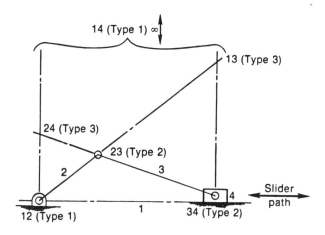

Figure 5.7   Instant centers for a slider-crank mechanism. The location of type 1 and type 2 instant centers can be determined by inspection, whereas the location of type 3 instant centers requires additional analysis.

The slider-crank mechanism shown in Figure 5.7 depicts these three types of instant centers. Usually, fixed and permanent instant centers can be readily identified by inspection. Typically, imaginary instant centers must be located by more detailed analysis. Generally, the circle diagram method (described in Section 5.6) is used to determine the imaginary centers.

Note that the instant center 14 (read "one-four") for the path of slider 4 on frame 1 is indeterminate because it lies at infinity. This is because the slider path is a straight line, and therefore the slider can be considered a body that actually turns about a point located at infinity. An instant center that lies at infinity can be located along an infinite number of lines perpendicular to a straight path.

5.6 LOCATING INSTANT CENTERS

Obvious Instant Centers

Obvious instant centers are those that can be readily located (by inspection) on the mechanism. These may be of either fixed or permanent type. There are four types of obvious instant centers:

1. Instant center for pin-connected links (see Figure 5.8).
2. Instant center for sliding body (see Figure 5.9).
3. Instant center for rolling body (see Figure 5.10).
4. Instant centers for direct contact mechanisms.
    a. For sliding contact between 2 and 4, instant center 24 lies at the intersection of the common normal through the contact point and the line of centers [see Figure 5.11(a)].
    b. For rolling contact between bodies 2 and 4, both instant center 24 and the contact point are coincident and lie on the line of centers [see Figure 5.11(b)].

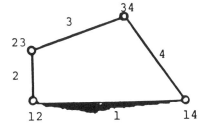

Figure 5.8   Instant center for pin-connected links.

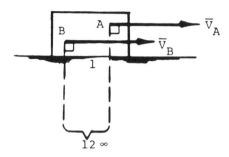

Figure 5.9   Instant center for a sliding body.

Circle Diagram Method

The circle diagram method is based on Kennedy's theorem, which states that **any three bodies having plane motion relative to one another have three instant centers, and they lie on a straight line.**

PROOF

Consider any three bodies 1, 2, and 3 having plane relative motion as shown in Figure 5.12. If we assume that bodies 2 and 3 are pinned to body 1 and, therefore, that instant centers 12 and 13 are known, the problem is to show that the third instant center of 2 and 3 must be on the straight line connecting 12 and 13. First, suppose that the instant center of 2 and 3 was at P'. Then, as a point in body 2, P' must move at right angles to line 12-P' [$\overline{V}_{23}^{(2)}$], and as a point in body 3, P' must move at right angles to line 13-P' [$\overline{V}_{23}^{(3)}$]. This means that point P' moves

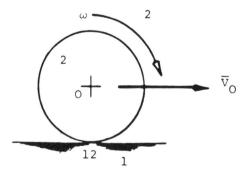

Figure 5.10   Instant center for a rolling body.

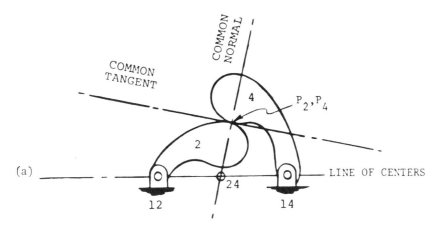

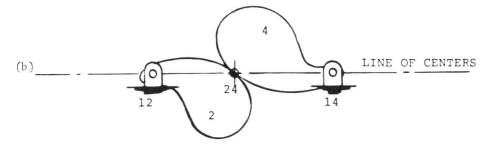

Figure 5.11   (a) Sliding contact; (b) rolling contact.

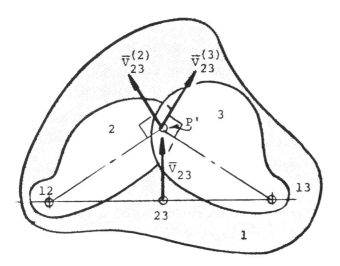

Figure 5.12   Three bodies in relative motion.

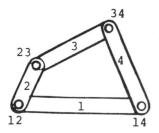

Figure 5.13 Mechanism.

in two different directions at the same time, which is impossible. Therefore, in order for the instant center of 2 and 3 to have the same velocity
direction in both 2 and 3, it must lie on the straight line joining 12 and
13. This proof also applies if all three bodies are moving.

Consider the four-bar linkage shown in Figure 5.13. Let it be required to find all instant centers of this mechanism.

PROCEDURE

1. Locate all fixed and permanent instant centers on the mechanism
    in Figure 5.13. These centers are usually found by inspection. For
    example, instant center 12 is located where link 1 joins link 2, and
    instant center 23 is located where link 2 joins link 3.
        Note that the order of the digits used to designate the instant
    centers is not important. That is, either 23 or 32 may be used to
    designate the same instant center.
2. Lay out points 1, 2, 3, and 4, approximately equally spaced and in
    sequence on a circle (see Figure 5.14), each point representing a
    link on the mechanism (Figure 5.13). Any straight line drawn to
    connect any two or these points represents an instant center on the
    mechanism.

Point represents
link number.

Figure 5.14   Step 2.

Dashed line represents
instant center to be
found.

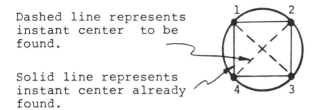

Solid line represents
instant center already
found.

Figure 5.15   Steps 3 and 4.

3. Using **solid lines**, indicate on the diagram (Figure 5.15) the instant centers that have been located so far on the mechanism (i.e., the fixed and permanent centers) by connecting the points that represent those centers. Lines 12, 23, 34, and 14 should therefore be drawn in as solid.

4. Connect all other points in the diagram (Figure 5.15) using **dashed lines** to indicate the instant centers that remain to be found. For example, lines 13 and 24 indicate those outstanding centers.

5. To locate these centers, examine the diagram in Step 4 (Figure 5.15) to find one dashed line that, if it were solid, will complete two solid triangles. For example, the dashed line 13 completes triangles 123 and 341 (Figure 5.15). Using these triangles, we can now locate instant center 13, noting that the sides of triangle 123 represent three instant centers—12, 23, and 13—which by Kennedy's theorem must lie on a straight line within or outside the mechanism. Hence, the instant center 13 lies somewhere on a line joining points 12 and 23, or its extension in the mechanism (see Figure 5.16).

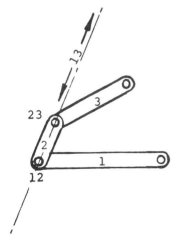

Figure 5.16   Step 5.

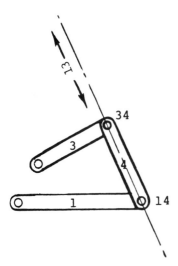

Figure 5.17   Step 5.

Similarly, the sides of triangle 341 represent three instant centers—34, 14, and 13—which again by Kennedy's theorem must lie on a straight line within or outside the mechanism. Hence, instant center 13 must lie on a line joining points 34 and 14, or its extension in the mechanism (see Figure 5.17). Since by both triangles, instant center 13 is given to be somewhere along lines 12-23 and 34-14 (Figures 5.16 and 5.17), it must be located at the intersection of these lines.

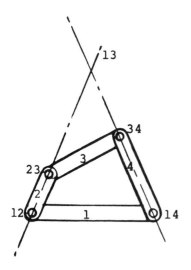

Figure 5.18   Step 6.

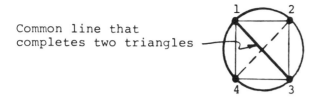

Common line that
completes two triangles

Figure 5.19   Step 6.

6. Accordingly, on the mechanism (Figure 5.18) draw two straight
   lines, 12-23-13 and 14-34-13, and at their intersection locate instant
   center 13. Then, on the diagram, change line 13 to a solid line
   before proceeding to find the other instant center 24 (Figure 5.19).

*It is important that after an instant center has been located on the
mechanism, it be immediately drawn in as a solid line on the diagram.* Otherwise, when locating the remaining centers, it may not be
possible to find additional pairs of *triangles are solid except for a
common dashed line*.

5.7   VELOCITY PROPERTIES OF THE
INSTANT AXIS

A link is considered to be rotating when all points in that link remain
at fixed distances from an axis of rotation. In a machine, each member
is rotating about a fixed axis or moving axis whose location varies from
one instant to another. For purposes of analysis, this moving axis may
be thought of as a stationary axis having properties similar to those of
a fixed axis, at any given instant. Also, it should be noted that

1. There is one instant axis of velocity for each floating link of a
   machine.
2. There is no single instant axis of velocity for all links of a machine.
3. Instant centers are also axes of rotation when they have no absolute
   motion. That is, their velocity is zero with respect to the frame.
   Such instant centers can be readily identified because their numerical designation includes the number representing the frame.
   For instance, in the slider-crank example, link 1 represents the
   frame and instant centers 12 and 13 are axes of rotation for links
   2 and 3, respectively. Instant 14 is also an axis of rotation but is
   indeterminate and lies at infinity. Instant centers 24, 23, and 34 are

not axes of rotation and may or may not have absolute motion with respect to the frame. Note that 12 is a fixed axis of rotation, whereas 13 is a moving axis of rotation.

4. The instant center of velocity is not an instant center of acceleration, although it moves as the link moves and may have an actual zero acceleration. Neither does it necessarily have zero acceleration as does the fixed center.

5.8 VELOCITY ANALYSIS BY INSTANT CENTERS

The determination of velocities in mechanisms by instant centers is based on three principles:

1. The velocity magnitude at a point in a rotating body is directly proportional to the radius of rotation of that point.
2. The velocity direction at a point in a rotating body is perpendicular to its radius of rotation.
3. An instant center is a point common to two bodies and has the same linear velocity (both magnitude and direction) in both bodies.

Consequently, if the linear velocity of any point in a body relative to an instant center is known, the velocity of any other point in that body, relative to the same instant center, can be determined graphically by proportions, as illustrated by the *radius of rotation method* in Figure 5.20.

Note that instant centers must be located first before velocities can be determined. Also, in mechanism analysis, velocities are usually determined about fixed instant centers (either normally fixed or momentarily fixed) having no velocity with respect to the frame.

To illustrate how these principles are applied, let us now consider two important concepts as they apply to a four-bar linkage (Figure 5.21): (1) fixed axis (or center) and (2) link extensions.

The term *fixed axis* is used to refer to an instant center that is normally fixed or one that is momentarily fixed (an instant axis). To identify the fixed axis of a mechanism, we look for those centers whose numerical designations include the fixed link of the mechanism, that is, the link whose velocity is zero. Since in this example the fixed link is 1, it immediately becomes evident that the fixed axes are (12), (13), and (14).

Note also that a fixed axis typically has a numerical designation that includes both the frame and link that rotates about the axis. For

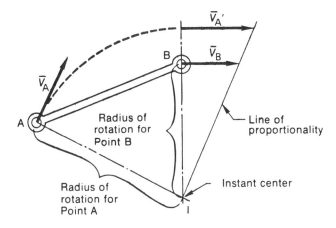

Figure 5.20   Velocity is proportional to radius of rotation. The velocity of any point in a link is proportion to its radius of rotation about an instant center. As a result, if the velocity of one point (such as B) is known, the velocity of any other point (such as A) in that link can be determined graphically by proportions.

example, the fixed axes (12), (13), and (14) have, respectively, links 2, 3, and 4 rotating about them. The other centers—(24), (34), and (23)—can be considered joints on the extended mechanism. These centers may or may not have a velocity.

Link extensions are imaginary bodies defined by three instant centers, one of which typically is a fixed axis, and by a common link that rotates about this fixed axis. To identify the link extension for a given link, we simply locate the fixed axis (or center) about which that link rotates and the other centers whose numerical designations include that link. These centers, together, comprise the link extension for the given link.

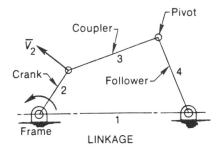

Figure 5.21   Four-bar linkage.

The four-bar mechanism has three rotating members: links 2, 3, and 4. Link extensions for these rotating links are defined by the following sets of centers:

(12) (23) (24)   where link 2 is the common link and center (12) the fixed axis of rotation. This link extension is depicted as triangular plate ② in Figure 5.22(a).

(13) (23) (34)   where link 3 is the common link and center (13) the fixed axis of rotation. This link extension is depicted as triangular plate ③ in Figure 5.22(b).

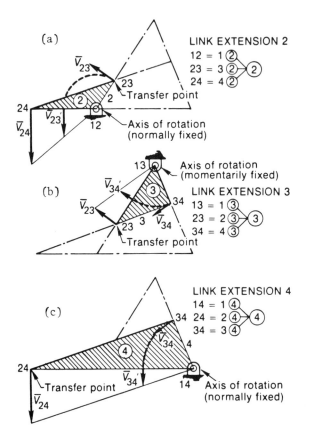

Figure 5.22   Link extensions for a four-bar linkage. Link extensions 2(a) and 4(c) have fixed axes of rotation. Link extension 3(b) has a moving axis rotation, denoted by the imaginary instant center 13, which for analytical purposes is assumed to be momentarily fixed.

(14) (24) (34)   where link 4 is the common link and center (14) the fixed axis of rotation. This link extension is depicted as triangular plate ④ in Figure 5.22(c).

For purposes of analysis, link extensions essentially reduce complex mechanisms to several simpler mechanisms, thereby reducing the computation of velocities to simple graphics. Also, link extensions help to illustrate such concepts as axes of rotation, common links, and transfer points—all of which are essential elements of velocity analysis.

5.9   VELOCITY ANALYSIS OF A SIMPLE MECHANISM

Consider again the four-bar linkage shown in Figure 5.21. For the position shown, the velocity $\bar{V}_{23}$ is given and the velocity $\bar{V}_{34}$ is required.

PROCEDURE

First, we determine all instant centers for the mechanism. The total number (N) of instant centers of a mechanism is given by the equation

$$N = \frac{n(n-1)}{2}$$

where n is the number of links in the mechanism. To locate these centers, the circle diagram method has been found to be most useful, particularly for the more complex mechanisms. Instant centers for the four-bar linkage are shown in Figure 5.23.

Next, we examine the figure to identify the *fixed centers* and *link extensions* that tend to have rotation about these centers. Again, these link extensions are shown in Figure 5.22.

Finally, having defined the link extensions and their respective axes of rotation, we are ready to apply principles 1 and 2, using the radius of rotation method (shown in Figure 5.20) to determine the required velocities of points in the linkage. In the present example, the velocity of link 4 ($\bar{V}_4$) can be determined by proportions using the link extension (13) (23) (34) and known velocity $\bar{V}_{23}$. For instance, velocity

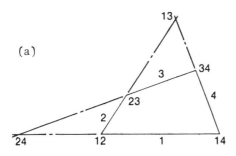

(a)

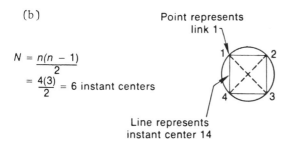

(b)

Point represents
link 1

$$N = \frac{n(n-1)}{2}$$

$$= \frac{4(3)}{2} = 6 \text{ instant centers}$$

Line represents
instant center 14

Figure 5.23   Instant centers for a four-bar linkage. Instant centers 24 and 13 are both imaginary. However, center 13 is also an axis of rotation for floating link 3. (a) Instant center diagram; (b) circle diagram.

$\overline{V}_{34}$ at 34 is simply proportional to the radius of rotation of 34 about 13. Therefore,

$$\overline{V}_{34} = \overline{V}_{23} \frac{13\text{-}34}{13\text{-}23}$$

where 13-34 and 13-23 denote the radii of rotation of 34 and 23, respectively. Note that $\overline{V}_{34}$ is a velocity common to both links 3 and 4.

This observation leads to the concept of a transfer point. A **transfer point** is an instant center that has the same velocity in two different links or link extensions. In the four-bar linkage, 34 is a transfer point for links 3 and 4 and for link extensions 3 and 4. Also, 23 is a transfer point for links 2 and 3, and 24 is a transfer point for links 2 and 4.

The concept of a transfer point provides a powerful tool for velocity analysis, whereby the velocity of any point in a link can be determined without knowing the velocity of another point in the same link. For example, suppose that only $\overline{V}_{23}$ is known; then the velocity at any point in link 4, or its extension, can be determined without knowing the velocity at 34. This is accomplished by first finding the velocity at 24, using link extension 2, which contains the given velocity $\overline{V}_{23}$, then transferring the velocity found at 24 ($\overline{V}_{24}$) to link extension 4 to find the velocity of any other point on that extension. This means that the velocity at point 34, which is a point on link extension 4, can be determined as follows:

1. The velocity at 24, if we use link extension 2, is

$$\overline{V}_{24} = \overline{V}_{23} \frac{12\text{-}24}{12\text{-}23}$$

2. The velocity at 34, when we use link extension 4, is

$$\overline{V}_{34} = \overline{V}_{24} \frac{14\text{-}34}{14\text{-}24}$$

EXAMPLE 5.2

Figure 5.24 shows a quick-return mechanism whose crank 2 turns at 1 rad/sec (clockwise). (a) Locate all instant centers and (b) determine the angular velocity of the follower link 3.

SOLUTION

1. Number of instant centers for this four-link mechanism is given by

$$N = \frac{n(n-1)}{2}$$

$$= \frac{4(4-1)}{2} = 6$$

a. Locate the obvious instant centers on the mechanism. They are 12, 13, 24, and 34.
b. With the aid of the circle diagram, locate the remaining centers, namely, 14 and 32.
2. The required angular velocity of link 3 is determined as follows:
a. Solve for the linear velocity of transfer center 23 (or $\overline{V}_{23}$). This is done by considering link extension 2, which is defined by

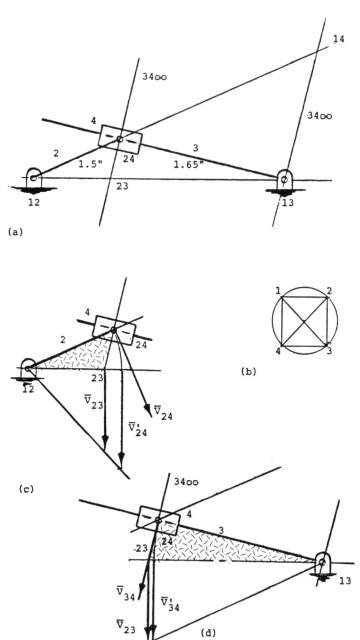

Figure 5.24   Quick-return mechanism: (a) Instant center locations; (b) circle diagram; (c) link extension 2; (d) link extension 3.

the centers 12, 24, and 23, where 12 is the instant axis and $\overline{V}_{24}$ is determinable from

$$V_{24} = (12\text{-}24)\omega_2 = 1.5 \times 1 = 1.5 \text{ in./sec}$$

Then, from proportionality,

$$\overline{V}_{23} = \frac{12\text{-}24}{12\text{-}23} V_{24}$$

b. Consider the link extension 3, defined by the instant centers 13, 23, and 34 (at the slider), where 13 is the instant axis, $\overline{V}_{23}$ is known from above, and determine $\overline{V}_{34}$ from the proportionality

$$\overline{V}_{34} = \frac{13\text{-}23}{13\text{-}34} \overline{V}_{23}$$

$$= 1.25 \text{ in./sec}$$

c. Calculate the angular velocity of link 3 from the relationship

$$\omega_3 = \frac{V_{34}}{13\text{-}34}$$

$$= \frac{1.25}{1.65}$$

$$= 0.75 \text{ rad/sec}$$

5.10   VELOCITY ANALYSIS OF A COMPOUND MECHANISM

Consider a steam locomotive driven by a compound mechanism having six links and 15 instant centers (Figure 5.25a). As indicated by the circle diagram, nine of the instant centers were identified by inspection, whereas the remaining six imaginary instant centers had to be located by detailed analysis. Given a velocity $\overline{V}_A$ at instant center 36 in link 6, find the velocity at point B in link 2, using extended links to simplify the analysis (Figure 5.25b).

First, the velocity at instant center 26 must be determined. Instant center 26 is the transfer point between links 2 and 6. Therefore, 26 has the same velocity in both link extensions 2 and 6.

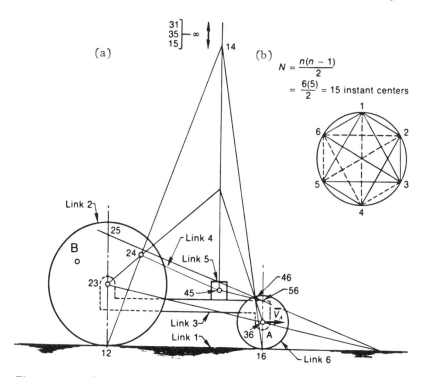

Figure 5.25   Determining velocities by link extensions: (a) Mechanism; (b) circle diagram.

For link extension 6 the axis of rotation is 16 and the common link is 6. By inspection it can be determined that link extension 6 consists of instant centers 16, 26, and 36 (Figure 5.26a). Thus, the velocity at the transfer point 26 is determined by proportions as

$$\overline{V}_{26} = \overline{V}_{36} \frac{16\text{-}26}{16\text{-}36}$$

For link extension 2 the axis of rotation is 12, the common link is 2, and the constituent instant centers are 12, 23, and 26 (Figure 5.26b). Note that transfer point 26 is common to both link extensions. Given velocity $\overline{V}_{26}$, the velocity 23 is obtained from

$$\overline{V}_{23} = \overline{V}_{26} \frac{12\text{-}23}{12\text{-}26}$$

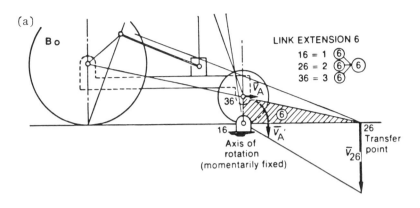

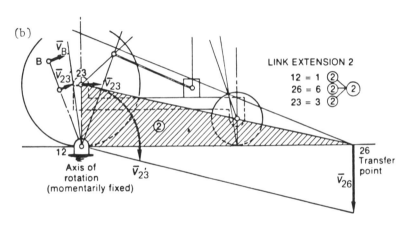

Figure 5.26   Velocity analysis by link extension. Given a velocity in link 6, velocities in link 2 can be determined once the transfer point (instant center 26) is identified and the transfer point velocity ($\overline{V}_{26}$) is determined. (a) Link extension 6; (b) link extension 2.

Finally, the velocity at point B is determined graphically by proportions as

$$\overline{V}_B = \overline{V}_{23} \frac{12\text{-}B}{12\text{-}23}$$

It should be noted that $\overline{V}_B$ could have been determined more directly by recognizing that the velocity at 23 is the same as that of 36. However, transfer points were used deliberately in the analysis, for instructive purposes.

Velocity Analysis of a Shaper Mechanism

EXAMPLE 5.3

Figure 5.27(a) shows a shaper mechanism whose crank 2 turns at 1 rad/ sec. Determine (a) all instant centers of the mechanism and (b) the linear velocity of the cutting tool link 6.

SOLUTION

1. As before, the number of instant centers for this six-link mechanism is obtained by

$$N = \frac{n(n - 1)}{2}$$

$$= \frac{6(6 - 1)}{2}$$

$$= 15$$

   a. Locate the obvious instant centers on the mechanism. They are 12, 13, 16, 24, 34, 45, and 56. See Figure 5.27(b).
   b. With the aid of the circle diagram, locate the remaining centers, namely, 14, 15, 25, 35, 36, 46, and 26. See Figure 5.27(b).
2. The required velocity $\overline{V}_{56}$ may be determined by first solving for the velocity of transfer point 23 on link 2 (or $\overline{V}_{23}$), then transferring this velocity to link 3 to solve for the velocity of center 35, also a point on link 3, and finally, using $\overline{V}_{35}$ on link 5 to solve for $\overline{V}_{56}$. The procedure is as follows:
   a. Find $\overline{V}_{23}$. This velocity is found by considering the link extension 2, Figure 5.27(c), containing the centers (12) (24) (23), where (12) is the instant axis, and $\overline{V}_{24}$ is determinable from

$$\overline{V}_{24} = (12\text{-}24)\omega_2 = (1.0)1 = 1 \text{ in./sec}$$

Then, by proportionality,

$$\overline{V}_{23} = \frac{(12\text{-}23)}{(13\text{-}24)}\overline{V}_{24}$$

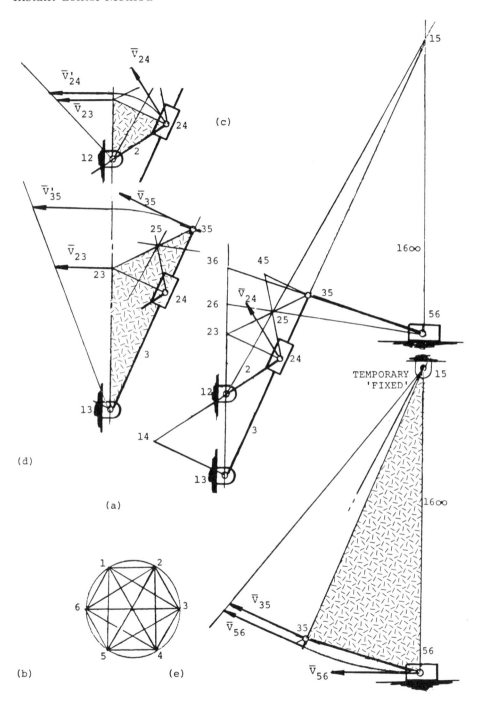

Figure 5.27   Shaper analysis: (a) Shaper with instant centers; (b) circle diagram; (c) link extension 2; (d) link extension 3; (e) link extension 5.

b. Find $\overline{V}_{35}$. This velocity is found by considering the link exten-
    sion 3, Figure 5.27(d), containing the centers (13) (23) (35),
    where (13) is the instant axis, and $\overline{V}_{23}$ is known. Then, by
    proportionality,

$$\overline{V}_{35} = \frac{(13\text{-}35)}{(13\text{-}23)} \overline{V}_{23}$$

c. Find $\overline{V}_{56}$. This velocity is found by considering the link exten-
    sion 5, Figure 5.27(e), defined by the centers (15) (35) (56),
    where (15) is the instant axis and $\overline{V}_{35}$ is known. Then, by pro-
    portionality,

$$\overline{V}_{56} = \frac{(15\text{-}56)}{(15\text{-}35)} \overline{V}_{35}$$

$$= 1.4 \text{ in./sec} \quad \text{(scaled)}$$

Note that the same result would be obtained if the following link
extensions and centers are considered:

Link extension 2 ... (12) (24) (25)

Link extension 5 ... (15) (25) (56)

5.11 VELOCITY ANALYSIS OF ROLLING ELEMENTS

The method of instant centers can be effectively applied to mechanisms
consisting of rolling elements as in planetary gear trains and cam-follower mechanisms. For example, consider the cam-follower mechanism
shown in Figure 5.28, where the cam 2 rotates counterclockwise at an
angular velocity of ω_2 rad/sec. Let it be required to determine the linear
velocity of the follower 4.

PROCEDURE

1. Label all instant centers as shown in Figure 5.28a.
2. Select the appropriate transfer point and links or link extensions
    needed for the analysis. Since the center 23 is a point in link 2 as
    well as link 3, it is logical to select this point as the transfer point.

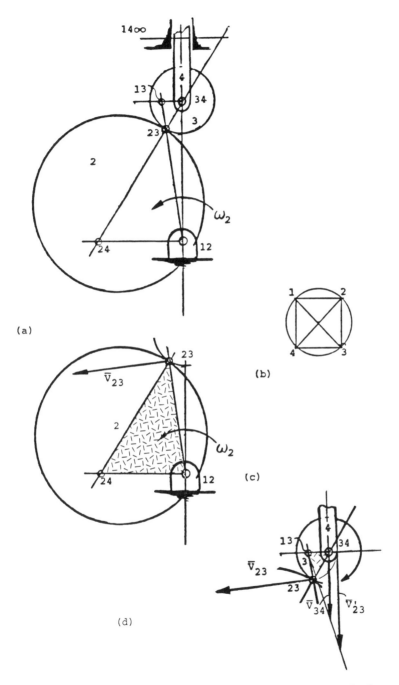

Figure 5.28 (a) Cam-follower with instant centers; (b) circle diagram; (c) link extension 2; (d) link extension 3.

Then link 2, whose axis is known, can be used to determine the velocity of the center 23, whereas link extension 3, defined by 13 (the instant axis), 23 (found from link 2), and 34, can be used to determine the velocity of 34, as required.

3. Lay out vector $\overline{V}_{23}$ from

$$V_{23} = (12\text{-}23)\omega_2$$

4. Determine the velocity $\overline{V}_{34}$ from proportionality

$$\overline{V}_{34} = \frac{12\text{-}23}{12\text{-}34} \overline{V}_{23}$$

Rotate vector $\overline{V}_{23}$ so that its radial line falls on 13-34 and draw a line of proportionality from instant axis 13 through the head of rotated vector $\overline{V}'_{23}$. Then from center 34, draw a vector parallel to $\overline{V}'_{23}$ terminating at the proportionality line. This vector will represent the required velocity $\overline{V}_{34}$.

EXAMPLE 5.4

Consider the planetary gear train shown in Figure 5.29, where gear 1 is fixed and the angular velocity of the planet arm 4 is ω_4 (clockwise). Let it be required to determine the linear and angular velocities in the ring gear.

SOLUTION

1. Label all instant centers as shown in Figure 5.29.
2. Select the appropriate transfer point(s) and link(s) or link extension(s) needed for the analysis. Since center 24 is a point in link 2 as well as link 4, and the velocity of this point is determinable from link 2 data, this point may be considered a transfer point. Then link extension 2, defined by centers 12, 24, and 23, can be used to determine the velocity of center 23.
3. Lay out vector $\overline{V}_{24}$. This velocity is given by

$$V_{24} = (R_1 + R_2)\omega_4$$

4. Determine the velocity $\overline{V}_{23}$ using link extension 2, where 12 is the instant axis and the velocity of 24 is known. From proportionality,

$$\overline{V}_{23} = \frac{12\text{-}24}{12\text{-}23} \overline{V}_{24}$$

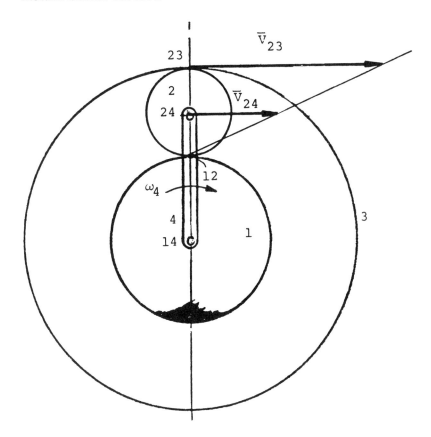

Figure 5.29   Example 5.4.

Draw a line of proportionality from instant axis 12 through the head
of $\overline{V}_{24}$. Then from center 12 draw a vector parallel to $\overline{V}_{24}$ terminating
at the proportionality line. This vector will represent the required
velocity $\overline{V}_{23}$.

EXAMPLE 5.5

Consider the gear set in Figure 5.30. Given that the angular velocities
for gear 5 and arm 4 are ω_5 (CW) and ω_4 (CW) respectively, let it be
required to find $\overline{V}_6$.

SOLUTION

1.   Label all instant centers, as shown in Figure 5.30.
2.   Select the appropriate links (or gears) along with the centers needed
     for the analysis. To determine $\overline{V}_6$, it is useful to note that center 36

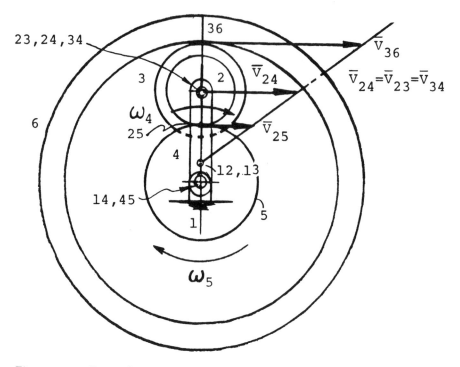

Figure 5.30   Example 5.5.

can be considered a point in gear 2 as well as gear 3, since both gears are fastened together. Hence, we consider gear 2 and its centers 24, 25, and 12. Here, the velocities $\overline{V}_{24}$ and $\overline{V}_{25}$ are known, but the location of 12, the instant axis, is not. This axis, however, can be found from rotational principles, where

$$\overline{V}_{24} = \frac{12\text{-}24}{12\text{-}25} \, \overline{V}_{25}$$

a.   Lay out velocities $\overline{V}_{25}$ and $\overline{V}_{24}$

$$V_{25} = \omega_5 R_5$$

and

$$V_{24} = \omega_4 (R_5 + R_2)$$

Note that

$$\overline{V}_{24} = \overline{V}_{23} = \overline{V}_{34}$$

   b. Draw a straight line touching the heads of $\overline{V}_{24}$ and $\overline{V}_{25}$ and locate
       12 where this line intersects the arm centerline. This center is
       also the instant axis of gear 3 (or 13) since gears 2 and 3 are
       one body.
3. With axis 13 located, we next consider gear 3 and its centers 13,
   34, and 36, where the velocity of 34 (same as 24) is known and
   that of 36 is to be determined. Then, we determine $\overline{V}_{36}$ from

$$\overline{V}_6 = \overline{V}_{36} = \frac{13\text{-}36}{13\text{-}34}\,\overline{V}_{34}$$

From 13, construct a line of proportionality touching the head of
$\overline{V}_{34}$, and from 36, draw a vector parallel to $\overline{V}_{34}$, terminating at the
line of proportionality. This vector will represent $\overline{V}_{35}$, or the required ring gear velocity $\overline{V}_6$.

EXAMPLE 5.6

Consider the planetary wheel arrangement in Figure 5.31. Given the
angular velocity of wheel 2 is ω_2 rad/sec, find the angular velocity of
arm 5.

SOLUTION

1. Label all obvious instant centers as shown in Figure 5.31.
2. Select the appropriate links (or gears) and link centers needed to
   be analyzed. In this problem it is noted that the instantaneous axes
   for all moving bodies are known except that of body 3. They are
   12 for body 2, 14 for body 4, and 15 for body 5. In order to determine
   the angular velocity of arm 5, we need to consider wheel 5 and
   centers 35, 45, and 15, where 15, the instant axis, is defined and
   the linear velocity $\overline{V}_{35}$ is required. But, since $\overline{V}_{45}$ is not known, we
   cannot obtain directly $\overline{V}_{35}$ from wheel 5 alone. However, noting
   that center 35 is also a point in wheel 3, we then consider wheel
   3 and centers 35, 23, and 13, where 13 is yet an undefined instant
   axis, but the velocity of center 23 (or $\overline{V}_{23}$) is determinable from the

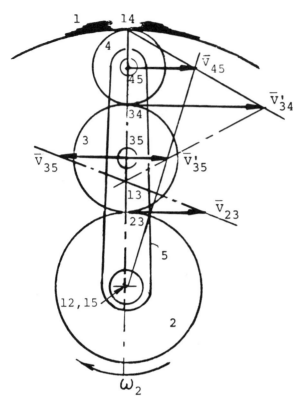

Figure 5.31   Example 5.6.

problem data. At this point, it should be clear that if we locate 13, we can determine $\overline{V}_{35}$ immediately from

$$\overline{V}_{35} = \frac{13\text{-}35}{13\text{-}23}\ \overline{V}_{23}$$

3. Axis 13 can be located by assuming a velocity of wheel 4 and relating this velocity to wheels 5 and 3, as follows:
    a. Assume a value for the linear velocity of center 34 (or $\overline{V}'_{34}$) and lay out this vector.
    b. Consider wheel 4. Since $\overline{V}'_{34}$ is the velocity of a point in wheel 4 and instant axis 14 is known, determine the assumed velocity of 45 (or $\overline{V}'_{45}$) from proportionality.
    c. Consider arm 5. Since $\overline{V}'_{45}$ is also the velocity of a point on arm 5 and instant axis 15 is known, determine the assumed velocity of 35 (or $\overline{V}'_{35}$) from proportionality.

d.   Consider wheel 3. With two velocities $\overline{V}'_{45}$ (or $\overline{V}'_{35}$) defined on body 3, the axis of rotation 13 can now be located by drawing a line of proportionality connecting the heads of $\overline{V}'_{45}$ and $\overline{V}'_{35}$, then extending this line to intersect the arm centerline at the required instant axis 13.

4.   With instant axis 13 defined and the velocity of center 23 (also a point in 3) determinable, the true velocity of 35 (or $\overline{V}_{35}$) can be found as follows:

a.   Lay out the vector $\overline{V}_{23}$.

$$V_{23} = R_2\omega_2$$

b.   Draw a line of proportionality from the head of $\overline{V}_{23}$ through 13 and from center 35 draw a velocity vector perpendicular to arm 5 and terminating at this line. This vector will represent the required linear velocity $\overline{V}_{35}$.

5.   Finally, determine the angular velocity of arm 5 from the expression

$$\omega_5 = \frac{\overline{V}_{35}}{R_2 + R_3}$$

Note that it would have made no difference in the above results what the magnitude or direction of velocity $\overline{V}_{45}$ was assumed, since the instant axis 13 depends only on the proportionality between $\overline{V}'_{34}$ and $\overline{V}'_{35}$.

Summary of Steps for Example 5.6

1.   Use center 14 and $\overline{V}'_{34}$ to obtain $\overline{V}'_{45}$.
2.   Use center 15 and $\overline{V}'_{45}$ to obtain 35.
3.   Use $\overline{V}'_{34}$ and $\overline{V}'_{35}$ to obtain 13.
4.   Use 13 and $\overline{V}_{23}$ to obtain $\overline{V}_{35}$.
5.   Use $\overline{V}_{35}$ and 13-15 to obtain ω_5.

Note

An advantage of the instant center method in the planetary wheel or gear train application is its power in aiding the visualization of the motion of the planetary system. The method also ensures that the proper direction of the output rotation is obtained without risk of error.

5.12   SUMMARY

To find the linear velocity of a point in a link, we first identify the link (or its extension) in which the point occurs and its axis of rotation. If the velocity of any point in this link (or its extension) is known, the velocity of that point can readily be determined from the principles of rotation. However, if there is no known velocity in this link (or its extension), we (1) select a point (i.e., an instant center) common to this link and a second link (or its extension), where the velocity of such a point can be determined from rotation principles; (2) determine this velocity; and (3) transfer the velocity found to the first link (or its extension) to find the required velocity, again using the principles of rotation.

6
Relative Velocity Method

6.1 INTRODUCTION

The relative velocity method is probably the most common among the graphical methods used in velocity analysis. Compared to the other graphical methods, it readily provides solutions not only for absolute velocities, but also for relative velocities of points in a mechanism without requiring the location of instant centers. This singular feature makes it most desirable to use when determining the relative velocities needed for acceleration analysis. The relative velocity method is based on two important concepts: (1) relative motion and (2) the rigid-body principle, considered earlier.

6.2 RELATIVE MOTION CONCEPT

Relative motion has been defined as the motion of a body with respect to another body that is itself moving. If the motion of the body is with respect to a stationary frame of reference or fixed point such as the earth, the motion is defined as **absolute motion**.

If we think of two bodies, A and B, having independent or absolute motion, the velocity of A relative to B is the velocity that A appears to have to an observer traveling on B. To illustrate the concept of relative

121

motion, let us consider three ways in which we can observe a car A traveling at 50 mph while seated in a second car, B.

1. Our car (B) is parked on the shoulder of the road. Then, in the adjacent lane, car A comes speeding past at 50 mph and immediately we have the experience of seeing the vehicle moving at 50 mph.
2. Our car (B) is moving at 50 mph in one lane, and in another lane, car A is moving in the opposite direction with a speed of 50 mph. As it speeds past our car, car A appears to be moving at a speed of 100 mph.
3. Our car (B) is now moving at 40 mph, and along comes car A, still in another lane and moving at 50 mph, but in the same direction as our car. We now have the experience of seeing car A moving at 10 mph.

Note that the velocity of car A, although the same in each case, does in fact vary, from our point of view, and that we realize the true velocity only when we make our observation from a stationary position, as in case 1. This proves that the velocity that a moving body appears to have is always dependent on the observation point or frame of reference.

Consider two ships, A and B, whose velocities, $\overline{V}_A$ and $\overline{V}_B$, in still water are in the direction shown in Figure 6.1(a). It is desired to find the velocity that B appears to have to an observer on A.

Since the required velocity is to be relative to A, we must consider a method of bringing A to rest. To do this, we imagine the water to be a stream moving with a velocity equal and opposite to that of A (i.e., with a velocity in $-\overline{V}_A$). This means that as fast as A moves forward, the stream moves backward with the same velocity, and consequently A makes no progress as far as the earth is concerned. The same effect would be produced if one were to run forward on an endless belt moving with a velocity equal and opposite to that of the runner. As far as the surroundings are concerned, the runner would be stationary.

Now, consider what is happening to B. Not only is it moving with its original velocity $\overline{V}_B$, but it has added to it the velocity of the stream, which is $-\overline{V}_A$. Therefore,

$$\text{Velocity of } B \text{ relative to } A = \text{velocity of } B + \text{velocity of stream}$$

$$= \text{velocity of } B + (-\text{velocity of } A)$$

$$= \text{velocity of } B - \text{velocity of } A$$

$$= \text{difference of velocities}$$

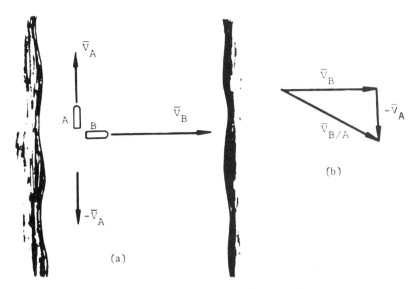

Figure 6.1   Vectorial representation of relative velocity.

Hence, we can conclude that **the velocity of B relative to A is the vectorial difference between the velocities of A and B**. Symbolically,

$$\overline{V}_{B/A} = \overline{V}_B - \overline{V}_A \tag{6.1}$$

Note the order of subscripts on both sides of the equation. An alternative form of Eq. (6.1) is

$$\overline{V}_B = \overline{V}_A + \overline{V}_{B/A}$$

It should be noted that whereas **the relative velocity is the vectorial difference between two velocities, the resultant of two velocities is the vectorial sum of those velocities**. Figure 6.1(b) illustrates the vector polygon that represents Eq. (6.1).

6.3   THE VELOCITY POLYGON

The velocity polygon is an alternative form of the vector polygon presented earlier. To illustrate the procedure, let us consider the same ships A and B discussed in Section 6.2.

PROCEDURE (refer to Figure 6.2)

1. Define a point o, called the pole, as the origin for the construction. This is a point of zero velocity and it represents all the fixed points on the mechanism. All absolute velocities originate from this point. By *"absolute velocity"* we mean the **real and true velocity** of a body, as observed from a stationary frame of reference such as the earth.
2. Since the velocities $\overline{V}_A$ and $\overline{V}_B$ are absolute velocities, draw $\overline{V}_A$ and $\overline{V}_B$ from point o and define their respective termini, a and b.
3. From point a, the terminus of $\overline{V}_A$, draw a third vector to terminate at point b, the terminus of $\overline{V}_B$, thereby closing the polygon. This vector defines the relative velocity $\overline{V}_{B/A}$.

Note the agreement in the results between the vector polygon and the velocity polygon given in Figure 6.2(a) and (b), particularly the fact that $\overline{V}_{B/A}$ is the same in both cases. This agreement is also validated by the vector equations shown.

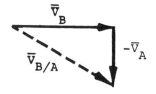

VECTOR POLYGON

$$\overline{V}_B + (-\overline{V}_A) - \overline{V}_{B/A} = 0$$

$$\overline{V}_B - \overline{V}_A = \overline{V}_{B/A}$$

(a) TRIANGULAR METHOD

VELOCITY POLYGON

$$\overline{V}_B - \overline{V}_{B/A} - \overline{V}_A = 0$$

$$\overline{V}_{B/A} = \overline{V}_B - \overline{V}_A$$

(b) PARALELLOGRAM METHOD

Figure 6.2 Vectorial representations of relative velocity: (a) Vector polygon; (b) velocity polygon.

6.4 THE VELOCITY POLYGON CONVENTION

The convention just employed in developing the velocity polygon [Figure 6.2(b)] can be summarized as follows:

$\overline{oa}$ represents the velocity of A or $\overline{V}_A$.
$\overline{ob}$ represents the velocity of B or $\overline{V}_B$.
$\overline{ab}$ represents the velocity of B relative to A or $\overline{V}_{B/A}$.

Note that the letter to which the arrow points indicates the velocity under consideration. In the vectors $\overline{oa}$ and $\overline{ob}$, the arrows point toward a and b, respectively. Hence, the vectors represent the velocities of A and B, respectively. Similarly, in the vector $\overline{ab}$, the arrow points toward b and away from a; hence, the vector represents the velocity of B relative to A. If the arrow were reversed, pointing toward a and away from b, the vector would represent the velocity of A relative to B.

6.5 VELOCITY POLYGON: LINKAGE APPLICATION

Application of the velocity polygon to linkage analysis is based on the following two important principles covered earlier:

1. If two points lie on the same rigid body, their relative velocity is the vectorial difference between their absolute velocities.
2. If two points lie on the same rigid body, their relative velocity is perpendicular to the line connecting the two points.

6.6 RELATIVE VELOCITY OF TWO POINTS ON A RIGID BODY

Consider floating link AB in Figure 6.3. Imagine that we were seated at end A looking toward the other end, B. Since the distance between A and B does not change (i.e., it is a rigid body) and the link has motion, B would appear to move about us in a circular path. This is the only motion that B could have.

Therefore, the linear velocity of B relative to A ($\overline{V}_{B/A}$) must act perpendicular to link AB and proportional to the distance from A to B, or

$$\overline{V}_{B/A} = AB \times \omega \quad \text{(in one direction)}$$

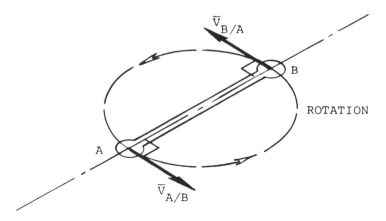

Figure 6.3   Relative velocity of points on a link.

Similarly, if we were seated at end B looking toward A, A would also appear to have the same circular motion about us. Therefore, the linear velocity of A relative to B ($V_{A/B}$) must also act perpendicular to link AB and proportional to the distance from A to B, or

$$\overline{V}_{A/B} = AB \times \omega \quad \text{(in the opposite direction)}$$

As a result, it is easy to see that the linear velocity of B relative to A is equal and opposite to the linear velocity of A relative to B, or

$$\overline{V}_{B/A} = -\overline{V}_{A/B}$$

In summary, if two points, A and B, lie on the same rigid body, and the velocity of B relative to A is required, we must assume that A is fixed and B rotates about it. Similarly, if the velocity of A relative to B is required, we must assume that B is fixed and A rotates about it.

6.7   VELOCITIES OF END POINTS ON A FLOATING LINK

If the velocity of one end of a floating link is completely known (in direction and magnitude) and that of the other end is known only in direction, the magnitude of the latter velocity can be determined from a velocity polygon. For example, consider link AB of Figure 6.4(a).

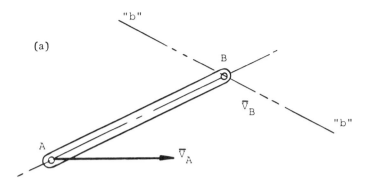

(a)

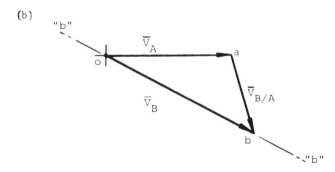

(b)

Figure 6.4   Velocities of end points on a link: (a) Link; (b) velocity polygon.

Given the velocity of point A (both magnitude and direction) and of point B (direction only), determine the complete velocity $\overline{V}_B$.

Since point A has motion, the velocity of point B is obtained from the vector equation

$$\overline{V}_B = \overline{V}_A + \overline{V}_{B/A}$$

where the vector $\overline{V}_{B/A}$ is the vectorial difference between the completely known velocity $\overline{V}_A$ and partially known velocity $\overline{V}_B$. Also, since points A and B are on the same link, the velocity of B relative to A ($\overline{V}_{B/A}$) must act perpendicularly to the line joining A to B. Hence, by laying out the velocity of A, which is known, and the velocity directions of B and $\overline{V}_{B/A}$, the magnitudes of both $\overline{V}_B$ and $\overline{V}_{B/A}$ can be determined.

PROCEDURE

1. Define a point o as a pole for the velocity polygon.
2. From point o, lay out the vector $\overline{V}_A$, terminating at a point a.
3. Through point o, draw a line "b"–"b" parallel to the given direction of vector $\overline{V}_B$. The magnitude and orientation of this vector are defined by a point b on this line, which is as yet unknown.
4. To define $\overline{V}_B$, we must relate the velocity of point A, which is completely known, to that of B, which is only partially known. This means that we must seek to determine the velocity of B relative to A (or $\overline{V}_{B/A}$). Referring to the link, we know that the velocity vector $\overline{V}_{B/A}$ must be perpendicular to link AB; and from the velocity polygon convention, we also know that this velocity must be directed from point a to point b on the polygon. Therefore, draw a vector in the direction perpendicular to link AB, originating from point a and terminating at a point b in line "b"–"b". This line, $\overline{ab}$, represents the vector $\overline{V}_{B/A}$, and the line ob represents the required velocity vector $\overline{V}_B$j.

A quick check of the completed velocity polygon [Figure 6.4(b)] reveals the balanced vector equation

$$\overline{V}_B = \overline{V}_A + \overline{V}_{B/A}$$

Note that it would have made no difference in the results obtained for $\overline{V}_B$ if we had considered the velocity of A relative to B ($\overline{V}_{A/B}$) instead of the velocity of B relative to A ($\overline{V}_{B/A}$). However, as far as the velocity polygon is concerned, since $\overline{V}_{A/B}$ is directed opposite to $\overline{V}_{B/A}$, the direction of vector $\overline{V}_{B/A}$ would have to be reversed, in order to be consistent with the vector equation

$$\overline{V}_B = \overline{V}_A - \overline{V}_{A/B}$$

6.8 VELOCITY OF ANY POINT ON A LINK

Point on Centerline of Two Known Velocities

If the velocities of two points on a link are completely known, the velocity of any other point on that link can easily be determined from the velocity triangle using the method of proportionality. Let us consider again link AB in Figure 6.5(a), where the velocities $\overline{V}_A$ and $\overline{V}_B$ are both known from the velocity triangle oab [Figure 6.5(b)]. It is required to find the velocity of a point C in the link.

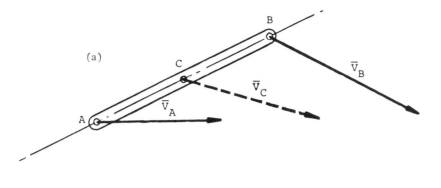

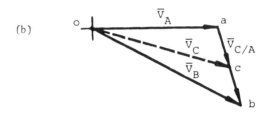

Figure 6.5   Velocity of any point on a link: (a) Link; (b) velocity polygon.

The velocity of point C is given by the vector equation

$$\overline{V}_C = \overline{V}_A + \overline{V}_{C/A}$$

where $\overline{V}_{C/A}$, the velocity of C relative to A, must be (1) perpendicular to AC and (2) proportional to AC. For example, if point C were located at the midpoint of the link between A and B, vector $\overline{V}_{C/A}$ would be defined by a distance $\overline{ac}$ on the velocity polygon, where point c is the midpoint between a and b. Hence, for any point C on link AB, the velocity $\overline{V}_{C/A}$ is obtained by selecting a point c between a and b on the velocity polygon such that

$$\frac{ac}{ab} = \frac{AC}{AB}$$

The distance $\overline{ac}$ will then satisfy both conditions of direction and magnitude of vector $\overline{V}_{C/A}$. Therefore, a line drawn from o to c, that is, $\overline{oc}$, will represent $\overline{V}_C$, the vector sum of $\overline{V}_A$ and $\overline{V}_{C/A}$.

PROCEDURE

1. Determine a point c on line $\overline{ab}$ of the triangle such that

$$\frac{ac}{ab} \text{ (on the polygon)} = \frac{AC}{AB} \text{ (on the link)}$$

2. Join $\overline{oc}$.
3. Now since all vectors originating from the pole o represent absolute velocities, line $\overline{oc}$ represents magnitude and direction of velocity of C $(\overline{V}_C)$. A quick check of the velocity polygon [Figure 6.5(b)] will verify the vector equation $\overline{V}_C = \overline{V}_A + \overline{V}_{C/A}$.

Point Offset From Centerline of Two Known Velocities

In the preceding section, the point of unknown velocity considered was located on the centerline connecting two points of known velocity. Let us now consider the expanded link ABD in Figure 6.6(a), where the velocities of points A and B are both given as shown, and let it be required to find the velocity of point D, located away from center-line AB.

The velocity of D $(\overline{V}_D)$ is given by the vector equation

$$\overline{V}_D = \overline{V}_A + \overline{V}_{D/A}$$

or

$$\overline{V}_D = \overline{V}_B + \overline{V}_{D/B}$$

where $\overline{V}_{D/A}$ and $\overline{V}_{D/B}$ can be obtained from a velocity polygon [Figure 6.6(b)].

PROCEDURE

1. Develop the velocity polygon, assuming a simple link AB. (Note that this polygon is the same as that of the preceding example.)
2. Through point a on the polygon, draw a line perpendicular to side AD on the link to indicate the direction of $\overline{V}_{D/A}$. Since A is rigidly

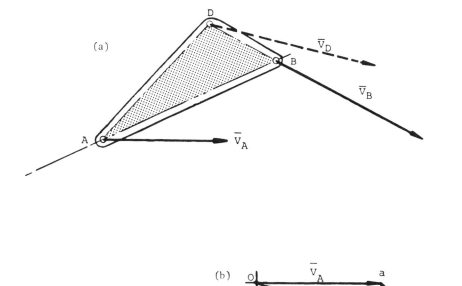

Figure 6.6   (a) Link; (b) velocity polygon.

   connected to D, the only direction that the velocity of D relative to
   A could have is one perpendicular to a line joining A to D.
3. Similarly, to indicate the direction of $\overline{V}_{D/B}$, draw a line through
   point b perpendicular to DB.
4. Define the point where the two perpendiculars in Steps 2 and 3
   intersect as point d. This point defines the terminus of vector $\overline{V}_D$.
5. Draw a line to connect point d to the polar origin o. This line defines
   the magnitude of the required vector $\overline{V}_D$.

The Velocity Image: Alternative Approach

In the example of Figure 6.6, we note that the relative velocity $V_{D/A}$ is
given by the vector extending from point a to point d, and the relative
velocity $V_{D/B}$ is given by the vector extending from b to d. We also note
that triangle abd is similar to triangle ABD, since it was produced by
lines drawn perpendicular to corresponding sides of the link. Because

of this similarity (or proportionality), triangle *abd* is often referred to as the ***velocity image*** of link *ABD*.

The velocity image is a useful concept in velocity analysis. If the velocities of any two points on a link are known in the velocity polygon, the velocity of a third point on that link can readily be determined by constructing the velocity image, making sure that the lines which define the image are perpendicular to the corresponding lines that form the link. Also, the letters used to designate both the link and image must run in the same cyclic order.

6.9   VELOCITY ANALYSIS OF A SIMPLE MECHANISM

Consider the four-link mechanism shown in Figure 6.7(a), in which the angular velocity of link 2 or ω_2 is 10 rad/sec. Determine the angular velocities of links 4 and 3.

Since point C is on link 4, the angular velocity of link 4 is the same as that of point C and is given by

$$\omega_4 = \omega_C = \frac{V_C}{CD}$$

where V_C is the magnitude of the linear velocity of point C. Hence, we must first determine $\overline{V}_C$, which is given by the vector equation

$$\overline{V}_C = \overline{V}_B + \overline{V}_{C/B}$$

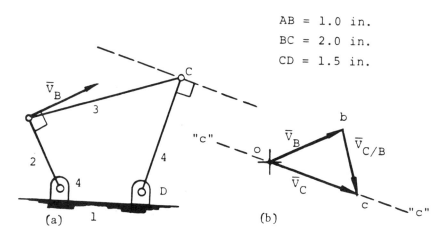

AB = 1.0 in.
BC = 2.0 in.
CD = 1.5 in.

Figure 6.7   Velocity analysis of a four-bar mechanism: (a) Four-bar mechanism; (b) velocity polygon.

Similarly, the angular velocity of link 4 is found from

$$\omega_3 = \omega_{CB} = \frac{V_{C/B}}{CB}$$

PROCEDURE [see Figure 6.7(b)]

1. Define a point o as the pole or origin for a velocity polygon.
2. Starting from point o and using a convenient scale, lay out vector $\overline{ob}$ perpendicular to link 2 to represent velocity $\overline{V}_B$, whose magnitude is computed as

$$V_B = \omega_2 \times AB = 10 \text{ in./sec}$$

3. Through point o, draw a line "c"–"c", perpendicular to link 4, to indicate the direction of velocity $\overline{V}_C$. Note that $\overline{V}_C$ must be perpendicular to CD since link 4 could have circular motion only about the fixed axis D. However, the orientation and magnitude are as yet unknown.
4. To define the magnitude of $\overline{V}_C$, we must find the velocity $\overline{V}_{C/B}$, which relates point b (known) to point c (unknown) on the polygon. By definition, this velocity must be perpendicular to link BC and, by convention, must be directed to point c from point b on the polygon. Therefore, draw a line perpendicular to BC extending from point b and terminating at line "c"–"c". This point of termination defines point c or the magnitude of $\overline{V}_C$; and it also defines the magnitude of $\overline{V}_{C/B}$.
5. Scale velocity vectors $\overline{V}_C$ and $\overline{V}_{C/B}$ to obtain the magnitudes.

$$V_C = 12.0 \text{ in./sec}$$

$$V_{C/B} = 8.5 \text{ in./sec}$$

6. Finally, as required angular velocities of links 4 and 3 are given as

$$\omega_4 = \frac{V_C}{CD} = \frac{12.0}{1.5} = 8.0 \text{ rad/sec}$$

and

$$\omega_3 = \frac{V_{C/B}}{CB} = \frac{8.5}{2.0} = 4.25 \text{ rad/sec}$$

6.10   VELOCITIES OF SLIDING CONTACT MECHANISMS

When one body slides on another, the difference in their absolute velocites (or their relative velocity) is defined as the **velocity of sliding**. The velocity of sliding is always directed along the common tangent drawn through the contact point.

To illustrate this concept, consider the cam-and-follower mechanism shown in Figure 6.8(a), where P is the point of contact between the two bodies. The angular velocity of the cam ω_C is known and $\overline{V}_{P(C)/P(F)}$, the velocity of sliding, is required.

By definition, the velocity of sliding is the velocity of P on C relative to the velocity P on F, or

$$\overline{V}_{P(C)/P(F)} = \overline{V}_{P(C)} - \overline{V}_{P(F)}$$

The velocity of sliding must, by definition, be along the common tangent $T–T$. Therefore, $\overline{V}_{P(C)/P(F)}$ can be determined as follows:

PROCEDURE

1. Define polar origin o to start the velocity polygon [Figure 6.8(b)].
2. From origin o, lay out known velocity $\overline{V}_{P(C)}$ (magnitude and direction) perpendicular to OP.

$$\overline{V}_{P(C)} = OP \times \omega \quad \text{(directed perpendicular to } OP\text{)}$$

3. Define the terminus of $\overline{V}_{P(C)}$ as $p(C)$.
4. Through the origin draw line "$p(F)$"–"$p(F)$" parallel to the follower axis to indicate the direction of $\overline{V}_{P(F)}$.
5. To define $\overline{V}_{P(C)/P(F)}$, draw a line starting from $p(C)$ extending toward line "$p(F)$"–"$p(F)$" in the known direction of sliding (i.e., parallel to tangent line $T–T$). The point at which the two lines intersect defines the terminus of $\overline{V}_{P(F)}$, and line $p(F)–p(C)$ defines the magnitude of $\overline{V}_{P(C)/P(F)}$. Vector $\overline{V}_{P(C)/P(F)}$ is pointed toward $p(C)$, in accordance with convention. A quick check of the completed polygon reveals the vector equation

$$\overline{V}_{P(C)/P(F)} = \overline{V}_{P(C)} - \overline{V}_{P(F)}$$

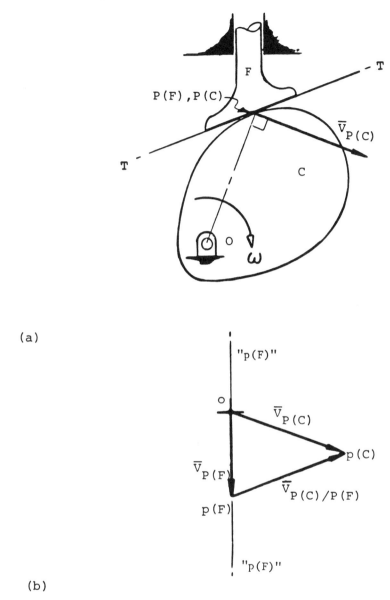

(a)

(b)

Figure 6.8   Velocities of points on a cam mechanism: (a) Mechanism; (b) velocity polygon.

EXAMPLE 6.1

Consider the quick-return mechanism shown in Figure 6.9(a). Crank
OP rotates at 2 rad/sec and slider S slides on follower F. Determine
the angular velocity of the follower and the velocity of sliding.

SOLUTION

To find the angular velocity of the follower, we must first find the linear
velocity of point P on F [$\overline{V}_{P(F)}$]. This unknown velocity is related to the
known velocity $\overline{V}_{P(S)}$ as follows:

$$\overline{V}_{P(F)} = \overline{V}_{P(S)} + \overline{V}_{P(F)/P(S)}$$

where $\overline{V}_{P(F)/P(S)}$ is the velocity of sliding (or the velocity of slip), which
is directed along the instantaneous path of the slider.

After determining the velocity of point P on F, we obtain the
required angular velocity from the relationship

$$\omega_F = \frac{V_{P(F)}}{QP}$$

Thus, we proceed as follows:

1. Define polar origin o to start the velocity polygon [Figure 6.9(b)].
2. Lay out the velocity of point P on S [$\overline{V}_{P(S)}$] perpendicular to OP.
    The magnitude of this velocity is obtained from

    $$V_{P(S)} = \omega(OP)$$

    $$= 2(1.5) = 3 \text{ in./sec}$$

3. Define the terminus of the vector $\overline{V}_{P(S)}$ as $p(S)$.
4. Through point o draw a line "$p(F)$"–"$p(F)$" to indicate the known
    direction of $\overline{V}_{P(F)}$.
5. Lay out the relative velocity (or velocity of slip) starting from $p(S)$
    and moving in the known direction of slip toward line "$p(F)$"–
    "$p(F)$". This relative velocity vector intersects line "$p(F)$"–"$p(F)$"
    and defines the magnitude of $\overline{V}_{P(F)}$ as well as that of the slip velocity
    $\overline{V}_{P(S)/P(F)}$ at that point. Label the point of intersection $p(F)$. Note that
    the orientation of the velocity of the sliding vector depends on
    whether we consider the velocity of the slider relative to the fol-

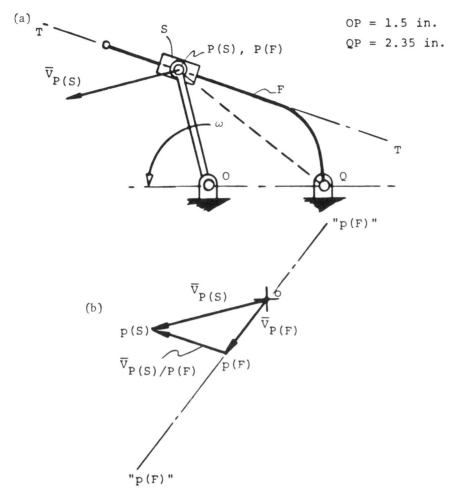

Figure 6.9   Velocity analysis of a quick-return mechanism: (a) Mechanism; (b) velocity polygon.

lower, where vector $\overline{V}_{P(S)/P(F)}$ is pointed toward p(S), or the velocity of the follower relative to the slider, where vector $\overline{V}_{P(F)/P(S)}$ would be pointed in the opposite direction, toward p(F).

6. Scale vectors $\overline{V}_{P(F)}$ and $\overline{V}_{P(S)/P(F)}$ to obtain their magnitudes as follows:

$$V_{P(F)} = 1.7 \text{ in./sec}$$

and

$$V_{P(S)/P(F)} = 2.0 \text{ in./sec}$$

7. Determine the angular velocity of F from the relationship

$$V_{P(F)} = \omega_{QP} \times QP$$

Therefore,

$$\omega_{QP} = \frac{V_{P(F)}}{QP}$$

$$= \frac{1.7}{2.35} = 0.72 \text{ rad/sec}$$

6.11   VELOCITIES OF A BODY WITH ROLLING CONTACT

Consider wheel W in Figure 6.10(a), which rolls, without slipping, on a track. As the wheel moves to the right, the center A is considered the moving frame of reference and the velocity of every other point on the wheel is relative to the velocity of this point. Therefore, we apply the relative motion concept to determine the velocities of points B and C as follows:

PROCEDURE

1. Define polar origin o [Figure 6.10(b)].
2. Locate the velocity vector $\overline{V}_A$. The direction of this vector is perpendicular to a line joining A to the instant center I (or AI), and the magnitude is given by

$$V_A = AI \times \omega$$

3. Define the terminus of $\overline{V}_A$ as a.
4. Lay out the direction of the vector $\overline{V}_B$ perpendicular to a line joining B to I (or BI). Therefore, draw a line "b"–"b", of undefined length, through pole o.
5. Determine the vector $\overline{V}_B$.
    a. Determine the vector $\overline{V}_{B/A}$. By definition, $\overline{V}_{B/A}$ assumes that point B rotates about point A and this velocity is perpendicular to a line connecting B to A (i.e., AB) on the wheel. Also, this

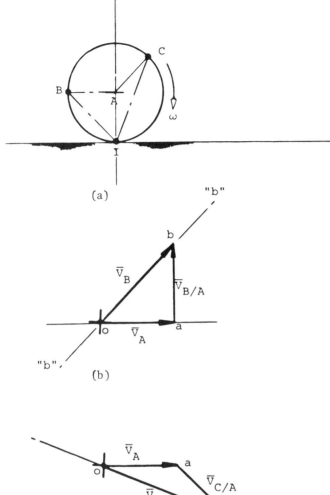

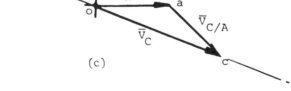

Figure 6.10   Velocities of a body with rolling contact.

velocity is represented by a vector headed from a to b on the polygon.
b.   Therefore from a, draw a line in a direction perpendicular to AB to intersect line "b"–"b". This point of intersection defines the termini of vectors $\overline{V}_{B/A}$ and $\overline{V}_B$.

6. Label the point of intersection b. A quick check of the polygon should immediately reveal the vector equation:

$$\overline{V}_B = \overline{V}_A + \overline{V}_{B/A}$$

Following the procedure described above, the velocities of point C $(\overline{V}_C)$ can be determined as shown in Figure 6.10(c).

6.12 VELOCITY ANALYSIS OF A COMPOUND MECHANISM

Consider the toggle mechanism in Figure 6.11(a). Wheel W rotates at 10 rpm. Find the velocities of slider S and point P.
  We first note the following:

1. The velocity of the slider is the same as that of point E, that is, $\overline{V}_E$.
2. Linkage may be analyzed in two parts:
   a. The four-bar section $ABCD$ to determine $\overline{V}_C$.
   b. The slider-crank section DCE, using the value found for $\overline{V}_C$ in part (a) to find $\overline{V}_E$.
3. $\overline{V}_E$ is found using the velocity image of link CE.

   PROCEDURE

1. Define polar origin o to start the velocity polygon [Figure 6.11(b)].
2. Lay out, using a convenient scale, velocity vector $\overline{V}_B$. The magnitude of this vector is given as

$$V_B = AB \times \omega_{AB}$$

Therefore,

$$\omega_{AB} = 10 \left(\frac{2\pi}{60}\right) = 1.05 \text{ rad/sec}$$

3. Define the terminus of vector $\overline{V}_B$ as point b.
4. Through the origin draw a line "c"–"c" perpendicular to CD to indicate the orientation of velocity vector $\overline{V}_C$.

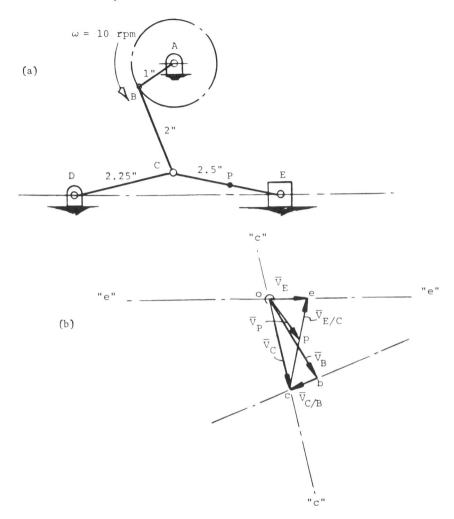

Figure 6.11   Velocities of points on a toggle mechanism: (a) Mechanism; (b) velocity polygon.

5.   Lay out relative velocity $\overline{V}_{C/B}$. Starting from point b, draw a line heading toward the "c"–"c" in a direction perpendicular to link BC. This line should meet "c"–"c" at point c, which defines the termini of both $\overline{V}_C$ and $\overline{V}_{B/C}$.
6.   Scale the magnitude of $\overline{V}_C$, that is, the line $\overline{oc}$.

$$V_C = 1.06 \text{ in./sec}$$

7. Again, through the origin, draw the line "e"–"e" to indicate the orientation of the absolute velocity vector $\overline{V}_E$ (horizontal).
8. Lay out relative velocity vector $\overline{V}_{E/C}$. Starting from point c, draw a line heading toward line "e"–"e" in a direction perpendicular to link CE. This line should meet line "e"–"e" at point c to define the magnitudes of both $\overline{V}_E$ and $\overline{V}_{E/C}$.
9. Scale the magnitude of $\overline{V}_E$, that is, the line $\overline{oe}$.
10. To find the velocity of point P ($\overline{V}_P$), locate a point p in line $\overline{ec}$ on the polygon such that the ratio of $\overline{cp}$ to $\overline{pe}$ is the same as ratio of CP to PE on the link, or

$$\frac{cp}{pe} = \frac{CP}{PE}$$

11. Connect point p with a vector originating from point o. This vector $\overline{op}$ then defines the velocity $\overline{V}_P$.
12. Scale the magnitude of $\overline{V}_P$.

$$V_P = 0.58 \text{ in./sec}$$

II.B

GRAPHICAL TECHNIQUES: ACCELERATION ANALYSIS

Acceleration is a very important property of motion to a machine designer. By Newton's law, $F = ma$, the force imposed on machine members is directly proportional to the acceleration. Although fundamentally kinematics is not concerned with forces, the study of acceleration is vitally important to the dynamic analysis, because of its influence on stresses, bearing loads, vibration, and noise.

In high-speed machines, the accelerations and resulting inertial forces can be very large compared to static forces that do useful work. For example, in modern automobile and aircraft engines, the stresses imposed on the connecting rods as a result of these accelerations are considerably greater than those produced on the piston by gas pressure. Also, a small imbalance of the rotating members can produce forces that are many times larger than the weights of the members. Therefore, a complete acceleration analysis is a prerequisite to stress analysis and proper design of machine members.

7

Linear Acceleration Along Curved Paths

7.1 INTRODUCTION

A body moving along a curved path is always subjected to an acceleration since its velocity changes from one position to the next. If the body moves with a constant angular speed, the velocity changes is one direction only, and the resultant acceleration is a **normal acceleration**. If the body moves with a variable angular speed, the velocity change is one of direction and magnitude, and the resultant acceleration is a summation of the **normal acceleration** (due to the direction change) and a **tangential acceleration** (due to the magnitude change).

7.2 NORMAL ACCELERATION

Consider a point A on a body that moves in a circular path of radius R with constant angular speed through an angle $\Delta\theta$ [Figure 7.1(a)]. The initial velocity at point A_i is $\overline{V}_A^i$ and the final velocity at point A_f is $\overline{V}_A^f$.

   The velocity polygon representing this motion is given in Figure 7.1(b). Here the resultant change in velocity of point A ($\Delta\overline{V}_A = \overline{V}_A^f - \overline{V}_A^i$) is in direction only, since $\overline{V}_A^i$ and $\overline{V}_A^f$ are equal in magnitude. Also, for a small angular displacement $\Delta\theta$, $\Delta\overline{V}_A$ is oriented normal or perpendicular to the instantaneous linear velocity.

145

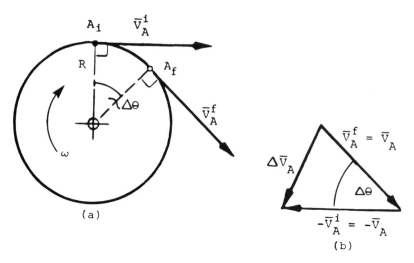

Figure 7.1   Disk with uniform rotation.

The normal acceleration magnitude of point A or A_A^N is obtained from

$$\overline{A}_A^N = \frac{\overline{V}_A^f - \overline{V}_A^i}{\Delta T} = \frac{\Delta \overline{V}_A}{\Delta T}$$

where

$$A_A^N = \frac{\Delta V_A}{\Delta T}$$

$$\Delta V_A = V_A \, \Delta\theta \quad \text{(for small } \Delta\theta)$$

$$A_A^N = \frac{V_A \, \Delta\theta}{\Delta T}$$

However,

$$\frac{\Delta\theta}{\Delta T} = \omega$$

$$A_A^N = V_A\omega$$

$$= R\omega \times \omega = R\omega^2$$

Also,

$$\omega = \frac{V}{R}$$

$$A_A^N = V_A \frac{V_A}{R} = \frac{V_A^2}{R} \qquad (7.1)$$

This relationship states that the magnitude of the acceleration of any point on a body rotating at a constant speed is equal to the square of the angular velocity of the body multiplied by the distance of the point from the center of rotation, or the square of the linear velocity divided by the distance of the point from the axis of rotation. This acceleration, called the normal acceleration, is always directed radially toward the center of rotation perpendicular or normal to the instantaneous linear velocity.

A familiar example of radial or normal acceleration is that of a car traveling around a circular track at a constant speed. Although the speedometer reading remains the same, there is acceleration because the direction of velocity is continually changing. Whenever there is a change of direction of velocity, there is acceleration. Furthermore, it is the normal acceleration that keeps the car on its circular path. If it ceases, the car will at once move in the direction tangent to the circular track.

It should be noted from Eq. 7.1 that if the point has rectilinear motion, the normal acceleration is always zero. That is,

$$A_A^N = \frac{V_A^2}{R_\infty} = \frac{V_A^2}{\infty} = 0$$

7.3   TANGENTIAL ACCELERATION

Consider a point A on a body that moves in a circular path of radius R with increasing angular speed ω through an angle $\Delta\theta$ (Figure 7.2). As before, $\overline{V}_A^i$ is the initial velocity at point A_i and $\overline{V}_A^f$ the final velocity at point A_f.

Here, for convenience, we consider the velocity change due to magnitude only. This gives the vector equation

$$\Delta\overline{V}_A = \overline{V}_A^f - \overline{V}_A^i \quad \text{(in tangential direction only)}$$

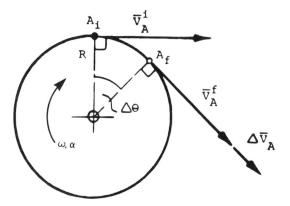

Figure 7.2   Disk with nonuniform rotation.

Therefore, the magnitude of the tangential acceleration of point A (A_A^T) is

$$A_A^T = \frac{\Delta V_A}{\Delta T}$$

$$= \frac{R\omega^f - R\omega^i}{\Delta T}$$

$$= \frac{R\,\Delta\omega}{\Delta T}$$

However,

$$\frac{\Delta\omega}{\Delta T} = \alpha$$

Therefore,

$$A_A^T = R\alpha \tag{7.2}$$

This relationship states that the magnitude of the tangential acceleration of any point on a rotating body whose angular velocity is changing is equal to the angular acceleration multiplied by the distance of the point from the center of rotation. (*Note:* This is the same relationship as that derived by an alternative method in Section 2.3.)

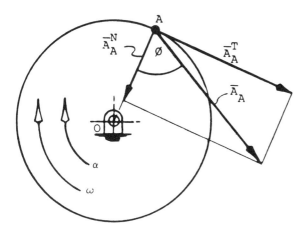

Figure 7.3   Resultant acceleration of a point with nonuniform rotation.

7.4   RESULTANT ACCELERATION

The total or resultant acceleration of a point A (or $\overline{A}_A$) (Figure 7.3) is the vectorial sum of the normal or radial acceleration ($\overline{A}_A^N$) and the tangential acceleration ($\overline{A}_A^T$). Symbolically,

$$\overline{A}_A = \overline{A}_A^N + \overline{A}_A^T$$

where the magnitude is given by

$$A_A = |\overline{A}_A| = \sqrt{(A_A^N)^2 + (A_A^T)^2}$$
$$= \sqrt{(R\omega^2)^2 + (R\alpha)^2}$$

The direction of this acceleration or the angle ϕ that the vector makes with the radius R is obtained from

$$\phi = \tan^{-1} \frac{A_A^T}{A_A^N}$$

EXAMPLE 7.1

A, B, and C are points on a rigid body that rotates about center O as shown in Figure 7.4. Given that the angular velocity ω is 2 rad/sec and the angular acceleration α is 3 rad/sec², calculate and check graphically

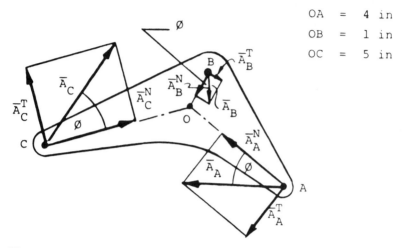

OA	=	4 in
OB	=	1 in
OC	=	5 in

Figure 7.4   Example problem.

the magnitudes and phase angles of the resultant acceleration of A, B, and C.

SOLUTION

The normal acceleration magnitude A^N is given by

$$A^N = r\omega^2$$

Therefore,

$$A_A^N = 4(2)^2 = 16 \text{ in./sec}^2$$
$$A_B^N = 1(2)^2 = 4 \text{ in./sec}^2$$
$$A_C^N = 5(2)^2 = 20 \text{ in./sec}^2$$

The tangential acceleration magnitude A^T is given by

$$A^T = r\alpha$$

Therefore,

$$A_A^T = 4(3) = 12 \text{ in./sec}^2$$
$$A_B^T = 1(3) = 3 \text{ in./sec}^2$$
$$A_C^T = 5(3) = 15 \text{ in./sec}^2$$

The resultant acceleration magnitude A is given by

$$A = \sqrt{(A^N)^2 + (A^T)^2}$$

Therefore,

$$A_A = \sqrt{(16)^2 + (12)^2} = 20 \text{ in./sec}^2$$

$$A_B = \sqrt{(4)^2 + (3)^2} = 5 \text{ in./sec}^2$$

$$A_C = \sqrt{(20)^2 + (15)^2} = 25 \text{ in./sec}^2$$

The phase angle ϕ is given by

$$\phi = \tan^{-1} \frac{A^T}{A^N}$$

Therefore,

$$\phi_A = \tan^{-1} \frac{12}{16} = 36.8°$$

$$\phi_B = \tan^{-1} \frac{3}{4} = 36.8°$$

$$\phi_C = \tan^{-1} \frac{15}{20} = 36.8°$$

Figure 7.4 also shows the required graphical solution.

Note that this problem could have been solved using the fact that both normal and tangential components are proportional to their distances from the center of rotation. That is, having found the values of components for point A, we could have determined the values of the corresponding components for points A and B by simple ratio.

Also, note that the phase angle ϕ is the same for all points on the link. This is because this angle depends only on the angular velocity and angular acceleration of the link, as can be seen from the following derivation:

$$\phi = \tan^{-1} \frac{A^T}{A^N}$$

$$= \tan^{-1} \frac{r\alpha}{r\omega^2}$$

$$= \tan^{-1} \frac{\alpha}{\omega^2}$$

EXAMPLE 7.2

In Figure 7.5, pulleys P_2 (2 in. in diameter) and P_3 (6 in. in diameter) are driven by pulley P_1 (4 in. in diameter) via a continuous belt. At this instant, P_1 has an angular velocity of 1 rad/sec counterclockwise and an angular acceleration of 5 rad/sec². Determine the acceleration of points A, B, C, and D on the belt.

SOLUTION

For point A on pulley P_1

  Normal acceleration:

  $$\overline{A}_A^N = r_1\omega^2 \quad \text{(directed toward center of pulley)}$$

  $$= 2(1)^2 = 2 \text{ in./sec}^2 \quad \text{(directed toward center of pulley)}$$

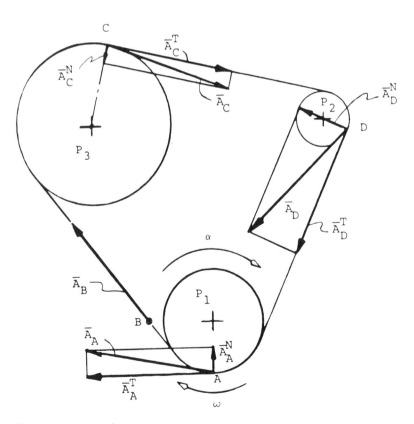

Figure 7.5   Resultant acceleration of points on a belt.

Tangential acceleration:

$$\overline{A}_A^T = r_1\alpha \quad \text{(directed along the belt)}$$

$$= 2(5) = 10 \text{ in./sec}^2 \quad \text{(directed along the belt)}$$

Resultant acceleration:

$$\overline{A}_A = \overline{A}_A^N + \overline{A}_A^T \quad \text{(vectorial sum)}$$

$$\text{where } A_A = \sqrt{10^2 + 2^2}$$

Therefore, $\overline{A}_A = 10.2 \text{ in./sec}^2$ (directed as shown)

For point B on the belt

$$\overline{A}_B = \overline{A}_A^T \quad \text{(same as tangential acceleration of point A)}$$

$$= 10 \text{ in./sec}^2 \quad \text{(directed along the belt)}$$

For point C on pulley P_3

Normal acceleration:

$$\overline{A}_C^N = r_3\omega^2 \quad \text{(directed toward center of pulley)}$$

$$= 3\left(\frac{2}{3}\right)^2 = 1.33 \text{ in./sec}^2 \quad \substack{\text{(directed toward} \\ \text{center of pulley)}}$$

Tangential acceleration:

$$\overline{A}_C^T = \overline{A}_B \quad \text{(same as belt acceleration)}$$

$$= 10 \text{ in./sec}^2 \quad \text{(directed along the belt)}$$

Resultant acceleration:

$$\overline{A}_C = \overline{A}_C^N + \overline{A}_C^T \quad \text{(vectorial sum)}$$

$$\text{where } A_C = \sqrt{10^2 + 1.33^2}$$

Therefore, $\overline{A}_C = 10.1 \text{ in./sec}^2$ (directed as shown)

For point D on pulley P_2

Normal acceleration:

$$\overline{A}_D^N = r_2\omega^2 \quad \text{(directed toward center of pulley)}$$

$$= 1\left(\frac{2}{1}\right)^2 = 4 \text{ in./sec}^2 \quad \substack{\text{(directed toward} \\ \text{center of pulley)}}$$

Tangential acceleration:

$$\overline{A}_D^T = \overline{A}_B \quad \text{(same as belt acceleration)}$$

$$= 10 \text{ in./sec}^2 \quad \text{(directed along the belt)}$$

Resultant acceleration:

$$\overline{A}_D = \overline{A}_D^N + \overline{A}_D^T \quad \text{(vectorial sum)}$$

where $A_D = \sqrt{10^2 + 4^2}$

Therefore, $A_D = 10.8 \text{ in./sec}^2$ (directed as shown)

Figure 7.5 also shows the graphical solution.

7.5 PROPORTIONALITY OF ACCELERATIONS

It was shown earlier that if a body rotates about a fixed point, the normal and tangential accelerations of a point located on that body at a distance r from the center of rotation are given by

$$\overline{A}^N = r\omega^2 \quad \text{(directed along r toward the center)}$$

$$\overline{A}^T = r\alpha \quad \text{(directed perpendicular to r)}$$

Therefore, if we consider points A and B on link OAB in Figure 7.6, it is easy to see that

$$\frac{A_B^T}{A_A^T} = \frac{r_B}{r_A}$$

and

$$\frac{A_B^N}{A_A^N} = \frac{r_B}{r_A}$$

which indicates that both normal and tangential components of any two points along the radial line OA are proportional. Similarly, since the acceleration components of the points are proportional, the sum of these components, or the total acceleration, must be proportional.

A graphical representation of the proportionality of accelerations $\overline{A}_B$ and $\overline{A}_A$ and their components can be seen in Figure 7.6, where the

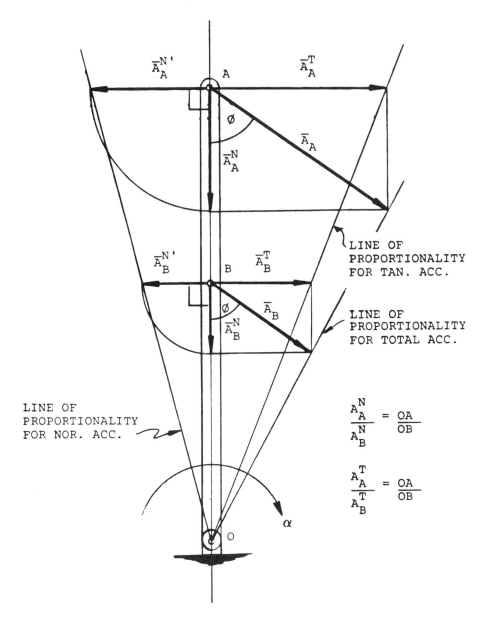

Figure 7.6   Proportionality of accelerations: Link OBA.

termini of these vectors lie on respective straight lines (termed **lines of proportionality**) that pass through the center of rotation O. Note that the proportionality line for the normal components $\overline{A}_B^N$ and $\overline{A}_A^N$ is a straight line drawn through the center O to touch the termini of these vectors when rotated 90° from the radial line.

7.6 RELATIVE ACCELERATION OF TWO POINTS ON A RIGID BODY

Just as in the velocity case, if the absolute accelerations of two points on a body are known, the relative acceleration between these points is the vectorial difference of their absolute accelerations. For example, if in link AC shown in Figure 7.7(a) the accelerations of both points A and C are known, then [see Figure 7.7(b)]

$$\overline{A}_{A/C} = \overline{A}_A - \overline{A}_C \tag{7.3}$$

or

$$\overline{A}_{C/A} = \overline{A}_C - \overline{A}_A \tag{7.4}$$

Normal and Tangential Accelerations

Using the vector $\overline{A}_{A/C}$, we can resolve this vector into its normal and tangential components simply by dropping two perpendiculars from the terminus of this vector: one to the line AC and the other to a line

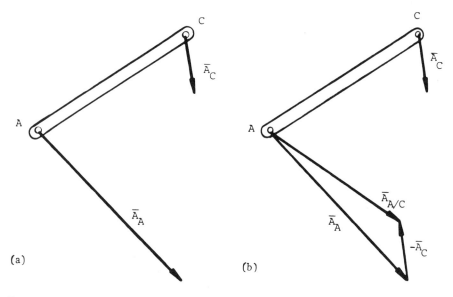

Figure 7.7   Relative acceleration of points on a rigid body.

normal to AC passing through point A. This construction yields $A^N_{A/C}$ along AC and $\overline{A}^T_{A/C}$ along the line normal to AC (see Figure 7.8).

For the vector $\overline{A}_{C/A}$, a similar construction can be made at point C. Consequently, the components for $\overline{A}^N_{C/A}$ and $\overline{A}^T_{C/A}$ will have directions opposite to those of $\overline{A}^N_{A/C}$ and $\overline{A}^T_{A/C}$.

Normal and tangential acceleration relationships for a point relative to another point on a link are similar to those for a point that undergoes pure rotation. For example, in Figure 7.8, where the acceleration of point A relative to point C (which is assumed to have a nonzero acceleration) is known, the following relationships apply:

1. Normal acceleration of A relative to C

$$\overline{A}^N_{A/C} = AC \times \omega^2_{AC} \quad \text{(directed from } A \text{ to } C\text{)}$$

2. Tangential acceleration of A relative to C

$$\overline{A}^T_{A/C} = AC \times \alpha_{AC} \quad \text{(directed perpendicular to } AC\text{)}$$

3. Resultant acceleration of A relative to C

$$\overline{A}_{A/C} = \overline{A}^N_{A/C} + \overline{A}^T_{A/C} \quad \text{(vectorially)}$$

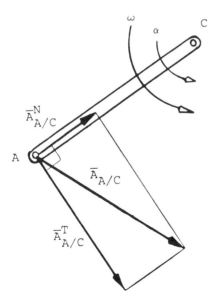

Figure 7.8   Relative acceleration.

4. Magnitude of acceleration of A relative to C

$$|\overline{A}_{A/C}| = \sqrt{(A_{A/C}^N)^2 + (A_{A/C}^T)^2}$$

5. Direction of resultant acceleration $\overline{A}_{A/C}$

$$\phi = \tan^{-1} \frac{A_{A/C}^T}{A_{A/C}^N}$$

It should be evident that if the acceleration of point C were zero, Eq. (7.4) would become

$$\overline{A}_{A/C} = \overline{A}_A$$

and all relative relationships would be reduced to absolute relationships, as in pure motion.

EXAMPLE 7.3

In link BC shown in Figure 7.9(a), the angular velocity of point C relative to point B is 1 rad/min (counterclockwise) and the relative angular acceleration is 0.5 rad/min^2 (clockwise). If the absolute linear acceleration of point B is 5 in./min, as shown, determine the absolute linear acceleration of point C.

SOLUTION

The absolute linear acceleration of point C can be represented in terms of the acceleration of point B and the relative motion of point C to point B, by the vectorial relationship

$$\overline{A}_C = \overline{A}_B + \overline{A}_{C/B} \tag{7.5}$$

where

$$\overline{A}_{C/B} = \overline{A}_{C/B}^N + \overline{A}_{C/B}^T \tag{7.6}$$

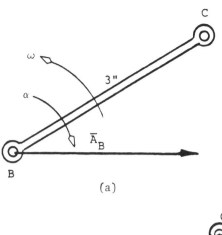

(a)

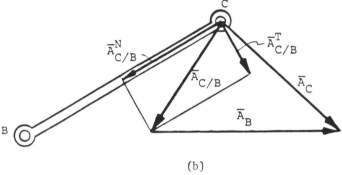

(b)

Figure 7.9   Example problem: (a) Floating link; (b) acceleration diagram.

However,

$$A_{C/B}^N = BC \times \omega^2$$

$$= 3(1.0)^2$$

$$= 3 \text{ in./min}^2$$

and

$$A_{C/B}^T = BC \times \alpha$$

$$= 3(0.5)$$

$$= 1.5 \text{ in./min}^2$$

Therefore, graphical application of Eqs. (7.5) and (7.6) yields

$$\overline{A}_{C/B} = 3.4 \text{ in./min}^2 \quad [\text{directed as shown in Figure 7.9(b)}]$$

$$\overline{A}_C = 4.2 \text{ in./min}^2 \quad [\text{directed as shown in Figure 7.9(b)}]$$

7.7   ACCELERATION OF ANY POINT IN A LINK

Point on Centerline of Two Known Accelerations

Normal and Tangential Component Method

Consider floating link AB shown in Figure 7.10(a). Suppose that the absolute accelerations of two points, A and B, are known and the absolute acceleration of a third point, C, is required.

The acceleration of point C can be found by applying the relationship

$$\overline{A}_C = \overline{A}_A + \overline{A}_{C/A}$$

where $\overline{A}_{C/A}$ is unknown but can be determined from proportionality with $\overline{A}_{B/A}$, which is obtainable from

$$\overline{A}_{B/A} = \overline{A}_B - \overline{A}_A$$

The proportionality between $\overline{A}_{C/B}$ and $\overline{A}_{B/A}$ is based on the fact that A, B, and C are points on the same link and therefore have the same angular velocity and acceleration. Hence,

$$\frac{A_{C/A}^N}{A_{B/A}^N} = \frac{CA}{BA}$$

$$\frac{A_{C/A}^T}{A_{B/A}^T} = \frac{CA}{BA}$$

PROCEDURE

1. Determine the acceleration of point B relative to point A [Figure 7.10(b)], using the vector equation

    $$\overline{A}_{B/A} = \overline{A}_B - \overline{A}_A$$

2. Resolve vector $A_{B/A}$ into its normal and tangential components, $\overline{A}_{B/A}^N$ along the link and $\overline{A}_{B/A}^T$ perpendicular to the link [Figure 7.10(c)].

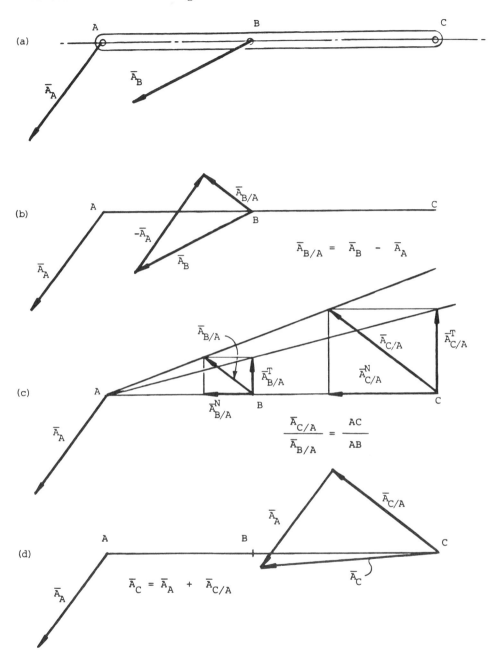

Figure 7.10   Normal and Tangential Component Method.

3. Determine $A_{C/A}$ [Figure 7.10(c)]. Since $A_{C/A}$ is proportional to $\overline{A}_{B/A}$, complete the following steps:
    a. From point A, draw a line proportionality for the tangential accelerations, touching the terminus of $\overline{A}^T_{B/A}$.
    b. From point C, construct a vector perpendicular to AB, terminating at the proportionality line in Step a. This vector defines $\overline{A}^T_{C/A}$.
    c. Again from point A, draw a second line of proportionality for the relative accelerations, touching the terminus of $\overline{A}_{B/A}$.
    d. From point C, construct a vector parallel to $\overline{A}_{B/A}$, terminating at the second line of proportionality in Step c. This vector defines $\overline{A}_{C/A}$.
4. Determine $\overline{A}_C$ [Figure 7.10(d)] using the vector equation

$$\overline{A}_C = \overline{A}_A + \overline{A}_{C/A}$$

Orthogonal Component Method

Consider again floating link AB in Figure 7.11(a), where the accelerations of points A and B are known and the acceleration of point C is required. In this method we make use of the fact that since A, B, and C are on the rigid body, their orthogonal (rotational and translational) components of acceleration must be proportional to each other.

   If we consider the rotational components for all points on the link centerline, the termini of these vectors must lie on a straight line. This line may be termed the *line of proportionality for the rotational components* of acceleration.

   Similarly, if we consider the translational components, the termini of these vectors, when rotated through a common angle away from the link centerline, must lie on a straight line. This line may be termed the *line of proportionality for the translational components* of acceleration.

   Therefore, to determine the acceleration of point C, we must first determine the rotational and translational components of acceleration at point C, using their respective lines of proportionality. Then, by adding these components, we obtain the required absolute acceleration of point C. Following is the procedure for construction.

   PROCEDURE

1. Resolve vectors $\overline{A}_A$ and $\overline{A}_B$ into their translational and rotational components (along the link and perpendicular to the link).

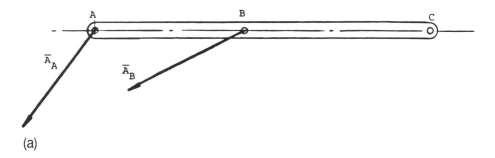

(a)

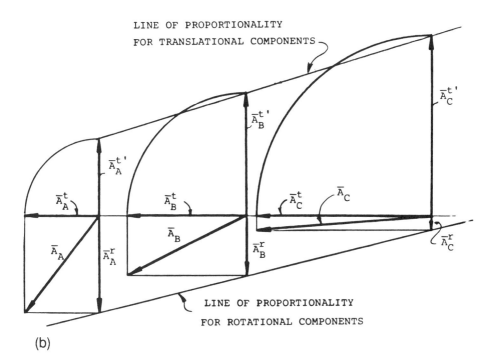

(b)

Figure 7.11   Orthogonal Component Method.

2. Establish the line of proportionality for rotational components of acceleration. This is a straight line drawn to touch vectors $\overline{A}_A^r$ and $\overline{A}_B^r$.
3. Determine $\overline{A}_C^r$. This is a vector drawn from point C perpendicular to the link and terminating at the proportionality line.
4. Establish the line of proportionality for translational components of acceleration. This is obtained by first rotating vectors $\overline{A}_A^t$ and $\overline{A}_B^t$ in a direction perpendicular to the link, then joining the termini

with a straight line. Note that the line of proportionality is used to obtain the magnitudes *only* (not directions) of the translational components of accelerations of all points along the link.

5. Determine the magnitude of $\overline{A}_C^t$. This is obtained by drawing an equivalent vector $\overline{A}_C^{t'}$, which extends from point C perpendicular to the link and terminates at the proportionality line. To obtain the true direction of $\overline{A}_C^t$, the equivalent vector $\overline{A}_C^{t'}$ must be rotated $90°$ to act in the direction consistent with $\overline{A}_A^t$ and $\overline{A}_B^t$ along the link.

6. Determine $\overline{A}_C$. This is obtained by summing the tangential and translational components $\overline{A}_C^r$ and $\overline{A}_C^t$ obtained above.

$$\overline{A}_C = \overline{A}_C^r + \overline{A}_C^t$$

Note also that if the directions of the given translational components of the two points oppose each other, this will indicate that the vectors are proportional with respect to a point that lies between the two points. Hence, these vectors must be rotated to opposite sides of the link centerline to establish the line of proportionality. This line of proportionality must therefore intersect the line joining the two points.

Instant Center of Acceleration Method

It has been established that the components of acceleration of a point on a rotating body are proportional to the distance of that point from the center of rotation. Therefore, if we consider any floating link, such as AB in Figure 7.12(a), having the same conditions as in methods 1 and 2, by locating the center of acceleration of the two given points A and B, we can readily determine the acceleration of any other point, such as point C, by proportion.

Here it should be noted that the center of acceleration or point of zero acceleration is the center of rotation. This should not be confused with the instant center used for velocity analysis. Although the instant center as defined for velocity analysis does have zero velocity, it may or may not have zero acceleration.

To determine the center of acceleration we employ the following construction, called the *four-circle method*:

The Four-Circle Method [Figure 7.12(b)]

1. Extend the lines of action of vectors A and B until they intersect at a point K.
2. Construct a circle to circumscribe the triangle formed by points A, B, and K.

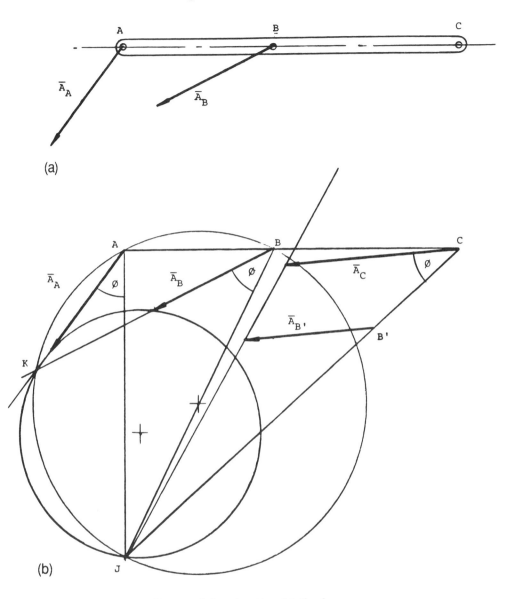

Figure 7.12   Instant Center of Acceleration Method.

3. Construct a second circle to pass through point K and the termini of vectors $\overline{A}_A$ and $\overline{A}_B$. This circle will intersect the first circle at another point, J. This point locates the center of acceleration of points A and B.
4. Connect points A and B to point J, using straight lines. Lines AJ and BJ are, therefore, the radii of rotation for points A and B.

Note the proportionality between the absolute acceleration vectors $\overline{A}_A$ and $\overline{A}_B$ and their respective radii of rotation. Also note the equality between phase angles JAK and JBK.

PROCEDURE (for finding $\overline{A}_C$)

Having determined the center of acceleration, complete the following steps:

1. Draw a straight line to connect point C and center J.
2. Rotate vector $\overline{A}_B$ to a point B' on radius CJ.
3. Using the rotated vector $\overline{A}_{B'}$ as a gauge, construct a proportionality line from point J to pass through the terminus of $\overline{A}_{B'}$.
4. From point C, construct a vector parallel to $\overline{A}_{B'}$ and terminating at the line of proportionality. This vector defines the required acceleration of point C $(\overline{A}_C)$.

Note that although called the four-circle method, the method requires only two circles to locate the center of acceleration. The method takes its name from the **four-circle theorem** in geometry, which states that *the four circles that circumscribe each side of a quadrilateral and the apex formed by extending their adjacent sides intersect at a point.*

Polygon Method

Consider again floating link AB where, as before, the accelerations of points A and B are known and the acceleration of point C is required [Figure 7.13(a)]. This method employs the **acceleration image concept**, which is similar to the velocity image concept considered earlier. Based on the geometric similarities between points on the link and their relative accelerations on the polygon, we can use proportions to determine the relative acceleration of any other point on the link and, hence, the absolute acceleration of that point.

PROCEDURE [Figure 7.13(b)]

1. Define a point o', called the pole, from which all absolute acceleration vectors originate.
2. Lay out given acceleration vectors $\overline{A}_A$ and $\overline{A}_B$ originating from the pole and label the respective termini a' and b'.
3. Draw a straight line to connect terminus a' to terminus b'. This line represents the magnitude of the relative acceleration between

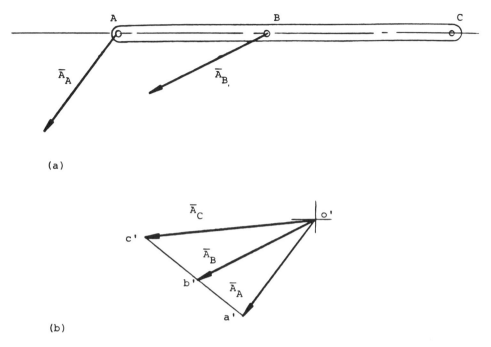

(a)

(b)

Figure 7.13   Polygon Method.

points B and A (or A and B) and, as in the velocity case, is called the acceleration image of link AB.

4.   Locate point c on line $a'b'$ in Step 3 such that

$$\frac{a'c'}{a'b'} = \frac{AC}{AB}$$

This point defines the terminus of vector $\overline{A}_C$, the required acceleration vector.

5.   Construct the vector $\overline{A}_C$ extending from the pole and terminating at point c' to determine its magnitude and direction.

Point Offset from Centerline of Two Known Accelerations

In the preceding section, the point of unknown acceleration was located on the centerline connecting two points of known acceleration. Let us now consider an expanded link BCD, in Figure 7.14, where the accelerations of points B and D are given as shown, and that of point D,

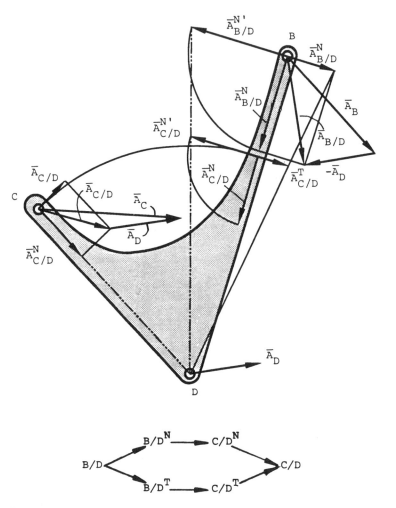

Figure 7.14　Acceleration of point offset from centerline.

which is offset from the centerline of the known acceleration points, is required. Also, let finding the angular velocity and acceleration of the link be required. Using the normal and tangential component method, or method 1, this problem can be solved as follows:

PROCEDURE

A: *Determining the Acceleration of Point C*

1. Determine $\overline{A}_{B/D}$ vectorially at B

$$\overline{A}_{B/D} = \overline{A}_B - \overline{A}_D$$

2. Join B to D with a straight line and resolve vector $A_{B/D}$ into its normal and tangential components along BD and perpendicular to centerline BD, respectively. Vectorially,

$$\overline{A}_{B/D} = \overline{A}^N_{B/D} + \overline{A}^T_{B/D}$$

3. Determine $\overline{A}^T_{C/D}$ as follows:
   a. Join C to D with a straight line and rotate point C about point D to meet BD at E.
   b. Draw the line of proportionality for the tangential acceleration vectors from D to the tip of $\overline{A}^T_{B/D}$.
   c. From E, draw vector $\overline{A}^T_{C/D}$ parallel to $\overline{A}^T_{B/D}$ and terminating at the proportionality line.
   d. Finally, rotate vector $\overline{A}^{T'}_{C/D}$ back to C and perpendicular to CD.
4. Determine $\overline{A}^N_{C/D}$ as follows:
   a. Draw the line of proportionality for the normal acceleration vectors from D to the tip of $\overline{A}^{N'}_{B/D}$ or $\overline{A}^N_{B/D}$ rotated.
   b. From E, draw vector $\overline{A}^{N'}_{C/D}$ parallel to $\overline{A}^{N'}_{B/D}$ and terminating at the proportionality line.
   c. Rotate vector $\overline{A}^{N'}_{C/D}$ first about E to centerline BD and then about D back to C along centerline CD.
5. Determine $\overline{A}_{C/D}$ by vector summation

$$\overline{A}_{C/D} = \overline{A}^N_{C/D} + \overline{A}^T_{C/D}$$

6. Determine $\overline{A}_C$. This is obtained from the vectorial relationship

$$\overline{A}_C = \overline{A}_D + \overline{A}_{C/D}$$

Therefore to vector $\overline{A}_{C/D}$ found in Step 5, add $\overline{A}_D$ to obtain $\overline{A}_C$.

B: *Determining the Angular Velocity of the Link*

We may use either $\overline{A}^N_{B/D}$ or $\overline{A}^N_{C/D}$, since all points in the link have the same motion relative to any other reference point. Choosing $\overline{A}^N_{C/D}$,

$$A^N_{C/D} = CD \times \omega^2_{CD}$$

or

$$\omega_{C/D} = \left(\frac{A^N_{C/D}}{CD} \right)^{1/2}$$

C: *Determining the Angular Acceleration*

Again, as in the angular velocity case above, we may employ

$$\overline{A}_{B/D}^{T} \qquad \text{or} \qquad \overline{A}_{C/D}^{T}$$

If we choose $\overline{A}_{B/D}^{T}$, then we can write the expression

$$A_{B/D}^{T} = BD \times \alpha$$

from which

$$\alpha = \frac{A_{B/D}^{T}}{BD}$$

7.8   CORIOLIS ACCELERATION

If one link slides radially on another link that is rotating, the sliding link (or slider) experiences an acceleration perpendicular to the radial line joining it to the center of rotation. Part of this acceleration is the effect of the changing distance of the slider from the center, and part is the effect of rotation of the radial sliding velocity vector. The resulting tangential acceleration of the slider relative to the other link is called the *Coriolis acceleration*.

Consider rotating link AD in Figure 7.15, which turns at a constant velocity ω about a fixed axis A as slider S freely slides (radially outward) on it. B and C are two coincident points at a distance r from A. Point C is on link AD directly beneath point B, which is on the slider.

First Change (Figure 7.16)

Consider the relative velocity of point B to point C radially along the link as the slider rotates from B, C to B', C' (a change in the direction of $\overline{V}_{B/C}$ due to rotation). If we assume that $\Delta\theta$ is small, the magnitude of ΔV can be expressed as

$$\Delta V = V_{B/C}\, \Delta\theta$$

and

$$\frac{\Delta V}{\Delta T} = V_{B/C}\frac{\Delta\theta}{\Delta T}$$

Therefore,

$$A_B^{(1)} = V_{B/C}\omega$$

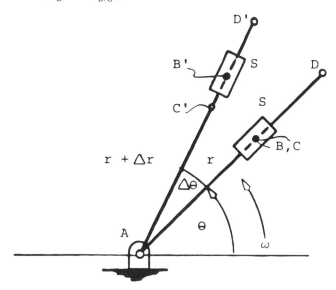

Figure 7.15   Rotating link with slider.

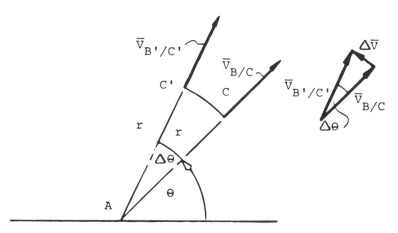

Figure 7.16   First change.

or

$$\overline{A}_B^{(1)} = V_{B/C}\omega \quad \text{(in the direction of } \Delta\overline{V})$$

Second Change (Figure 7.17)

Consider the tangential velocity of slider S as it moves outward from the center (a change in $\overline{V}_B$ due to the change in slider distance from A). Here the magnitude of $\Delta\overline{V}$ is given by

$$\Delta V = V_{B'} - V_B$$

$$= \Delta r\,\omega$$

and

$$\frac{\Delta V}{\Delta T} = \frac{\Delta r}{\Delta T}\,\omega$$

Therefore,

$$A_B^{(2)} = V_{B/C}\omega$$

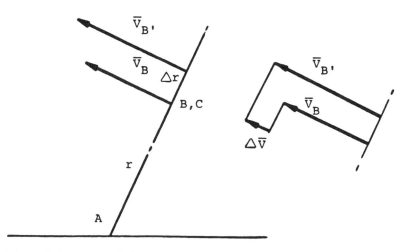

Figure 7.17   Second change.

or

$$\overline{A}_B^{(2)} = V_{B/C}\omega \quad \text{(in the direction of } \Delta\overline{V})$$

Summary

Combining the results of the two changes, we obtain the Coriolis acceleration

$$\overline{A}^{\text{Cor}} = \overline{A}_B^{(1)} + \overline{A}_B^{(2)}$$

or

$$\overline{A}^{\text{Cor}} = 2V_{B/C}\omega \quad \text{(in the direction of } \Delta\overline{V})$$

The Coriolis acceleration vector is always directed perpendicular to the rotating link and pointed as if it had been rotated about its tail through an angle 90° with vector $\overline{V}_{B/C}$ in the direction of ω, the angular velocity of the rotating link on which the sliding occurs.

EXAMPLE 7.4 (Figure 7.18)

Consider a block B that slides on a rotation rod AD and has a velocity of 5 ft/sec relative to a point C on the rod ($V_{B/C} = 5$ ft/sec). The angular velocity of the rod is 4 rad/sec. Determine the Coriolis acceleration in terms of both magnitude and direction for the following cases:

1. Block B moves outward and toward D as AD rotates counterclockwise [Figure 7.18(a)].
2. Block B moves inward and toward A as AD rotates counterclockwise [Figure 7.18(b)].
3. Block B moves inward and toward A as AD rotates clockwise [Figure 7.18(c)].
4. Block B moves outward and toward D as A rotates clockwise [Figure 7.18(d)].

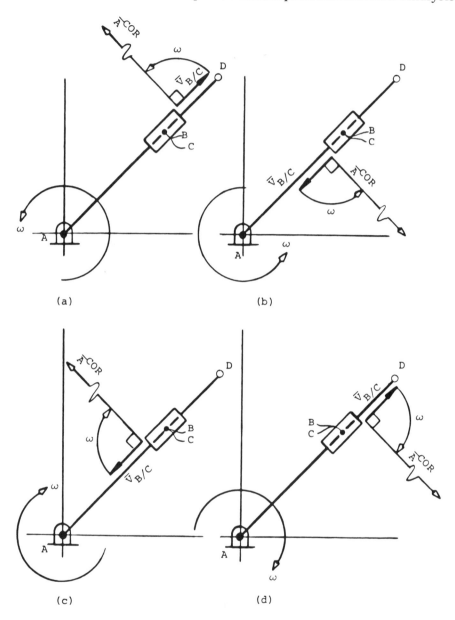

Figure 7.18   Coriolis acceleration for the four cases of Example 7.4: (a) Case
1; (b) case 2; (c) case 3; (d) case 4.

SOLUTION

The magnitude of the Coriolis acceleration in each case is given by

$$A^{Cor} = 2V_{B/C}\omega_{AD}$$

$$= 2(5)(4)$$

$$= 40 \text{ ft/sec}^2$$

Figure 7.18 shows the direction of the Coriolis acceleration in each case. Note that the direction of $\overline{A}^{Cor}$ is obtained from rotating the vector $\overline{V}_{B/C}$ through an angle 90° in the same direction (clockwise or counter-clockwise) as that of rotating link AD.

8

Effective Component of Acceleration Method

8.1 INTRODUCTION

The effective component of acceleration method is similar to that of the effective component of velocity method in that it is also based on the rigid-body principle. However, it is somewhat more complex for bodies with combined motion since it involves the concepts of relative acceleration and relative velocity. It will be seen that the effective component of acceleration along a line connecting two points, A and B, on a rigid body is composed of two parts:

1. The acceleration component of point A relative to point B due to rotation of A about B
2. The acceleration component of the reference point B due to translation

8.2 ACCELERATION OF END POINTS ON A LINK

Consider link AB shown in Figure 8.1. Both the velocity and acceleration of point A are known in magnitude and direction, whereas only the line of action of the acceleration of point B (or $\overline{A}_B$) is known. This line of action is the line "b"-"b". Determine the complete acceleration of point B ($\overline{A}_B$).

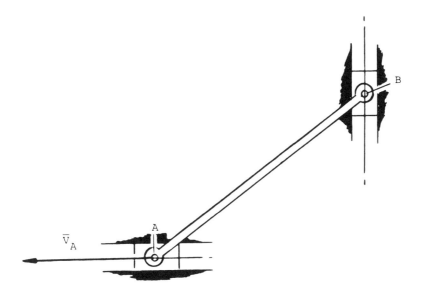

Figure 8.1   Linkage.

The acceleration of point B is related to that of point A by the vector equation

$$\overline{A}_B = \overline{A}_A + \overline{A}_{B/A}$$

and similarly, the effective components of $\overline{A}_B$ and $\overline{A}_A$ along link AB are related by

$$\overline{A}_B^{AB} = \overline{A}_A^{AB} + \overline{A}_{B/A}^{AB}$$

To determine $\overline{A}_{B/A}^{AB}$, we must consider the motion of point B relative to point A. That is, we assume that point A is fixed while B rotates about it. Therefore, the only effective component of acceleration that B can have along link AB is the normal or radial acceleration $(\overline{A}_{B/A}^N)$. In other words,

$$\overline{A}_{B/A}^{AB} = \overline{A}_{B/A}^N$$

where

$$A_{B/A}^N = \frac{V_{B/A}^2}{AB}$$

Therefore,

$$A_{B/A}^{AB} = \frac{V_{B/A}^2}{AB}$$

This means that we must find $V_{B/A}$ in order to find $\overline{A}_{B/A}^{AB}$. $V_{B/A}$ can be determined from a velocity polygon.

PROCEDURE [Figure 8.2(a)]

1. At point A, lay out the effective component of the given acceleration $\overline{A}_A$, that is, $\overline{A}_A^{AB}$. This is obtained, as in the velocity case, by dropping a perpendicular line from the terminus $\overline{A}_A$ to the extended line AB.
2. Find $\overline{A}_B^{AB}$ on link AB. This is obtained from the relationship

    $$\overline{A}_B^{AB} = \overline{A}_A^{AB} + \overline{A}_{B/A}^{AB}$$

    where $\overline{A}_{B/A}^{AB}$ is the radial or normal component of acceleration of point B relative to point A.
3. Determine the effective component of $\overline{A}_B$ on link AB ($\overline{A}_B^{AB}$).
    a. Determine $V_{B/A}$. This may be obtained from a velocity polygon, as in Figure 8.2(b).
    b. Calculate the magnitude of $\overline{A}_{B/A}^{AB}$.

    $$A_{B/A}^{AB} = A_{B/A}^N = \frac{V_{B/A}^2}{AB}$$

    c. From point B, lay out vector $\overline{A}_B^{AB}$ (the summation of vectors $\overline{A}_{B/A}^{AB}$ and $\overline{A}_A^{AB}$) along link AB.
4. If we know the path of acceleration $\overline{A}_B$ to be along line "b"-"b", this vector can now be defined by constructing a perpendicular line from the terminus of $\overline{A}_B^{AB}$ to meet line "b"-"b", similar to the procedure used in the velocity case.

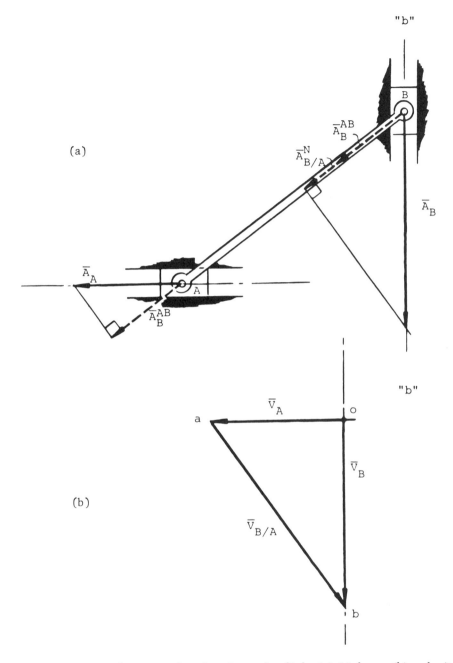

Figure 8.2   Acceleration of end points of a link: (a) Linkage; (b) velocity polygon.

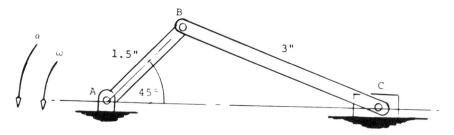

Figure 8.3   Slider-crank mechanism.

8.3   SLIDER-CRANK ANALYSIS

Crank AB of the slider-crank mechanism shown in Figure 8.3 rotates with an angular velocity of 2 rad/sec (counterclockwise) and angular acceleration of 2 rad/sec^2 (counterclockwise). Determine the acceleration of the slider.

The acceleration of slider C is related to that of point B by the vector equation

$$\overline{A}_C = \overline{A}_B + \overline{A}_{C/B}$$

and similarly, the effective components of these accelerations along BC are related by

$$\overline{A}_C^{BC} = \overline{A}_B^{BC} + \overline{A}_{C/B}^{BC}$$

where

$$A_{C/B}^{BC} = A_{C/B}^N$$

and

$$A_{C/B}^N = \frac{V_{C/B}^2}{BC}$$

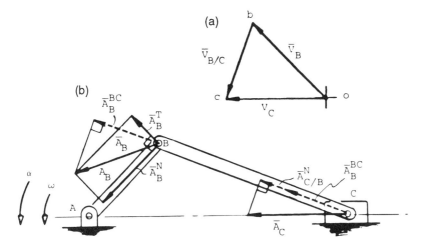

Figure 8.4   Acceleration of points on a slider-crank mechanism: (a) Acceleration diagram; (b) velocity polygon.

PROCEDURE [Figure 8.4(b)]

1.   From point B, lay out acceleration vector $\overline{A}_B$ to a convenient scale using the vector equation

$$\overline{A}_B = \overline{A}_B^N + \overline{A}_B^T$$

where

$$A_B^N = AB \times \omega^2$$

$$= 1.5(2)^2 = 6.0 \text{ in./sec}^2$$

$$A_B^T = AB \times \alpha$$

$$= 1.5(2) = 3 \text{ in./sec}^2$$

2.   Construct the effective component of $\overline{A}_B$ on BC $(\overline{A}_B^{BC})$. Drop a perpendicular from the terminus of $\overline{A}_B$ to extended link BC.
3.   Determine the effective component of $\overline{A}_C$ on link BC $(\overline{A}_C^{BC})$.
   a.   Determine the magnitude $V_{C/B}$ from a velocity polygon [Figure 8.4(a)].

$$V_{C/B} = 2.3 \text{ in./sec}$$

b. Lay out $\overline{A}_{C/B}^{BC}$ from point C along link BC. The magnitude of this vector is obtained

$$A_{C/B}^{BC} = \frac{(2.3)^2}{3} = 1.76 \text{ in./sec}^2$$

c. Lay out vector $\overline{A}_B^{BC}$ added to vector $\overline{A}_{C/B}^{BC}$ along BC.
4. Determine the absolute acceleration of point C $(\overline{A}_C)$. Since the path of the slider is a straight line, the direction of its acceleration must be along the same path. Therefore, to find the magnitude of this acceleration, drop a perpendicular from the effective component $\overline{A}_C^{BC}$ to intersect the line of action of $\overline{A}_C$. The point of intersection defines the acceleration $\overline{A}_C$.
5. Scale the vector $\overline{A}_C$ to determine its magnitude.

$$A_C = 7.5 \text{ in./sec}^2$$

8.4 FOUR-BAR LINKAGE ANALYSIS

Crank AB of the four-bar linkage in Figure 8.5 has an angular velocity of 2 rad/sec (counterclockwise) and is accelerating at the rate of 1 rad/sec². Determine the acceleration of point C.

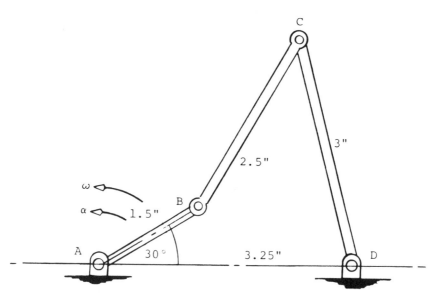

Figure 8.5 Four-bar mechanism.

In the example of Section 8.4, the acceleration of point C was related to point B by the vector equation

$$\overline{A}_C = \overline{A}_B + \overline{A}_{C/B}$$

and the effective component of the acceleration of point C along link BC was obtained from

$$\overline{A}_C^{BC} = \overline{A}_B^{BC} + \overline{A}_{C/B}^{BC}$$

In that example, since the direction of acceleration of point C was known, only one effective component of that acceleration was required to define the acceleration completely. In this example, however, the direction of acceleration of point C is not known. Therefore, another effective component is needed to define $\overline{A}_C$. This leads us to consider the effective component of point C along link CD, which is given by the equation

$$\overline{A}_C^{CD} = \overline{A}_D^{CD} + \overline{A}_{C/D}^{CD}$$

where

$$\overline{A}_D^{CD} = 0 \quad \text{(since point } D \text{ is fixed)}$$

$$\overline{A}_{C/D}^{CD} = \overline{A}_{C/D}^{N}$$

where

$$\overline{A}_{C/D}^{N} = \frac{V_C^2}{CD} \quad \text{(directed from } C \text{ to } D\text{)}$$

Therefore,

$$\overline{A}_C^{CD} = \frac{V_C^2}{CD} \quad \text{(directed from } C \text{ to } D\text{)}$$

With the two effective components $\overline{A}_C^{BC}$ and $\overline{A}_C^{CD}$ determined, the required acceleration of point C can readily be found.

PROCEDURE

1. From point B [Figure 8.6(b)], lay out acceleration vector $\overline{A}_B$ to a convenient scale, applying the relationship.

$$\overline{A}_B = \overline{A}_B^N + \overline{A}_B^T \quad \text{(vectorial sum)}$$

where

$$A_B^N = AB + \omega^2$$

$$= 1.5(4) = 6 \text{ in./sec}^2$$

$$A_B^T = AB \times \alpha$$

$$= 1.5(1) = 1.5 \text{ in./sec}^2$$

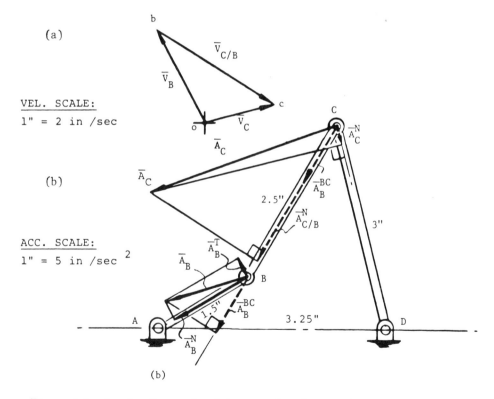

(a)

VEL. SCALE:
1" = 2 in /sec

(b)

ACC. SCALE:
1" = 5 in /sec ^{2}

(b)

Figure 8.6   Accelerations of points on a four-bar mechanism: (a) Velocity polygon; (b) acceleration diagram.

2. Construct the effective component of $\overline{A}_B$ on BC $(\overline{A}_B^{BC})$. Drop a perpendicular from the terminus of $\overline{A}_B$ to extended link BC.
3. Find the effective component of $\overline{A}_C$ on link BC $(\overline{A}_C^{BC})$.

$$\overline{A}_C^{BC} = \overline{A}_B^{BC} + \overline{A}_{C/B}^{BC}$$

where

$$A_{C/B}^{BC} = A_{C/B}^{N} = \frac{V_{C/B}^2}{BC}$$

a. Therefore, construct a velocity polygon [Figure 8.6(a)] to find $\overline{V}_{C\ B}$.

$$V_{C/B} = 4.15 \text{ in./sec}$$

b. Determine $\overline{A}_{C/B}^{BC}$.

$$A_{C/B}^{BC} = \frac{4.15^2}{2.5} = 6.8 \text{ in./sec}^2$$

Therefore,

$$\overline{A}_{C/B}^{BC} = 6.8 \text{ in./sec}^2 \quad \text{(directed along } BC \text{ toward } B)$$

c. From point C, add vector $\overline{A}_{C/B}^{BC}$ to vector $\overline{A}_B^{BC}$ along link BC to obtain $\overline{A}_C^{BC}$.
4. Determine the effective component of $\overline{A}_C$ along link CD $(\overline{A}_C^{CD})$.

$$\overline{A}_C^{CD} = \overline{A}_C^{N} \quad \text{(radial acceleration of point } C)$$

where

$$A_C^{N} = \frac{V_C^2}{CD} \quad (V_C \text{ from velocity polygon})$$

$$= \frac{2.15^2}{3} = 1.54 \text{ in./sec}^2$$

Therefore,

$$\overline{A}_C^{CD} = 1.54 \text{ in./sec}^2 \quad \text{(directed from } C \text{ to } D)$$

5. From point C, lay out vector $\overline{A}_C^{CD}$ on link CD directed toward the fixed axis D.
6. With the two effective components of point C now known, the required acceleration $\overline{A}_C$ is obtained by projecting perpendiculars from the termini of these two components until they intersect. This point of intersection defines the magnitude and direction of required vector $\overline{A}_C$.
7. Scale the magnitude of vector $\overline{A}_C$.

$$A_C = 14.5 \text{ in./sec}^2$$

Therefore,

$$\overline{A}_C = 14.5 \text{ in./sec}^2 \quad \text{(directed as shown)}$$

8.5 SLIDING CONTACT MECHANISMS

The effective method of acceleration for sliding contact mechanisms is similar to that employed in the velocity analysis of the same mechanism type, where the unknown velocity is found from the principle which states that

the effective components of velocities of coincident points on two bodies in sliding contact are equal along a common normal to the sliding path.

Here, use is made of the principle, except that the effective components are the contact point accelerations, instead of velocities. In other words, it can be stated that

the effective components of acceleration of the contact points are equal in the direction perpendicular to the sliding path.

For example, in the quick-return mechanism $ABCD$ shown in Figure 8.7, where points B (on the slider) and C (on the rod) are coincident points in sliding contact, it can be shown that the effective component of acceleration $\overline{A}_C$ along NN is the same for the effective components of accelerations of $\overline{A}_B$ and $\overline{A}_{C/B}$ along NN, when added. Vectorially, this may expressed as

$$\overline{A}_C^{NN} = \overline{A}_B^{NN} + \overline{A}_{C/B}^{NN} \tag{8.1}$$

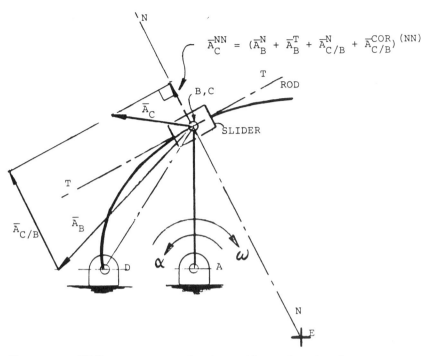

Figure 8.7   Sliding contact mechanism with acceleration diagram.

or

$$\overline{A}_C^{NN} = \overline{A}_B^{N(NN)} + \overline{A}_B^{T(NN)} + \overline{A}_{C/B}^{N(NN)} + \overline{A}_{C/B}^{T(NN)} + \overline{A}_{C/B}^{COR(NN)}$$ (8.2)

where

$\overline{A}_{C/B}^{N(NN)}$ = the effective component of the normal acceleration of C relative to B along NN

$\overline{A}_{C/B}^{T(NN)}$ = the effective component of the tangential acceleration of C relative to B along NN

$\overline{A}_{C/B}^{COR(NN)}$ = the effective component of the Coriolis acceleration component along NN

Note that since $\overline{A}_{C/B}^{T}$, by definition, is in the path of sliding, it has no effective component normal to that path. That is,

$$\overline{A}_{C/B}^{T(NN)} = 0$$

Hence, Eq. 8.2 is reduced to

$$\overline{A}_C^{NN} = \overline{A}_B^{N(NN)} + \overline{A}_B^{T(NN)} + \overline{A}_{C/B}^{N(NN)} + \overline{A}_{C/B}^{COR(NN)} \tag{8.3}$$

or

$$\overline{A}_C^{NN} = (\overline{A}_B^N + \overline{A}_B^T + \overline{A}_{C/B}^N + \overline{A}_{C/B}^{COR})^{(NN)} \tag{8.4}$$

EXAMPLE 8.1

To illustrate the method, let us consider the same quick-return linkage $ABCD$ shown in Figure 8.7 and repeated in Figure 8.8, where crank AB rotates with an angular velocity of 1 in./sec (clockwise) and angular acceleration of 1 in./sec^2 (counterclockwise). Points B and C are considered the contact points, located respectively on the crank and rod. Let it be required to find the acceleration of point C on the rod.

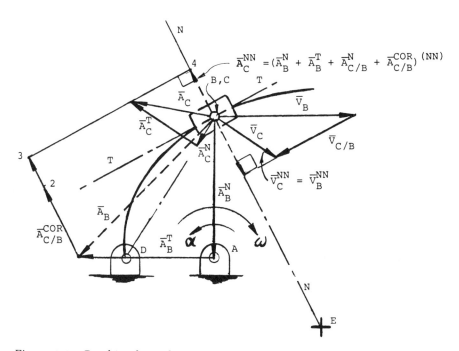

Figure 8.8 Combined acceleration and velocity diagram.

SOLUTION

1. From point B on the mechanism, draw vector $B1$ to represent the acceleration $\overline{A}_B$. This vector is determined from the vector equation

$$\overline{A}_B = \overline{A}_B^N + \overline{A}_B^T$$

where the magnitudes of A_B^N and A_B^T are given by

$$A_B^N = AB \times \omega^2$$

$$= 2.1 \text{ in./sec}^2$$

and

$$A_B^T = AB \times \alpha$$

$$= 2.1 \text{ in./sec}^2$$

2. Compute the magnitude of the Coriolis acceleration, which is given as

$$A_{C/B}^{COR} = 2V_{C/B} \times \omega_{CD}$$

where $V_{C/B}$ and ω_{CD} are determined from a velocity diagram obtained as follows:

a. Draw a straight line TT tangent to the curved rod at the contact point B or C, to indicate the path of sliding and through the same contact point draw another straight line NN normal to TT.
b. Plot the given velocity $\overline{V}_B$ perpendicular to AB. The magnitude of this vector is given by

$$V_B = AB \times \omega = 2.1(1) = 2.1 \text{ in./sec}$$

c. Plot the line of action of the unknown velocity V_C, noting that this line must be perpendicular to the line joining C to D.
d. Determining the effective component of the known velocity vector $\overline{V}_B$ along NN, or $\overline{V}_B^{NN}$, by dropping a perpendicular from the head of this vector to meet NN. Note that this perpendicular line meets the line of action of $\overline{V}_C$ in Step 2c. This meeting point will define the magnitudes of both vectors $\overline{V}_C$ and the

sliding velocity $\overline{V}_{C/B}$, the latter vector pointed toward the head of $\overline{V}_C$.

e. Scale vectors $\overline{V}_{C/B}$ and $\overline{V}_C$ to obtain

$$V_{C/B} = 1.3 \text{ in./sec}$$

$$V_C = 1.1 \text{ in./sec}$$

f. Compute the angular velocity of the rod CD from

$$\omega_{CD} = \frac{V_C}{CD}$$

$$= 0.44 \text{ rad/sec}$$

g. Finally,

$$A_{C/B}^{COR} = 2(1.3)(0.44)$$

$$= 1.14 \text{ in./sec}$$

3. From point 1, head of vector $\overline{A}_B$, draw vector $\overline{A}_{C/B}^{COR}$, the Coriolis acceleration vector, observing the orientation rule which requires that this vector be pointed in the direction of relative velocity $\overline{V}_{C/B}$, when rotated 90° about its tail, in the same angular direction as link CD. Label the head of this vector 2.

4. Compute the magnitude of the normal acceleration of point C relative to point B from

$$A_{C/B}^N = \frac{V_{C/B}^2}{CE}, \qquad \text{where } CE \text{ is the radius of the path of } C$$

$$= \frac{1.3}{3.5}$$

$$= 0.48 \text{ in./sec}^2$$

Lay out this vector from point 2 and pointed in its normal direction, parallel to NN. Note that

$$\overline{A}_{C/B}^N = -\overline{A}_{B/C}^N$$

Therefore, the vector points up. Label the head of this vector 3.

5. From point 3, the head of $\overline{A}_{C/B}^{N}$, drop a perpendicular to meet NN at point 4. This point will define the effective component of all acceleration vectors on the right-hand side of Eq. 6.4.
6. From point B, lay out the vector for the normal acceleration of point C, or $\overline{A}_{C}^{N}$, on link CD and label its terminus 5. The magnitude of this vector is given by

$$A_C^N = \frac{V_C^2}{CD}$$

$$= \frac{1.2^2}{2.5} = 0.48 \text{ in./sec}^2$$

7. Through 5, draw a perpendicular to $\overline{A}_C^N$ to represent the tangential acceleration of A_C, or $\overline{A}_C^T$. This perpendicular will meet the perpendicular drawn from point 3 (in Step 5) at point 6 to define the terminus of $\overline{A}_C$.
8. Finally, draw a vector extending from point B to point 6 to represent the required acceleration of point C on the rod. The magnitude of this vector when scaled gives

$$A_C = 1.3 \text{ in./sec}^2$$

Summary

In the above example, it is useful to observe the following:

1. The mechanism chosen represents the most general of the sliding contact type, since all possible components of the relative acceleration of the contact points ($\overline{A}_{C/B}^{N}$, $\overline{A}_{C/B}^{T}$, and $\overline{A}_{C/B}^{COR}$) are included. If the rod was straight, then the motion of C relative to B would be along the link and therefore $\overline{A}_{C/B}^{N}$ would not exist.
2. The choice of motion of C relative to B, rather than B relative to C, is intentional, as it ensures that all the vectors, except those related to the unknown acceleration $\overline{A}_C$, are pointed away from point B. In this way, the effective components of these vectors can be easily summed.
3. To obtain the velocity information needed for the acceleration analysis, a velocity analysis by any other method would be sufficient. However, by employing the effective component of velocity method

in this example, it is instructive to note that the complete analysis has been carried out for the first time by the effective component method.

4. Finally, the results obtained above may easily be verified using any one of the currently available methods, such as the relative acceleration method, equivalent linkage method, or velocity difference method.

9

Relative Acceleration Method

9.1 INTRODUCTION

The relative acceleration method is probably the fastest and most common among the graphical methods used for acceleration analysis. This is merely an extension of the relative velocity method used for velocity analysis and utilizes an acceleration polygon that is very similar to the velocity polygon. The method is based on the following principles:

1. All motions are considered instantaneous.
2. The instantaneous acceleration of a point A relative to another point B on a rigid link is obtained from the vectorial relationship

$$\overline{A}_A = \overline{A}_B + \overline{A}_{A/B}$$

3. The instantaneous motion of a point may be considered one of pure rotation in which the acceleration can be resolved into two rectangular components: one normal and one tangential to the path of rotation. Thus, the expression in Step 2 can be represented graphically as

$$\overline{A}_A^N + \overline{A}_A^T = \overline{A}_B^N + \overline{A}_B^T + \overline{A}_{A/B}^N + \overline{A}_{A/B}^T$$

4. The absolute as well as relative velocities of various points in the mechanism are known. This requirement makes it most desirable

to use the relative velocity or instant center method to determine the velocities involved.

9.2 THE ACCELERATION POLYGON

Let us consider the two points A and B on link AB shown in Figure 9.1. As we saw earlier, the acceleration of point B relative to point A is given by the vector expression

$$\overline{A}_B = \overline{A}_A + \overline{A}_{B/A}$$

Just as in the velocity case, an acceleration polygon can be developed to represent the vector expression. Furthermore, the construction procedure and convention employed for this development are the same in both cases.

PROCEDURE

1. Define a point o', called the pole, as the origin for the construction of the polygon. This is the point of zero acceleration. All absolute acceleration vectors originate from this point. By "absolute acceleration" we mean the real or true acceleration of a body as observed from a stationary frame of reference such as the earth.
2. Since the accelerations $\overline{A}_A$ and $\overline{A}_B$ are absolute accelerations, draw vectors $\overline{A}_A$ and $\overline{A}_B$ from point o' and define their respective termini as a' and b'.
3. From point a', the terminus of $\overline{A}_{A'}$, draw a third vector to terminate at point b', the terminus of $\overline{A}_{B'}$, thereby closing the polygon. This vector defines the relative acceleration $\overline{A}_{B/A}$.

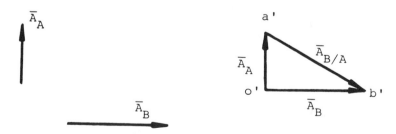

Figure 9.1 Acceleration polygon.

9.3 ACCELERATION POLYGON CONVENTION

The convention used to develop the acceleration polygon of Figure 9.1 can be summarized as follows:

$o'a'$ represents the acceleration vector $\overline{A}_A$.
$o'b'$ represents the acceleration vector $\overline{A}_B$.
$a'b'$ represents the acceleration vector $\overline{A}_{B/A}$.

As in the velocity polygon case (Section 6.4), note that the letter to which the vector is directed indicates the acceleration under consideration. For example, in the vectors $o'a'$ and $o'b'$, the arrows point toward a' and b', respectively. Hence, the vectors represent the accelerations of points A and B, respectively. Similarly, in the vector $a'b'$, the arrow points away from a' toward b'; hence, the vector represents the acceleration of point B relative to point A $(\overline{A}_{B/A})$. If the arrow were reversed, pointing toward a' and away from b', the vector $b'a'$ would represent the acceleration of A relative to B $(\overline{A}_{A/B})$.

9.4 GENERALIZED PROCEDURE

The construction procedure for developing the acceleration polygon of a mechanism will now be generalized using the four-bar mechanism as a model. Consider the four-bar mechanism $ABCD$ shown in Figure 9.2. In this mechanism, AB, the driven member, has a clockwise angular acceleration ω_{AB} and counterclockwise angular acceleration α_{AB}. It is required to find the acceleration of point C on the follower CD $(\overline{A}_C)$.

PROCEDURE

1. Determine the velocities of all points on the mechanism, including those with combined motion. This may be done using the instant center method, the effective component method, or as in the present case, the relative velocity method (Figure 9.3).
2. Define a starting point o', called the polar origin. *All absolute acceleration vectors originate from the polar origin.* By "absolute acceleration" we mean the real or true acceleration of the point as observed from a fixed frame of reference such as the earth.

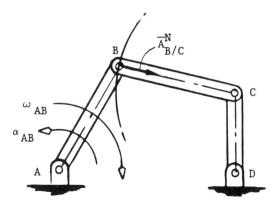

Figure 9.2   Four-bar mechanism.

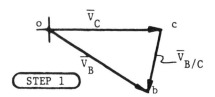

Figure 9.3   Velocity polygon.

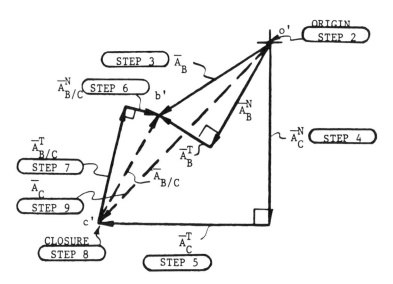

Figure 9.4   Acceleration polygon.

3. Lay out the known components of absolute acceleration of the drive member, starting with the normal acceleration, which has the magnitude $A_B^N = AB \times \omega_{AB}^2$ and is directed parallel to AB. Then add the tangential acceleration, which has the magnitude $A_B^T = AB \times \alpha_{AB}$ and is directed perpendicular to the normal acceleration, or $\overline{A}_B^N \perp \overline{A}_B^T$. The summation of these two components defines the absolute acceleration of point B $(\overline{A}_B = \overline{A}_B^N + \overline{A}_B^T)$ (Figure 9.4). If an angular acceleration α_{AB} is not given, $\overline{A}_B^T$ does not exist $(\overline{A}_B^T = 0)$, and $\overline{A}_B^N$ becomes the absolute acceleration of point B. Define the terminus of $\overline{A}_B$ as b'.

4. Select another point on the mechanism that has absolute motion and is rigidly connected to point B (call it point C) and lay out its normal acceleration component $\overline{A}_C^N$, noting that this vector is defined by the magnitude

$$A_C^N = \frac{V_{C/D}^2}{CD}$$

and is directed parallel to link CD. The normal components of acceleration are readily determined from the velocity data and link orientation.

5. Through the terminus of vector $\overline{A}_C^N$, construct a perpendicular to represent the tangential acceleration $(\overline{A}_C^T)$, whose direction is known but magnitude is as yet undefined. This line contains point c', the terminus of vector $\overline{A}_C$.

6. Consider next the relationship between point B, whose acceleration is **completely known**, and point C, whose acceleration is only **partially known**. This means that we must consider the acceleration of point B relative to point C

$$\overline{A}_{B/C} = \overline{A}_{B/C}^N + \overline{A}_{B/C}^T$$

where $\overline{A}_{B/C}^N$ can be completely determined from the velocity analysis and $\overline{A}_{B/C}^T$ is known in direction only. Therefore, lay out vector $\overline{A}_{B/C}^N$, knowing that

a. Its magnitude is given by

$$A_{B/C}^N = \frac{V_{B/C}^2}{BC}$$

b. Its direction is obtained by assuming that point C on link is fixed while point B rotates about it. $\overline{A}_{B/C}^N$ therefore lies on the link and is directed toward the center of rotation assumed.

c. Also, in accordance with the polygon convention, relative acceleration vectors normally *do not* originate at the pole o', but extend between the termini of the absolute acceleration vectors to which they relate. For example, the notation $\overline{A}_{B/C}$ means that polygon vector is directed from c' to b', and $\overline{A}_{C/B}$ means that polygon vector is directed from b' to c', where b' and c' are the termini of absolute vectors on the acceleration polygon, and B and C are the corresponding points in the linkage.

7. Through the tail* of $\overline{A}_{B/C}^N$, construct a perpendicular line of undefined length to represent the acceleration $\overline{A}_{B/C}^T$, whose direction is known but magnitude is yet to be determined.

8. Now since c' lies on both $\overline{A}_{B/C}^T$ (Step 7) and $\overline{A}_C^T$ (Step 5), it follows that the intersection of these two lines will define that point. Therefore, extend $\overline{A}_{B/C}^T$ and $\overline{A}_C^T$ until they intersect and label the point of intersection c'.

9. Finally, lay out a vector from origin o' to terminus c' to define the required absolute acceleration $\overline{A}_C$ and a vector from terminus c' to point b' to define the relative acceleration $\overline{A}_{B/C}$.

10. A quick check of the completed polygon should reveal the balanced vector equation

$$\overline{A}_B = \overline{A}_C + \overline{A}_{B/C}$$

or

$$\overline{A}_B^N + \overline{A}_B^T = \overline{A}_C^N + \overline{A}_C^T + \overline{A}_{B/C}^N + \overline{A}_{B/C}^T$$

Special Cases

Slider-Crank Mechanism

The slider-crank mechanism shown in Figures 9.5 to 9.7 may be considered a special case of the four-bar linkage the follower link—for

---

*Note that had the normal relative acceleration $\overline{A}_{C/B}^N$ been chosen instead of $\overline{A}_{B/C}^N$, the perpendicular drawn to represent $\overline{A}_{C/B}^T$ would have been constructed through the terminus of $\overline{A}_{C/B}^N$. This is because, by polygon convention, the normal acceleration vector $\overline{A}_{C/B}^N$ must head toward c' and away from b'.

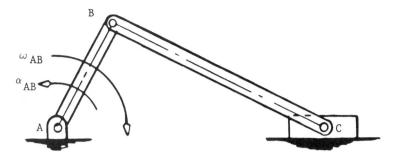

Figure 9.5   Slider-crank mechanism.

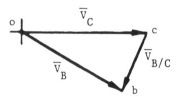

Figure 9.6   Velocity polygon: Step 1.

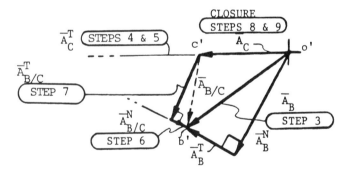

Figure 9.7   Acceleration polygon.

example, CD in Figure 9.2—is infinitely long. This means that the velocity of point C must be a straight line. Therefore, in Step 4 (Figure 9.7) the normal acceleration of the slider $\overline{A}_C^N$ is zero, and

$$\overline{A}_C = \overline{A}_C^T$$

In effect, the tangential component of acceleration $\overline{A}_C^T$ becomes the absolute acceleration of C and originates at o'.

Quick-Return Mechanism

Another special case is the quick-return mechanism shown in Figures 9.8 to 9.10. Here point B slides on link CD, which is itself rotating. This means that in addition to the normal acceleration of point B relative to a coincident point C on CD, we must consider an additional component of acceleration: the Coriolis acceleration.

Also, since the path of B relative to C is a straight line, the acceleration of B relative to C can only be tangential. That is, B can have *no normal component* of acceleration relative to C, or

$$\overline{A}_{B/C}^{N} = 0$$

Had the path of B on CD been a curve, there would have existed a normal acceleration component of B relative to C $(\overline{A}_{B/C}^{N})$, as will be seen later (see Section 9.7).

To determine the Coriolis acceleration, we note from earlier discussion that the magnitude of this vector is given by

$$A^{\text{Cor}} = 2V_{B/C}\omega_{CD}$$

where $V_{B/C}$ and ω_{CD} are obtained from the velocity analysis. Also, we note that in defining the direction of this vector, we always consider *(1) the linear velocity of the sliding body relative to that of the rotating body and (2) the angular velocity of the same body on which the sliding occurs*.

In this example, since body B slides on link CD, we consider the velocity of B relative to C or $\overline{V}_{B/C}$ (not $\overline{V}_{C/B}$) and the angular velocity of link CD or ω_{CD} (not ω_{AB}). Accordingly, in Step 6 (Figure 9.10), connect the vector $\overline{A}^{\text{Cor}}$ to terminus b' on the polygon, observing the following rules:

1. This vector has the orientation of vector $\overline{V}_{B/C}$ when rotated 90° about its tail in the direction of ω_{CD} (counterclockwise in this case).
2. Since point B relative to point C on the link is being considered, the polygon vector must go from c' to b' *(note the reversed letter sequence)*.

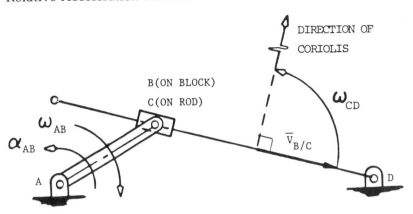

Figure 9.8   Quick-return mechanism.

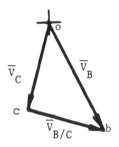

Figure 9.9   Velocity polygon: Step 1.

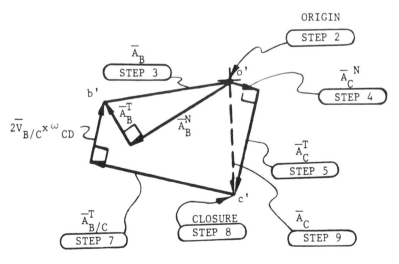

Figure 9.10   Acceleration polygon.

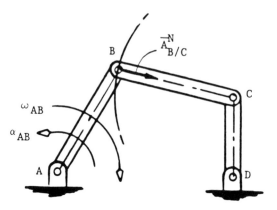

Figure 9.11   Four-bar mechanism.

EXAMPLE 9.1 (see Figure 9.11)

Let AB = 1.5 in., BC = 1.5 in., CD = 1 in., ω_{AB} = 1 rad/sec (clockwise),
and α_{AB} = 0.5 rad/sec^2 (counterclockwise). Determine $\overline{A}_C$.

SOLUTION

1. Determine the velocities. Construct a velocity polygon (Figure 9.12)
    and obtain

$$V_B = AB \times \omega_{AB} = 1.5(1) = 1.5 \text{ in./sec}$$

$$V_C = 1.45 \text{ in./sec}$$

$$V_{B/C} = 0.8 \text{ in./sec}$$

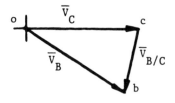

SCALE: 1 in. = $\dfrac{1 \text{ in.}}{\text{sec.}}$

Figure 9.12   Velocity polygon.

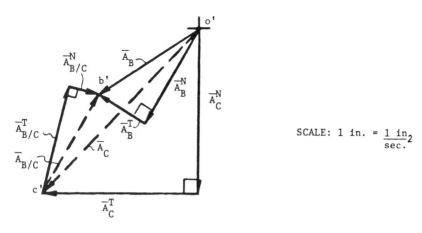

SCALE: 1 in. = $\dfrac{1 \text{ in}}{\text{sec.}^2}$

Figure 9.13   Acceleration polygon.

2. Define polar origin o' for the acceleration polygon (Figure 9.13).
3. Lay out the acceleration of the drive point $\overline{A}_B$.

$$A_B^N = AB \times \omega_{AB}^2 = 1.5(1^2) = 1.5 \text{ in./sec}^2$$

$$A_B^T = AB \times \alpha_{AB} = 1.5(0.5) = 0.75 \text{ in./sec}^2$$

$$\overline{A}_B = \overline{A}_B^N + \overline{A}_B^T$$

The vector originates at o' and terminates at b'.
4. Lay out the normal acceleration of the driven point $\overline{A}_C^N$. The magnitude of this vector is given as

$$A_C^N = \frac{V_C^2}{CD} = \frac{1.45^2}{1.0} = 2.13 \text{ in./sec}^2$$

The vector is directed parallel to link CD, originating from o'.
5. Add the tangential acceleration $\overline{A}_C^T$ (undefined length).

$$\overline{A}_C^T \perp \overline{A}_C^N$$

6. Lay out the normal acceleration of the drive point relative to the driven point $\overline{A}_{B/C}^N$.

$$\overline{A}_{B/C}^N = \frac{V_{B/C}^2}{BC} = \frac{0.8^2}{1.5} = 0.426 \text{ in./sec}^2$$

The vector is directed parallel to link BC, pointing to b' (or away from c').

7. Add the tangential acceleration $\overline{A}_{B/C}^T$ (undefined length).

$$\overline{A}_{B/C}^T \perp \overline{A}_{B/C}^N$$

8. Determine the magnitude of tangential accelerations $\overline{A}_C^T$ and $\overline{A}_{B/C}^T$. Extend undefined lines for these acceleration components in Steps 6 and 7, until the polygon is closed. $\overline{A}_C^T$ intersects $\overline{A}_{B/C}^T$ at c'.

9. Determine the required acceleration $\overline{A}_C$. Connect point c' to point o' and measure $o'c'$, or magnitude A_C.

$$\overline{A}_C = 3.2 \text{ in./sec}^2 \quad \text{(directed as shown in Figure 9.13)}$$

9.5 MECHANISM WITH EXPANDED FLOATING LINK

It was shown earlier that if the accelerations of two points on a link are known, the acceleration of a third point on that link can be found by proportion. In the following example, it will be seen that the polygon construction procedure can also be used to determine the acceleration of the third point.

EXAMPLE 9.2

Consider the mechanism $ABCE$ shown in Figure 9.14(a), where the crank AB rotates with an angular velocity of 2 rad/sec (clockwise). It is required to find the acceleration of point E on the expanded floating link BEC.

SOLUTION

1. Develop the velocity polygon for the mechanism as shown in Figure 9.14(b). From this polygon, the absolute and relative velocity magnitudes are obtained as follows:

$$V_B = 2.5 \text{ in./sec}$$

$$V_C = 2.8 \text{ in./sec}$$

$$V_{E/B} = 1.0 \text{ in./sec}$$

$$V_{E/C} = 1.5 \text{ in./sec}$$

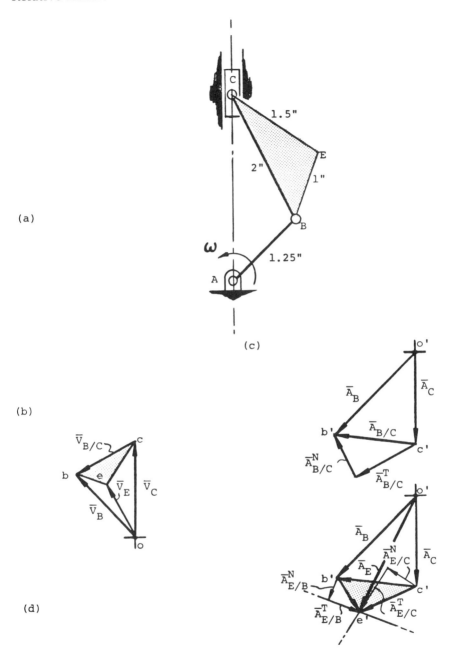

Figure 9.14   Mechanism with expanded floating link: (a) Mechanism; (b) velocity polygon; (c) basic acceleration polygon (for ABC); (d) complete acceleration polygon.

2. Develop the acceleration polygon for the basic mechanism ABC [Figure 9.14(c)] (ignoring point E for the time being), following the construction procedure discussed earlier. From this polygon, the accelerations of points B and C are obtained.

$$\overline{A}_B = 5 \text{ in./sec}^2 \quad \text{(directed as shown)}$$

$$\overline{A}_C = 3.9 \text{ in./sec}^2 \quad \text{(directed as shown)}$$

3. To determine the acceleration of point E ($\overline{A}_E$), complete the following steps [Figure 9.14(d)].

a. Lay out the vector for the normal acceleration of E relative to B ($\overline{A}^N_{E/B}$). The magnitude of this vector is obtained from

$$A^N_{E/B} = \frac{V^2_{E/B}}{EB} = \frac{(1)^2}{1}$$

$$= 1 \text{ in./sec}^2$$

This vector is directed from point E toward point B on the mechanism, but on the polygon, heads from point b' (which is defined) to point e' (which is undefined), **still maintaining its basic orientation**.

b. Through the terminus of vector $\overline{A}^N_{E/B}$, draw a line of undefined length perpendicular to this vector to indicate the direction of $\overline{A}^T_{E/B}$. This line contains point e', the terminus of vector $\overline{A}_E$.

c. From point c' on the polygon, lay out the vector for the normal acceleration of E relative to C. The magnitude of this acceleration is obtained from

$$A^N_{E/C} = \frac{V^2_{E/C}}{EC} = \frac{(1.5)^2}{1.5}$$

$$= 1.5 \text{ in./sec}^2$$

This vector is directed from point E to point C on the mechanism, but on the polygon, heads from point c' (which is defined) toward point e' (which is undefined), **still maintaining its basic orientation**.

d. Through the terminus of vector $\overline{A}^N_{E/C}$, draw a line of undefined length perpendicular to this vector to indicate the direction of $\overline{A}^T_{E/C}$. This line contains point e', the terminus of $\overline{A}_E$.

e. Since point e' is to be found on both lines $\overline{A^T_{E/B}}$ (Step c) and $\overline{A^T_{E/C}}$ (Step d), it must be located at the intersection of these lines. This point e' defines the terminus of the vector $\overline{A}_E$ drawn from the pole o'. Therefore,

$$\overline{A}_E = 5.6 \text{ in./sec}^2 \quad \text{(directed as shown)}$$

Acceleration Image

As in the relative velocity case, it should be observed from the polygon that the triangle formed by points b', e', c' is similar to link BEC, and for this reason, it is often referred to as the ***acceleration image*** of link BEC. Consequently, the acceleration of point E could have been determined more directly by constructing the acceleration image on link $b'c'$ of the polygon for the basic mechanism ABC. (Note that the letters used to designate both the links and the image must run in the same order and direction.)

9.6 COMPOUND MECHANISM

Let $ABCDEF$ represent a compound mechanism in the form of a shaper, where crank AB rotates with a constant angular velocity of 1 rad/sec (counterclockwise) [Figure 9.15(a)]. Determine the acceleration of point E on the slider.

PROCEDURE

1. Develop the velocity polygon [Figure 9.15(b)] and determine the velocities as follows:

$$V_B = AB \times \omega = 1.5(1) = 1.5 \text{ in./sec}$$

$$V_C = 1.2 \text{ in./sec}$$

$$V_{B/C} = 1.0 \text{ in./sec}$$

$$V_D = \frac{OD}{OC} \times V_C = \frac{6}{4}(1.2) = 1.8 \text{ in./sec}$$

$$V_E = 2.1 \text{ in./sec}$$

$$V_{D/E} = 0.75 \text{ in./sec}$$

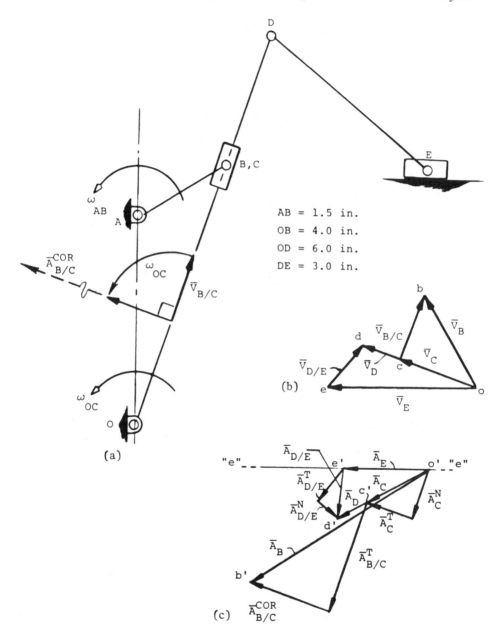

AB = 1.5 in.
OB = 4.0 in.
OD = 6.0 in.
DE = 3.0 in.

Figure 9.15   Acceleration analysis of a shaper mechanism: (a) Shaper mechanism; (b) velocity polygon; (c) acceleration polygon.

2. Define polar origin o' for the acceleration polygon [Figure 9.15(c)]
3. Lay out the acceleration of the driver at point B $(\overline{A}_B)$.

$$A_B^N = AB \times \omega^2 = 1.5(1)^2 = 1.5 \text{ in./sec}^2$$

$$A_B^T = AB \times \alpha = 0$$

$$\overline{A}_B = \overline{A}_B^N + \overline{A}_B^T = 1.5 \text{ in./sec}^2 \quad \text{(directed as shown)}$$

4. Lay out the normal acceleration of point-driven point C $(\overline{A}_C^N)$.

$$A_C^N = \frac{V_C^2}{OC}$$

$$A_C^N = \frac{(1.2)^2}{4}$$

$$\overline{A}_C^N = 0.36 \text{ in./sec}^2 \quad \text{(directed as shown)}$$

5. Add the tangential acceleration of point C (undefined length) to $\overline{A}_C^N$.

$$\overline{A}_C^T \perp \overline{A}_C^N$$

6. Determine the relative acceleration between drive point B and related point C $(\overline{A}_{B/C})$.

$$\overline{A}_{B/C} = \overline{A}_{B/C}^N + \overline{A}_{B/C}^T + \overline{A}_{B/C}^{Cor}$$

$$A_{B/C}^N = 0 \quad \text{(since the path of the slider is a straight line)}$$

and

$$\overline{A}_{B/C}^{Cor} = 2V_{B/C}\omega_{OC} \quad \text{(properly directed)}$$

or

$$A_{B/C}^{Cor} = 2(1)\left(\frac{1.2}{4}\right) = 0.6 \text{ in./sec}^2$$

   a. Lay out the vector $\overline{A}_{B/C}^{Cor}$ in accordance with convention. That is, the vector must have the same orientation as vector $\overline{V}_{B/C}$ when rotated 90° in the direction of ω_{OC}.

   b. Add the tangential relative acceleration $\overline{A}_{B/C}^{T}$ (undefined length) to $\overline{A}_{B/C}^{Cor}$.

$$\overline{A}_{B/C}^{T} \perp \overline{A}_{B/C}^{Cor}$$

7. Extend the tangential accelerations in Steps 4 and 5 until the polygon is closed. The acceleration of point C on the polygon is defined at the intersection of $\overline{A}_{C}^{T}$ and $\overline{A}_{B/C}^{T}$, that is, at point c'.

8. Draw vector $\overline{A}_{C}$ extending from the pole o' to c'. Note that

$$\overline{A}_{C} = \overline{A}_{C}^{N} + \overline{A}_{C}^{T}$$

and

$$\overline{A}_{B/C} = \overline{A}_{B/C}^{Cor} + \overline{A}_{B/C}^{T}$$

$$A_{C} = 0.5 \text{ in./sec}^2$$

9. Determine magnitude of the acceleration of point D (A_D). This is obtained from the proportion

$$\frac{A_D}{A_C} = \frac{OD}{OC}$$

$$A_D = \frac{6}{4}(0.5) = 0.75 \text{ in./sec}^2$$

10. Define the terminus of vector $\overline{A}_D$ as point d' on polygon.

11. Determine the acceleration of point E relative to point D.

$$\overline{A}_{D/E} = \overline{A}_{D/E}^{N} + \overline{A}_{D/E}^{T}$$

where

$$\overline{A}_{D/E}^{N} = \frac{V_{D/E}^2}{DE} \quad \text{(properly directed)}$$

$$= \frac{(0.75)^2}{3} = 0.1875 \text{ in./sec}^2 \quad \text{(properly directed)}$$

a.   Lay out vector $\overline{A}_{D/E}^{N}$. This vector must be pointed toward d' on the polygon and be parallel to DE on the mechanism.
b.   Add vector $\overline{A}_{D/E}^{T}$ to vector $\overline{A}_{D/E}^{N}$.

$$\overline{A}_{D/E}^{T} \perp \overline{A}_{D/E}^{N}$$

12.   Determine the acceleration of point E $(\overline{A}_E)$. Note that since the slider path is a straight line, there is no normal acceleration $(\overline{A}_E^{N} = 0)$, so $\overline{A}_E$ is equal to the tangential acceleration $\overline{A}_E^{T}$, which exists along the same path. Therefore,
a.   Lay out the direction of vector $\overline{A}_E$ by drawing a line "e"-"e" through pole o' parallel to the slider path.
b.   Extend vector $\overline{A}_{D/E}^{T}$ in Step 11b until it intersects line "e"-"e" (direction line). This point of intersection defines the magnitude of $\overline{A}_E$.
13.   Define the point of intersection in Step 12b as e'.
14.   Measure line o'-e', the vector $\overline{A}_E$.

$$\overline{A}_E = 0.625 \text{ in./sec}^2 \quad \text{(directed as shown)}$$

9.7   CAM-FOLLOWER MECHANISM

Consider the cam-follower mechanism shown in Figure 9.16(a). The cam (2) rotates counterclockwise at a constant angular velocity of 2 rad/sec. Find the acceleration of the follower (4).

At first glance, it would appear that to determine the acceleration of 4, it would be necessary first to determine the motion of the roller (3). However, since the path that 3 traces on 2 is generally not easily recognizable, because of the two curved surfaces in contact, a more direct approach is the following:

PROCEDURE

1.   Assume that the roller does not turn. Then the acceleration of contact point C is the same as that of any point in 4.
2.   Construct a pitch line to represent the locus of the center of the roller as it rolls on the cam.
3.   Designate the new point of contact between 2 and 4 as P_2 or P_4.
4.   Proceed with the analysis based on the assumption that the follower 4 actually slides on the expanded cam defined by the pitch line.

Figure 9.16(b) shows the velocity polygon for the mechanism. Note that since 4 rides on 2, the relative velocity $\overline{V}_{P4/P2}$ is considered for Coriolis

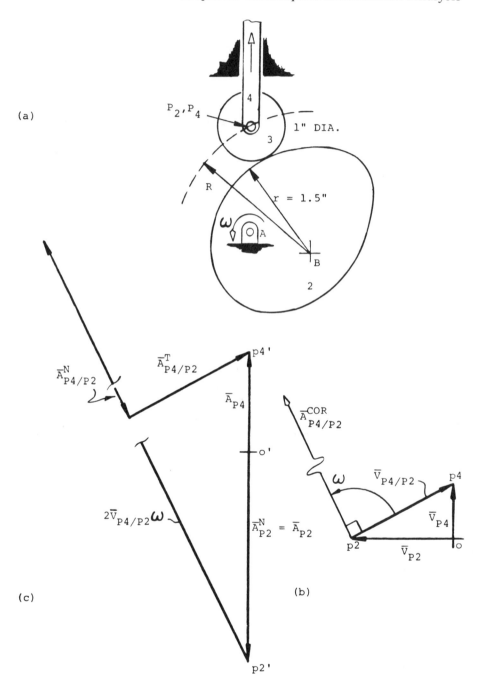

Figure 9.16 Acceleration analysis of a cam-follower mechanism: (a) Cam-follower mechanism; (b) velocity polygon; (c) acceleration polygon.

acceleration. This velocity has a direction tangent to the curvature of the cam at the contact point.

Figure 9.16(c) shows the acceleration polygon for the mechanism. Here it should be noted that the acceleration of P4 relative to P2 (or $\overline{A}_{P4/P2}$) consists of three components:

$$\overline{A}_{P4/P2} = \overline{A}^N_{P4/P2} + \overline{A}^T_{P4/P2} + \overline{A}^{Cor}_{P4/P2}$$

This equation, incidentally, is unlike those of earlier examples where $\overline{A}^N_{P4/P2}$ did not exist, because the sliding paths in those cases were straight lines. In the present case, since the path of P4 on P2 is a curve (or 4 is forced to move in a curved path as it rides on 2), there must exist a normal acceleration, $\overline{A}^N_{P4/P2}$, whose magnitude is given by

$$A^N_{P4/P2} = \frac{V_{P4/P2}}{r + 0.5}$$

where $r + 0.5$ or R is the radius of curvature of the path.

For the Coriolis acceleration $\overline{A}^{Cor}$, we consider, as before, the motion of P4 relative to P2, computing its magnitude from

$$A^{Cor}_{P4/P2} = 2V_{P4/P2}\omega$$

and defining its direction by rotating the vector $\overline{V}_{P4/P2}$ through an angle of 90° in the same angular direction (counterclockwise) as ω_2.

Following is a summary of calculations and results:

$$V_{P2} = 1.5(2) = 3 \text{ in./sec}$$

$$V_{P4} = 1.6 \text{ in./sec} \quad \text{(from the velocity polygon)}$$

$$V_{P4/P2} = 3.35 \text{ in./sec} \quad \text{(from the velocity polygon)}$$

$$A^N_{P2} = 1.5(2)^2 = 6 \text{ in./sec}^2$$

$$A^{Cor}_{P4/P2} = 2(3.35)(2) = 13.4 \text{ in./sec}^2$$

$$A^N_{P4/P2} = \frac{(3.4)^2}{2.0} = 5.6 \text{ in./sec}^2$$

$$A^T_{P4/P2} = 4.0 \text{ in./sec}^2 \quad \text{(from the acceleration polygon)}$$

$$A_{P4} = 2.8 \text{ in./sec}^2 \quad \text{(from the acceleration polygon)}$$

9.8 ACCELERATION OF BODIES IN ROLLING CONTACT

Rolling contact mechanisms, or mechanisms that contain at least one pair of rolling elements, belong to a special class of mechanisms characterized by the higher-paired elements they contain. Higher-paired elements are those whose surfaces are so shaped that only a line or point contact is possible between them. Included in this class are cams that are equipped with roller followers, roller bearings, and most gears, in which the paired elements roll without slipping on one another to produce in the process *cycloidal, epicycloidal, or hypocycloidal motion.**

Acceleration analysis of rolling contact mechanisms are typically not as straightforward as that of pin-jointed mechanisms such as the four-bar linkage. This is because, unlike pin-jointed mechanisms where the relative motion of the paired elements is purely rotational, the paired elements of rolling contact mechanisms move relative to each other with a form of combined motion that is partly rotational and partly linear motion. This combined motion often makes it difficult to visualize the absolute motions of the contact points, which are needed for a complete analysis.

In this section, we will present graphical solutions as well as mathematical expressions for the acceleration of the contact points of rolling members for two conditions, namely,

a. When one body rolls without slipping on another body which is fixed
b. When two bodies roll in plane motion without slipping about each other

The approach employs the principles of relative motion as well as those of pure rolling to develop the desired relationships.

Determining the acceleration characteristics of coincident points on bodies in rolling contact is of great importance to a designer for two reasons:

1. These bodies are commonly used to transmit power and motion. Analytic expressions obtained from this study can be applied to

---

Cycloidal motion is the path described by a point on a circle that rolls without slipping on a straight track. *Epicycloidal* motion is the path described by a point on a circle that rolls without slipping on the outside of another circle. *Hypocycloid* motion is the path described by a point on a circle that rolls without slipping on the inside of another circle.

investigate the possibility of slipping between the bodies, where power or motion is transmitted through friction contact, or to determine inertia loads on meshed teeth, where motion transmission occurs through gears.

2. The ability to visualize and compute the relative acceleration of the contact points provides the student with an alternative graphical approach to the more commonly used solutions for rolling contact problems. For example, it avoids the need to deal with the Coriolis acceleration that can be puzzling to many students.

Absolute Acceleration of the Contact Point (One Body Fixed)

It is well known that when two bodies are in pure rolling contact, there is no relative velocity between the contact points. However, that is not to say that there is no acceleration between these points. The fact is, as long as the contact points on the rolling bodies travel along different paths, there will always be at least a normal component of relative acceleration between these points. Note that there is no such component of acceleration when the path is a straight line, since the radius of a straight line has infinite length.

Consider the three cases below of one roller that rolls on another roller that is fixed. Points P on roller 2 and Q on roller 1 are the contact points. The angular velocity of the moving roller is ω_2 (clockwise) and angular acceleration α (clockwise). Let it be required to find the acceleration of P on roller 2 in each case.

Graphical Procedure

For the graphical approach to this problem, it is convenient to construct an acceleration polygon, based on the relationship

$$\overline{A}_P = \overline{A}_C + \overline{A}_{P/C} = \overline{A}_C^N + \overline{A}_C^T + \overline{A}_{P/C}^N + \overline{A}_{P/C}^T \quad \text{(vectorially)}$$

(9.1)

The general procedure may be summarized as follows:

CASE 1: ROLLER 2 ROLLS ON OUTER SURFACE OF ROLLER 1 (Figure 9.17)

1. Determine the velocities from a velocity diagram, using point C, in Figure 9.17(a) as the velocity pole. These velocities are determined as follows:

Velocity of point C $(V_C) = CP \times \omega_2$

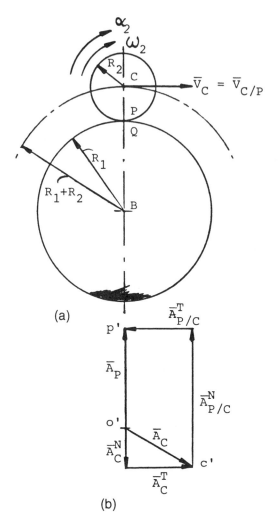

(a)

(b)

Figure 9.17   Case 1: (a) roller 1 on outer surface of roller 2; (b) acceleration diagram.

Velocity of point P (V_P) = 0

Velocity of P relative C = $CP \times \omega_2$ (same as V_C)

2.   Define the polar origin [see Figure 9.17(b)] and from this origin plot the vectors $\overline{A}_C^N$ and $\overline{A}_C^T$ where

$$\overline{A}_C^N = \frac{-V_C^2}{R_1 + R_2} \quad \text{(directed downward toward B)} \qquad (9.2a)$$

and

$$\overline{A}_C^T = R_2\alpha_2 \quad \text{(directed to the right)} \tag{9.2b}$$

Label the terminus of $\overline{A}_C$ as c'. *Note*: The path of C is a circle concentric to the surface of roller 1 with radius $R_1 + R_2$.

3. Plot the normal and tangential components of acceleration of P relative to C ($\overline{A}_{P/C}^N$ and $\overline{A}_{P/C}^T$), where

$$\overline{A}_{P/C}^N = \frac{V_C^2}{R_2} \quad \text{(directed upward toward C)} \tag{9.2c}$$

and

$$\overline{A}_{P/C}^T = -R_2\alpha_2 \quad \text{(directed to the left)} \tag{9.2d}$$

Draw vector $\overline{A}_{P/C}^N$ directed from c', and vector $\overline{A}_{P/C}^T$ perpendicular to it, directed in accordance with α_2. Define the terminus at $\overline{A}_{P/C}^T$ as p'.

4. From o', draw a vector to terminate at p'. This vector will define the required acceleration of point P.

CASE 2: ROLLER 2 ROLLS ON INNER SURFACE OF ROLLER 1 [Figure 9.18(a)]

If roller 1 (still fixed) is concave upward, the above procedure for finding $\overline{A}_P$ is the same, except the path radius of C is $R_1 - R_2$. Figure 9.18(b) shows the acceleration diagram for this case.

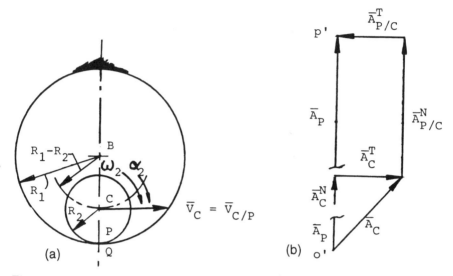

Figure 9.18   Case 2: (a) roller 1 on inner surface of roller 2; (b) acceleration diagram.

CASE 3:   ROLLER 2 ROLLS ON FLAT SURFACE [Figure 9.19(a)]

If roller 1 is infinitely large, which makes the rolling surface effectively flat, then the procedure to determine $\overline{A}_P$ is the same as in Case 1, except that point C has no normal acceleration (or $\overline{A}_C^N = 0$), since the path of C is linear. Figure 9.19(b) shows the acceleration diagram for this case.

Algebraic Expressions for $\overline{A}_P$

We start by recalling Eq. (9.1), namely,

$$\overline{A}_P = \overline{A}_C + \overline{A}_{P/C} = \overline{A}_C^N + \overline{A}_C^T + \overline{A}_{P/C}^N + \overline{A}_{P/C}^T \tag{9.1}$$

where

$$\overline{A}_C^N = \frac{-V_{C/B}^2}{CB} \quad \text{(directed downward toward } B)$$

$$\overline{A}_C^T = R_2\alpha_2 \quad \text{(directed toward the right)}$$

$$\overline{A}_{P/C}^N = \frac{V_{P/C}^2}{PC} \quad \text{(directed from } P \text{ toward } C)$$

$$\overline{A}_{P/C}^T = -R_2\alpha_2 \quad \text{(directed toward the left)}$$

Now, from the acceleration diagrams for the above cases, it is noted that since $\overline{A}_C^T$ and $\overline{A}_{P/C}^T$ are equal and opposite, they cancel each other. Therefore, Eq. (9.1) may be defined in terms of the normal acceleration components alone

$$\overline{A}_P = \overline{A}_C^N + \overline{A}_{P/C}^N \quad \text{(vectorially)} \tag{9.3}$$

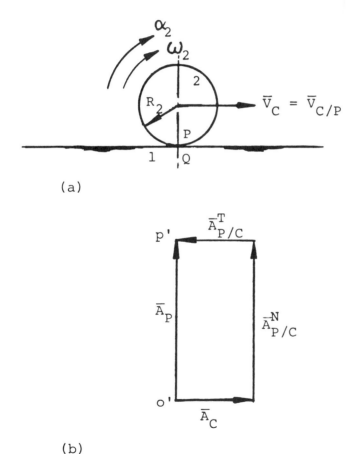

(a)

(b)

Figure 9.19   Case 3: (a) Roller 1 on flat surface; (b) acceleration diagram.

Hence, for Cases 1, 2, and 3, A_P may be deduced as follows:

CASE 1: ROLLER 2 ROLLS ON OUTER SURFACE OF ROLLER 1 [Figure 9.17(a)]

From Eq. 9.3, magnitude of vector $\overline{A}_P$ is given by

$$A_P = \frac{-V_C^2}{R_1 + R_2} + \frac{V_C^2}{R_2} \tag{9.4}$$

$$= V_C^2 \left[\frac{R_1}{R_2(R_1 + R_2)} \right] \tag{9.5}$$

$$= \omega_2^2 \left[\frac{R_1 R_2}{R_1 + R_2} \right] \tag{9.6}$$

CASE 2: ROLLER 2 ROLLS ON INNER SURFACE OF ROLLER 1 [Figure 9.18(a)]

Substituting $R_1 - R_2$ for $R_1 + R_2$ in Eq. (9.6) for the path radius of C, we get

$$A_P = \omega_2^2 \left[\frac{R_1 R_2}{R_1 - R_2} \right] \tag{9.7}$$

CASE 3: ROLLER 2 ROLLS ON FLAT SURFACE [Figure 9.19(a)]

Dividing the numerator and denominator of Eq. (9.6) by R_1 and letting R_1 equal infinity, we get

$$A_P = \omega_2^2 \times R_2 \tag{9.8}$$

Relative Acceleration of Contact Points (Both Bodies Moving)

Let us now investigate the relative acceleration between the coincident points P (on roller 2) and Q (on roller 1) when both bodies are moving.

To do this, we will consider the same roller arrangements as in the last section. However, for these cases, we let roller 1 be given in angular velocity ω_1 of 1 rad/sec (CW), while roller 2 rolls, as before, with an angular velocity ω_2 of 4 rad/sec (CW) and angular acceleration α_2 of 2 rad/sec (CW).

Graphical Procedure

The graphical procedure for finding this acceleration, based on this relationship, may be summarized as follows: From relative motion theory, we know that the acceleration of point P relative to point Q is the vectorial difference of the absolute accelerations of P and Q. That is,

$$\overline{A}_{P/Q} = \overline{A}_P - \overline{A}_Q \quad \text{(vectorially)} \tag{9.9}$$

where

$$\overline{A}_P = \overline{A}_C + \overline{A}_{P/C} \quad \text{(vectorially)}$$

Hence,

$$\overline{A}_{P/Q} = \overline{A}_C + \overline{A}_{P/C} - \overline{A}_Q \quad \text{(vectorially)} \tag{9.10}$$

$$= \overline{A}_{P/Q}^N + \overline{A}_{P/Q}^T$$

$$= \overline{A}_C^N + \overline{A}_C^T + \overline{A}_{P/C}^N + \overline{A}_{P/C}^T - \overline{A}_Q^N - \overline{A}_Q^T$$

$$\text{(vectorially)} \tag{9.11}$$

CASE 1: ROLLER 2 ROLLS ON OUTER SURFACE OF ROLLER 1 [Figure 9.20(a)]

1. Define the polar origin o' for the polygon.
2. Plot vectors $\overline{A}_C^N$ and $\overline{A}_C^T$, where the magnitudes are given by

$$A_C^N = \frac{V_{C/B}^2}{CB}$$

where $V_{C/B}$ is obtained from a velocity diagram [see Figure 9.20(b)] in accordance with the relationship

$$\overline{V}_{C/B} = \overline{V}_{C/Q} + \overline{V}_{Q/B} \tag{9.12a}$$

$$\text{or } V_{C/B} = R_2\omega_2 + R_1\omega_1 \tag{9.12b}$$

and

$$A_C^T = R_2 \times \alpha_2$$

Define the terminus of $\overline{A}$ as c'.

3. Plot vector $-\overline{A}_Q^N$, whose magnitude is given by

$$A_Q^N = R_1 \times \omega_1^2$$

Note that $\overline{A}_Q^T = 0$, since $\alpha_1 = 0$. Therefore, $\overline{A}_Q^N = \overline{A}_Q$. Define the terminus of $\overline{A}_Q$ as q'.

4. From point c', plot the normal and tangential acceleration vectors $\overline{A}_{P/C}^N$ and $\overline{A}_{P/C}^T$, whose magnitudes are given by

$$A_{P/C}^N = R_2\omega_2^2$$

$$A_{P/C}^T = R_2\alpha_2$$

Define the terminus of vector $\overline{A}_{P/C}$ as p'.

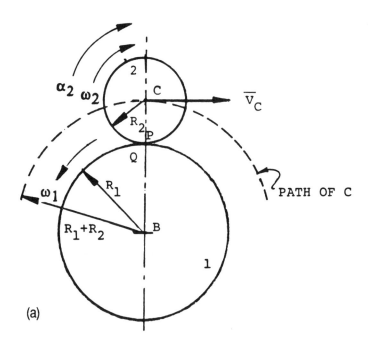

(a)

CASE 1

$$A_C^N = \frac{V_{C/B}^2}{CB} = \frac{(R_2\omega_2 + R_1\omega_1)^2}{R_1 + R_2}$$

$$= \frac{[2(4) + 5(1)]^2}{5 + 2}$$

$$= 24.14 \text{ in./sec}^2$$

$$A_C^T = R_2\alpha_2 = 2(2) = 4 \text{ in./sec}^2$$

$$A_{P/C}^N = R_2\omega_2^2 = 2(4^2) = 32 \text{ in./sec}^2$$

$$A_{P/C}^T = R_2\alpha_2 = 2(2) = 4 \text{ in./sec}^2$$

$$A_Q^N = R_1\omega_1^2 = 5(1^2) = 5 \text{ in./sec}^2$$

$$A_{P/Q} = 13 \text{ in./sec}^2 \quad \text{(scaled)}$$

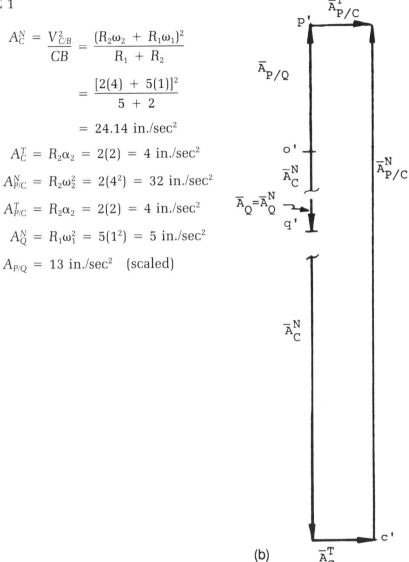

(b)

Figure 9.20   Case 1: (a) roller 2 on outer surface of roller 1; (b) acceleration diagram.

5. Draw a vector from q' to p' to complete the vector circuit. This vector will define the normal acceleration of P relative to Q (or $\overline{A}_{P/Q}^N$) or the required $\overline{A}_{P/Q}$.

CASE 2: ROLLER 2 ROLLS ON INNER SURFACE OF ROLLER 1 [Figure 9.21(a)]

For this case, the procedure is the same as in Case 1, except for the determination of $\overline{A}_C^N$ where the radius of the path of C is $R_1 - R_2$ and $V_{C/B} = R_2\omega_2 - R_1\omega_1$. Figure 9.21(b) shows the acceleration diagram for this case.

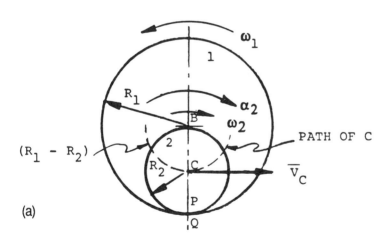

(a)

CASE 2

$$A_C^N = \frac{V_{C/B}^2}{CB} = \frac{(R_2\omega_2 - R_1\omega_1)^2}{R_1 - R_2}$$

$$= \frac{[2(4) - 5(1)]^2}{5 - 2}$$

$$= 3 \text{ in./sec}^2$$

$$A_C^T = R_2\alpha_2 = 2(2) = 4 \text{ in./sec}^2$$

$$A_{P/C}^N = R_2\omega_2^2 = 2(4^2) = 32 \text{ in./sec}^2$$

$$A_{P/C}^T = R_2\alpha_2 = 2(2) = 4 \text{ in./sec}^2$$

$$A_Q^N = R_1\omega_1 = 5(1^2) = 5 \text{ in./sec}^2$$

$$A_{P/Q} = 29 \text{ in./sec}^2 \quad \text{(scaled)}$$

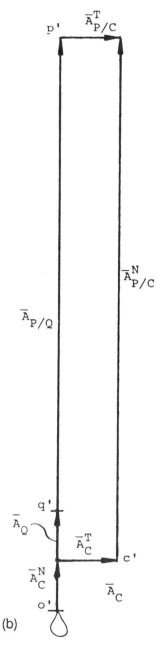

(b)

Figure 9.21   Case 2: (a) roller 2 on inner surface of roller 1; (b) acceleration diagram.

CASE 3: ROLLER 2 ROLLS ON FLAT SURFACE [Figure 9.22(a)]

For this case, the procedure is the same as in Case 1, except that $\overline{A}_C^N$ and $\overline{A}_Q^N$ are zero, since the paths of C and Q are both linear (for $R_1 = \infty$). Figure 9.22(b) shows the acceleration diagram for this case.

Algebraic Expressions for $\overline{A}_{P/Q}$ (Both Bodies Moving)

CASE 1: ROLLER 2 ROLLS ON OUTER SURFACE OF ROLLER 1 [Figure 9.20(a)]

We start by recalling Eq. (9.11)

$$\overline{A}_{P/Q} = \overline{A}_{P/Q}^N + \overline{A}_{P/Q}^T$$
$$= \overline{A}_C^N + \overline{A}_C^T + \overline{A}_{P/C}^N + \overline{A}_{P/C}^T - \overline{A}_Q^N - \overline{A}_Q^T \qquad (9.11)$$

where

$$\overline{A}_C^N = \frac{-V_{C/B}^2}{CB} \quad \text{(directed downward toward B)}$$

$$\overline{A}_C^T = R_2\alpha_2 \quad \text{(directed toward the right)}$$

$$\overline{A}_{P/C}^N = \frac{V_{P/C}^2}{PC} \quad \text{(directed from P toward C)}$$

$$\overline{A}_{P/C}^T = -R_2\alpha_2 \quad \text{(directed toward the left)}$$

$$\overline{A}_Q^N = -R_1\omega_1^2 \quad \text{(directed from Q toward B)}$$

$$\overline{A}_Q^T = R_1\alpha_1 = 0 \quad (\alpha_1 = 0)$$

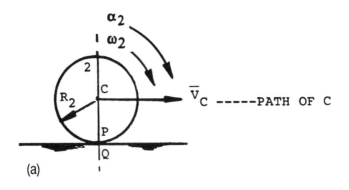

(a)

CASE 3

$A_C^N = R_2(0) = 0$ (path of C is straight line)

$A_C^T = R_2\alpha_2 = 2(2) = 4$ in./sec^2

$A_{P/C}^N = R_2\omega_2^2 = 2(4^2) = 32$ in./sec^2

$A_{P/C}^T = R_2\alpha_2 = 2(2) = 4$ in./sec^2

$A_Q^N = R_1(0) = 0$ (path of Q is straight line)

$A_{P/Q} = A_P = 32$ in./sec^2 (scaled)

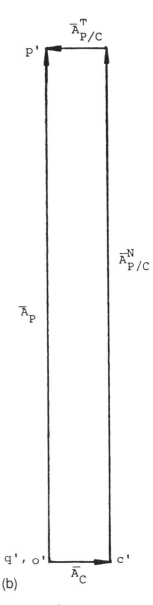

(b)

Figure 9.22   Case 3: (a) roller 2 on flat surface; (b) acceleration diagram.

Now, since the velocities of P and Q in the tangential direction are identical for no slipping to occur at these points, there can be no relative motion between P and Q in that direction. Hence, the tangential components of acceleration of P relative to Q (or $\overline{A}^T_{P/Q}$) must be zero.

Also, since $\overline{A}^T_C$ and $\overline{A}^T_{P/C}$ are equal and opposite, they cancel each other. Therefore, Eq. 9.11 may be written in terms of the normal component accelerations alone as follows:

$$\overline{A}_{P/Q} = \overline{A}^N_{P/Q} = \overline{A}^N_{P/C} + \overline{A}^N_C - \overline{A}^N_Q \tag{9.13}$$

$$A^N_{P/Q} = \frac{V^2_{P/C}}{PC} + \left(-\frac{V^2_{C/B}}{CB} \right) - \left(-\frac{V^2_{Q/B}}{QB} \right) \tag{9.14}$$

$$A^N_{P/Q} = \frac{V^2_{P/C}}{PC} - \frac{V^2_{C/B}}{CB} + \frac{V^2_{Q/B}}{QB} \tag{9.15a}$$

where

$$V_{P/C} = R_2\omega_2 \tag{9.15b}$$

$$V_{Q/B} = R_1\omega_1 \tag{9.15c}$$

$$V_{C/B} = R_2\omega_2 + R_1\omega_1 \quad \text{(from Eq. 9.12b)}$$

$$\frac{V^2_{C/B}}{C/B} = \frac{(R^2_2\omega^2_2 + 2R_1R_2\omega_1\omega_2 + R^2_1\omega^2_1)}{R_1 + R_2} \tag{9.16}$$

Substituting Eqs. (9.15b), (9.15c), and (9.12b) into Eq. (9.16) we get

$$A^N_{P/Q} = R_2\omega^2_2 - \frac{(R^2_2\omega^2_2 + 2R_1R_2\omega_1\omega_2 + R^2_1\omega^2_1)}{R_1 + R_2} + R_1\omega^2_1 \tag{9.17}$$

$$= \frac{R_1R_2(\omega^2_2 + \omega^2_1 - 2\omega_1\omega_2)}{R_1 + R_2} \tag{9.18}$$

$$= \left(\frac{R_1R_2}{R_1 + R_2} \right) (\omega_2 - \omega_1)^2 \tag{9.19}$$

$$= \left(\frac{R_1R_2}{R_1 + R_2} \right) (\omega_{2/1})^2 \tag{9.20}$$

or

$$A_{P/Q} = \text{ditto} \tag{9.21}$$

CASE 2: ROLLER 2 ROLLS ON INNER SURFACE OF ROLLER 1 [Figure 9.21(a)]

Since for this case, the radius of the path of C is $R_1 - R_2$, Eq. (9.21) is modified to give

$$A_{P/Q} = \left(\frac{R_1 R_2}{R_1 - R_2} \right) (\omega_{2/1})^2 \tag{9.22}$$

CASE 3: ROLLER 2 ROLLS ON FLAT SURFACE [Figure 9.22(a)]

Since for this case, roller 1 is assumed to be infinitely large, we then divide the numerator and denominator of Eq. (9.21), and by letting $R_1 = \infty$, we obtain

$$A_{P/Q} = R_2 \omega_{2/1}^2 \tag{9.23}$$

EXAMPLE 9.3

Find the acceleration of point C in the roller-cam mechanism shown in Figure 9.23(a), for the position shown, where the cam rotates with a counterclockwise angular velocity (ω_1) of 1 rad/sec and clockwise angular acceleration (α_1) of 0.5 rad/sec.

SOLUTION [See Figures 9.23(b) and 9.23(c)]

$$V_Q = AQ \times \omega_1 = 1.5(1) = 1.5 \text{ in./sec}$$

$$A_Q^N = AQ \times \omega_1^2 = 1.5(1^2) = 1.5 \text{ in./sec}^2$$

$$A_Q^T = AQ \times \alpha_1 = 1.5(0.5) = 0.75 \text{ in./sec}^2$$

$$A_C^N = \frac{V_C^2}{CD} = \frac{0.5^2}{2.25} = 0.11 \text{ in./sec}^2$$

$$A_{P/Q}^N = \frac{(R_1 R_2)}{(R_1 + R_2)} \omega_{2/1}^2 \quad \text{(from Eq. 9.20)}$$

$$= \frac{(1)(0.5)}{1.5} (\omega_2 - \omega_1)^2$$

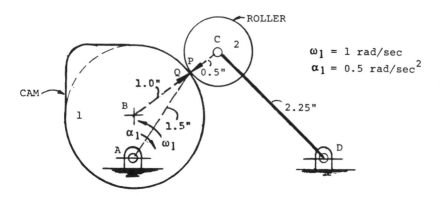

(a) ROLLER-CAM MECHANISM

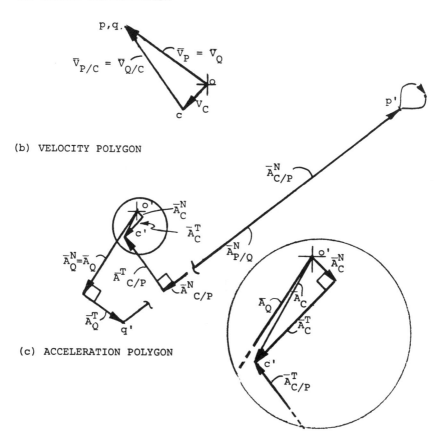

(b) VELOCITY POLYGON

(c) ACCELERATION POLYGON

Figure 9.23   Example 9.3: (a) Roller-cam mechanism; (b) velocity polygon; (c) acceleration polygon.

$$\omega_2 = \frac{V_{P/C}}{CP} = \frac{1.5}{0.5} = 3 \text{ rad/sec (CW)}$$

$$\omega_1 = 1.0 \text{ rad/sec (CCW)}$$

$$A_{P/Q}^N = 0.33[3 - (-1)]^2$$

$$= 0.33(4^2)$$

$$= 5.28 \text{ in./sec}^2$$

$$A_{C/P}^N = \frac{V_{C/P}^2}{CP} = \frac{1.5^2}{0.5} = 4.5 \text{ in./sec}^2$$

$$A_C = 0.45 \text{ in./sec}^2 \quad \text{(scaled)}$$

Note: It is important to note that in Eqs. (9.21) through (9.23), when the angular velocity ω_1 is zero, which means that roller 1 is fixed, the acceleration of P relative to Q becomes the absolute acceleration of point P, and these expressions revert, respectively, to Eqs. (9.6), (9.7), and (9.8).

Also note that the acceleration relationships derived here are not restricted to rolling circles, but are also applicable to noncircular bodies, as shown in the example above. In such cases, R_1 and R_2 represent the radii of the rolling curves at the point of contact.

9.9 SUMMARY

The generalized procedure for constructing the acceleration polygon may be summarized as follows:

1. Proceed from the "known" to the "unknown." That is,
   a. Lay out those absolute vectors whose magnitude and direction are known.
   b. Lay out those components of absolute and relative vectors that are known (magnitude and direction) or can be determined. These include normal accelerations and Coriolis acceleration.
   c. Add to the components in Step b their corresponding tangential accelerations (directions only) and extend these to close the polygon.
2. All absolute vectors on the acceleration polygon originate from pole o', whereas relative vectors extend between the termini of the absolute vectors.

3. a. A vector that originates from b' and terminates at c' represents the relative acceleration of point C to that of point B on the link.

   b. The choice between $\overline{A}_{B/C}$ and $\overline{A}_{C/B}$ makes no difference in the polygon configuration or the results, except that these vectors have opposite senses.

4. Two undefined vectors such as $\overline{A}^T_{B/C}$ and $\overline{A}^T_{C}$ will contain a common point c' on the polygon and define that point where they intersect.

10
Velocity-Difference Method

10.1   INTRODUCTION

The velocity-difference method of determining accelerations of points in a mechanism is probably the most straightforward of all the graphical methods used in kinematic analysis. Applicable to any type of mechanism—pin-connected, rolling contact, or sliding contact—the method does not rely on the use of sophisticated formulas, but instead employs the simple relationship

$$A = \frac{\Delta V}{\Delta T}$$

based on the fundamental definition, which states that **the acceleration (A) of a point is the rate of change of velocity (ΔV) of that point over the time interval (ΔT) in which the change occurs**. In terms of mechanism analysis, if the linear acceleration is required for a point in a mechanism at any given position of the mechanism, this acceleration may be found by first considering a small change in position ($\Delta\theta$) of the point, then determining the change in velocity (ΔV) resulting from the change in position, and finally dividing the change of velocity by the time interval (ΔT) during which the change has taken place.

   In applying the velocity-difference method, experience has shown that, generally, for reasonably accurate results, a position change of the mechanism based on a crank angular displacement of 1/10 rad is acceptable. This displacement is normally measured such that the given crank position is centrally located between its initial and final positions. That is, the initial and final crank positions are each 1/20 rad on either side of the given position.

The associated time interval ΔT depends on the angular velocity of the crank and its angular displacement. If, for example, the crank of a mechanism turns at a constant speed of 5 rad/sec, ΔT is the time it takes for the crank to move through 1/10 rad. In our earlier discussion it was noted that for constant angular speed, ΔT is obtained from

$$\omega = \frac{\Delta\theta}{\Delta T}$$

or

$$\Delta T = \frac{\Delta\theta}{\omega}$$

$$= \frac{1}{5 \text{ rad/sec}} \left(\frac{\text{rad}}{10}\right)$$

$$= 1/50 \text{ sec}$$

If, on the other hand, the crank turns with a uniform acceleration, which means that the angular velocity is changing, then again from our earlier discussion, ΔT is obtained from

$$\alpha = \frac{\Delta\omega}{\Delta T}$$

or

$$\Delta T = \frac{\Delta\omega}{\alpha}$$

where $\Delta\omega$ is the change in angular velocity between the initial and final positions of the mechanism. This uniform or constant acceleration case will be discussed in more detail in Section 10.4.

The following two sections illustrate the procedure for applying the velocity-difference technique to mechanisms in which the crank arm rotates with constant angular velocity.

Figure 10.1   Slider-crank mechanism.

10.2   SLIDER-CRANK MECHANISM ANALYSIS

Consider slider-crank mechanism ABC shown in Figure 10.1, where crank arm AB is rotating at a constant angular velocity of 1 rad/sec (counterclockwise). Let it be required to find the acceleration of point C for the position shown.

PROCEDURE

1.   Lay out the given mechanism in the position AB_iC_i (see Figure 10.2), where AB_i indicates the initial position of the crank, displaced

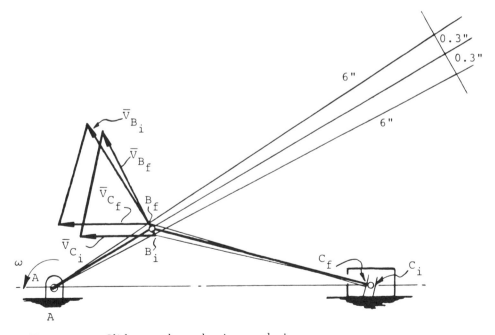

Figure 10.2   Slider-crank mechanism analysis.

1/20 rad (or 1/2 of 1/10 rad) clockwise from the given position. Note that the angular displacement 1/10 rad can be accurately laid out by applying the well-known relationship

$$S = R \, \Delta\theta$$

where S is the length of an arc that subtends angle $\Delta\theta$ and R is the radius of the arc. Hence, the angle 1/20 rad is obtained by laying out a segment of a circle having some convenient radius say, 6 in., in which case the arc length is

$$S = 6 \left(\frac{1}{20}\right) = 0.3 \text{ in.}$$

2. Determine the linear velocity of point C_i using any of the methods studied earlier. The relative velocity method, used in this example, yields

$$\overline{V}_{C_i} = 1.0 \text{ in./sec} \quad \text{(directed as shown in Figure 10.2)}$$

3. Lay out the given mechanism in position $AB_f C_f$ (see Figure 10.2), where AB_f indicates the final position of the crank, displaced 1/20 rad counterclockwise from the given position.
4. Determine the linear velocity from point C_f as in Step 2. This velocity has been determined to be

$$\overline{V}_{C_f} = 1.17 \text{ in./sec} \quad \text{(directed as shown in Figure 10.2)}$$

5. Determine the velocity difference ΔV_C, which is given by

$$\Delta\overline{V}_C = \overline{V}_{C_f} - \overline{V}_{C_i} \quad \text{(vectorial difference)}$$

This equation, which is represented graphically in Figure 10.2, yields

$$\Delta\overline{V}_C = 0.17 \text{ in./sec} \quad \text{(directed to the left)}$$

Note that the vectorial difference in this example yields the same results as the algebraic difference since $\overline{V}_{C_f}$ and $\overline{V}_{C_i}$ both have the same line of action.

6. Determine the acceleration of point C. The magnitude of this vector is given by

$$A_C = \frac{\Delta V_C}{\Delta T}$$

where

$$\Delta T = \frac{\Delta \theta}{\omega}$$

$$= 0.1 \text{ rad}/1.0 \text{ rad}/\text{sec} = 0.1 \text{ sec}$$

Therefore,

$$A_C = \frac{0.17}{0.1} \text{ in.}/\text{sec}^2$$

and the vector $\overline{A}_C = 1.70$ in./sec² (directed as shown in Figure 10.2). Note that the direction of acceleration is always the same as that for $\Delta \overline{V}_C$.

10.3 QUICK-RETURN MECHANISM ANALYSIS

Now consider the quick-return mechanism $ABCD$ in Figure 10.3. Crank AB turns with a constant angular speed of 30 rad/sec (counterclockwise) and the linear acceleration of point C on the rod is required.

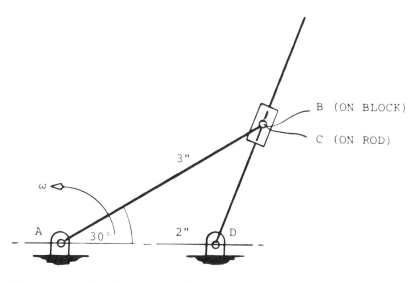

Figure 10.3 Quick-return mechanism.

PROCEDURE

We follow the same procedure as that given in Section 10.2.

1. Lay out the mechanism AB_iC_iD (Figure 10.4), showing the initial position of the mechanism rotated back 1/20 rad from the given position.
2. Determine the linear velocity of point C ($\overline{V}_{C_i}$) for this position (see Figure 10.4).

$$\overline{V}_{C_i} = 72 \text{ in./sec} \quad \text{(directed as shown)}$$

3. Lay out the mechanism AB_fC_fD (Figure 10.4), showing the final position of the mechanism rotated forward 1/20 rad from the given position.
4. Determine the linear velocity of point C ($\overline{V}_{C_f}$) for this position (see Figure 10.4)

$$\overline{V}_{C_f} = 69.5 \text{ in./sec} \quad \text{(directed as shown)}$$

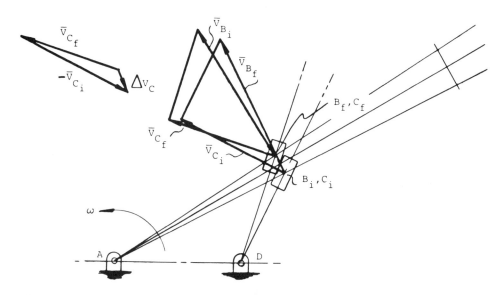

Figure 10.4   Quick-return mechanism analysis.

5. Determine the velocity difference $\Delta \overline{V}_C$ from the relationship

$$\Delta \overline{V}_C = \overline{V}_{C_f} - \overline{V}_{C_i} \quad \text{(vectorial difference)}$$

$$= 15.0 \text{ in./sec} \quad \text{(directed as shown in Figure 10.4)}$$

6. Determine the linear acceleration $\overline{A}_C$. The magnitude of this vector is given by

$$A_C = \frac{\Delta V_C}{\Delta T}$$

$$= \Delta V_C \frac{\omega}{\Delta \theta}$$

$$= 15.0 \left(\frac{30}{0.1} \right)$$

$$= 4500 \text{ in./sec}^2$$

Therefore,

$$\overline{A}_C = 4500 \text{ in./sec}^2 \quad \text{(directed like } \Delta \overline{V}_C \text{ as shown)}$$

7. To obtain the angular acceleration of CD (α_{CD}), apply the relationship

$$\alpha_{CD} = \frac{\Delta \omega_{CD}}{\Delta T}$$

where

$$\Delta \omega_{CD} = \omega_{CD_f} - \omega_{CD_i}$$

$$= \frac{V_{C_f}}{CD_f} - \frac{V_{C_i}}{CD_i}$$

$$= \frac{69.5}{1.7} - \frac{72}{1.52}$$

$$= 40.9 - 47.4 = -6.5 \text{ rad/sec}$$

$$\Delta T = \frac{\Delta \theta}{\omega}$$

$$= \frac{1}{10} \left(\frac{1}{30}\right) = 1/300 \text{ sec}$$

Therefore,

$$= \frac{-6.5}{1/300}$$

$$= -1950 \text{ rad/sec}^2$$

Note that in this analysis there was no need to determine the Coriolis acceleration, which is ordinarily the case in the relative acceleration method. In this respect, the velocity-difference method offers additional simplification to the solution of a problem that can otherwise be more complicated.

10.4 FOUR-BAR MECHANISM ANALYSIS

Consider the four-bar mechanism $ABCD$, where crank AB turns with a constant angular acceleration of 1 rad/sec^2 and for the position shown has an angular speed of 2 rad/sec (Figure 10.5). It is required to find the linear acceleration of point C on the follower.

The procedure for solving this problem is basically the same as that used in Sections 10.2 and 10.3, with the exception that in this case there is an angular acceleration, and therefore the magnitudes of the linear velocities of $\overline{V}_{B_i}$ and $\overline{V}_{B_f}$ are not the same. These velocity magnitudes are calculated from

$$V_{B_i} = AB\omega_i \tag{10.1}$$

and

$$V_{B_f} = AB\omega_f \tag{10.2}$$

where ω_i is the initial angular velocity of AB_i obtained from

$$\omega^2 = \omega_i^2 + 2\alpha \, \Delta\theta \tag{10.3}$$

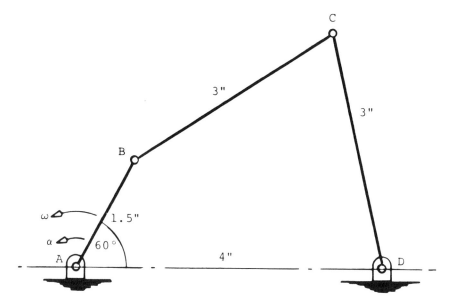

Figure 10.5   Four-bar mechanism.

and ω_f is the final angular velocity of AB_f obtained from

$$\omega_f^2 = \omega^2 + 2\alpha \, \Delta\theta$$

or

$$\omega^2 = \omega_f^2 - 2\alpha \, \Delta\theta \qquad\qquad (10.4)$$

Also, since the angular acceleration is constant, the angular speed is changing. ΔT must be expressed in terms of the angular acceleration α and changing speed ω. That is,

$$\Delta T = \frac{\Delta\omega}{\alpha} \qquad\qquad (10.5)$$

where $\Delta\omega = \omega_f - \omega_i$.

PROCEDURE

1. Figure 10.6 shows the mechanisms AB_iC_iD in the initial position of AB_fC_fD in the final position.

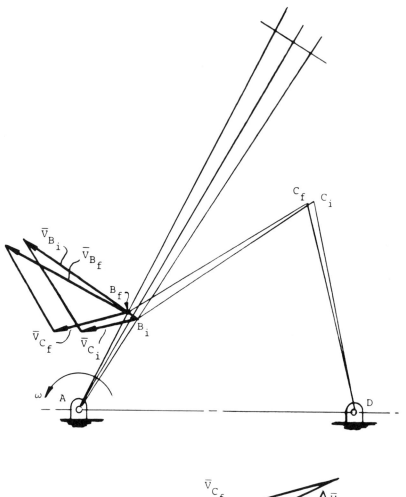

Figure 10.6 Four-bar mechanism analysis.

2. Determine ω_i. This is found by applying Eq. (10.3).

$$\omega^2 = \omega_i^2 + 2\alpha\ \Delta\theta$$

$$2^2 = \omega_i^2 + 2(1)\left(\frac{1}{20}\right)$$

$$\omega_i^2 = 4 - 0.1 = 3.9$$

$$\omega_i = 1.97 \text{ rad/sec}$$

3. Determine $\overline{V}_{B_i}$. From Eq. (10.1), the magnitude of this vector is computed as

$$V_{B_i} = 1.5(1.9)$$

$$= 2.95 \text{ in./sec}$$

Therefore,

$$\overline{V}_{B_i} = 2.45 \text{ in./sec} \quad \text{(directed as shown)}$$

4. Determine ω_f. Since α is constant, ω_f is easily determined by recognizing that its increment from ω at the given crank position is equal to that of the same ω from ω_i, due to the same angular displacement of $1/20$ rad in both cases. Therefore,

$$\omega_f = 2 + (2 - 1.97)$$

$$= 2.03 \text{ rad/sec}$$

5. Determine $\overline{V}_{B_f}$. The magnitude of this vector is computed as

$$V_{B_f} = 1.5(2.03)$$

$$= 3.04 \text{ in./sec}$$

Therefore,

$$\overline{V}_{B_f} = 3.04 \text{ in./sec} \quad \text{(directed as shown)}$$

6. Determine $\overline{V}_{C_f}$ using the mechanism AB_fC_fD, as before.

$$\overline{V}_{C_f} = 1.7 \text{ in./sec} \quad \text{(directed as shown)}$$

7. Determine $\overline{V}_{C_i}$ using the mechanism AB_iC_iD, as before.

$$\overline{V}_{C_i} = 1.25 \text{ in./sec} \quad \text{(directed as shown)}$$

8. Determine $\Delta\overline{V}_C$ from

$$\Delta\overline{V}_C = \overline{V}_{C_f} - \overline{V}_{C_i} \quad \text{(vectorial difference)}$$

$$= 0.46 \text{ in./sec} \quad \text{(directed as shown)}$$

9. Determine $\overline{A}_C$. The magnitude of this vector is computed as

$$A_C = \frac{\Delta V_C}{\Delta T}$$

where ΔT, from Eq. (10.5), is given by

$$\Delta T = \frac{\Delta\omega}{\alpha}$$

$$= \frac{2.03 - 1.97}{1}$$

$$= 0.06 \text{ sec}$$

Therefore,

$$A_C = \frac{0.46}{0.06}$$

$$= 7.7 \text{ in./sec}^2$$

Hence,

$$\overline{A}_C = 7.7 \text{ in./sec}^2 \quad \text{(directed like } \Delta\overline{V}_C \text{ as shown)}$$

Despite the simplicity of the velocity-difference method, it should be understood that because of the need to measure small changes in displacements as well as velocities, a high degree of drawing precision is required if sufficiently accurate results are to be achieved. In some cases, this requirement can impose practical limitations such that the results obtained may not be reliable.

11

Graphical Calculus Method

11.1 GRAPHICAL DIFFERENTIATION

A common method for obtaining velocity and acceleration curves for a mechanism consists of constructing a time-displacement curve and then by graphical differentiation developing the velocity and acceleration curves. *Graphical differentiation* consists of obtaining the slopes of various tangents along one curve and plotting these values as ordinates to establish a second curve. The method, as applied to the displacement and velocity curves, is based on the following principles:

1. The instantaneous velocity of a moving point can be represented graphically as the slope of the displacement curve at that instant, or

$$v = \frac{\Delta s}{\Delta t} \quad \text{where } \Delta t \text{ is very small}$$

2. The instantaneous acceleration can be represented as the slope of the velocity curve at that instant, or

$$a = \frac{\Delta t}{\Delta t} \quad \text{where } \Delta t \text{ is very small}$$

245

Graphical differentiation can be an effective tool for the designer, particularly when used during the preliminary stages of a mechanism design since it affords an overall picture of the velocity and acceleration throughout the motion cycle. For example, from motion curves such as the s-t, v-t, and a-t curves, one can readily determine

The maximum absolute values of displacement, velocity, and acceleration

Where and when the maximum values of angular displacement occur

Whether there is any abrupt change in displacement, velocity, and acceleration during the cycle

Also, there are many instances in which it is necessary to differentiate a curve for which an equation is difficult to obtain. In such instances, graphical differentiation is the most convenient approach.

Tangent Method

Consider the displacement-time curve shown in Figure 11.1. Suppose that it is required to develop the velocity-time curve from this curve.

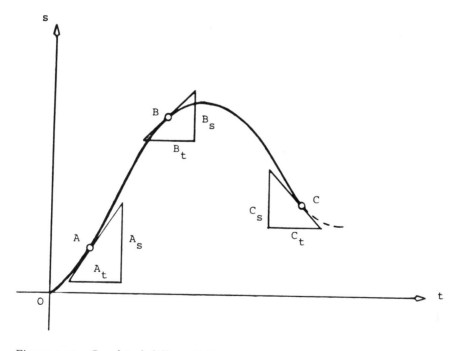

Figure 11.1   Graphical differentiation.

Since the instantaneous velocity at any point is represented by the slope of the displacement curve at that point, a tangent drawn through any point of the curve defines the velocity at that point. Therefore, at point A, the velocity v_A is obtained from

$$v_A = \text{slope of tangent at } A$$

$$= \frac{A_s k_s}{A_t k_t}$$

where

k_s = displacement scale, in./in.

k_t = time scale, sec/in.

Similarly,

$$v_B = \text{slope of tangent at } B = \frac{B_s(k_s)}{B_t(k_t)}$$

$$v_C = \text{slope of tangent at } C = -\frac{C_s(k_s)}{C_t(k_t)}$$

Here the negative sign indicates a negative slope of the curve and hence a negative velocity. Also, when the tangent to the curve is a horizontal line, this simply means that the velocity at the point of tangency is zero.

In addition, it should be clear that by making the time intervals A_t, B_t, C_t, and so on, equal, the velocities v_A, v_B, v_C, and so on, will be proportional to A_s, B_s, C_s, and so on. Hence, once the velocity v_A has been found, the velocities v_B, v_C, and so on, are readily obtained as follows:

$$v_B = \frac{B_s}{A_s} v_A$$

$$v_C = -\frac{C_s}{A_s} v_A$$

In this manner, the velocities of other points can be obtained to enable a smooth continuous curve to be drawn.

EXAMPLE 11.1

The curve shown in Figure 11.2(a) represents a typical displacement-
time curve for a cam mechanism. It is desired to obtain the velocity-
time curve from this displacement curve and, subsequently, the acceleration-time curve from the velocity curve.

SOLUTION

1. Draw tangents to various points on the given displacement-time
    curve: points A, B, C, D, and E.
2. Using the tangents as hypotenuses, construct right triangles at each
    point, making all the bases equal, that is,

$$A_t = B_t = C_t, \quad \text{and so on.}$$

3. Calculate the velocities at points A, B, C, D, and E as follows:

$$k_s = 2 \text{ in./in.}$$

$$k_t = 3 \text{ sec/in.}$$

$$k_v = 1 \text{ (in./sec)/in.}$$

$$v_A = \frac{A_s k_s}{A_t k_t} = \left(\frac{0.5}{1.0}\right)\frac{2}{3} = 0.33 \text{ in./sec}$$

$$v_B = \frac{B_s}{A_s} v_A = \left(\frac{1.3}{0.5}\right) 0.33 = 0.86 \text{ in./sec}$$

$$v_C = \frac{C_s}{A_s} v_A = \left(\frac{2.20}{0.5}\right) 0.33 = 1.45 \text{ in./sec}$$

$$v_D = \frac{D_s}{A_s} v_A = \left(\frac{0.5}{0.5}\right) 0.33 = 0.33 \text{ in./sec}$$

$$v_E = \frac{E_s}{A_b} v_A = \left(\frac{0}{0.5}\right) 0.33 = 0 \text{ in./sec}$$

Note that the remaining points on the velocity curve can be located
by inspection based on the symmetrical shape of the displacement
curve.

4. Draw the velocity-time axis using the same time scale as that used
    for the displacement-time axis [Figure 11.2(b)].

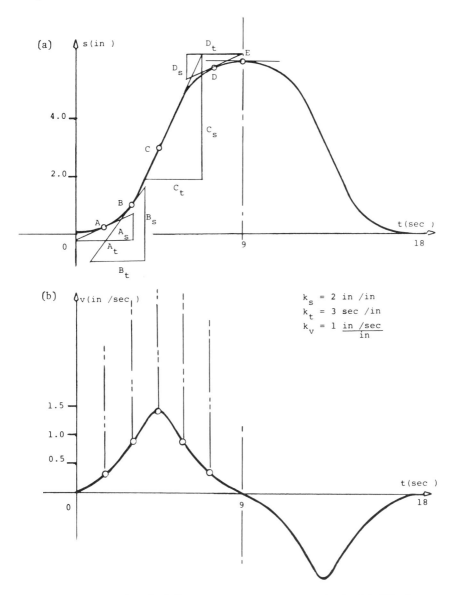

Figure 11.2 Graphical differentiation—tangent method: (a) Displacement curve; (b) velocity curve.

5. Plot the velocities calculated in Step 3 on the velocity-time axis.
6. Draw a smooth curve connecting the plotted points.
7. To obtain the acceleration curve, repeat Steps 1 through 6, replacing the displacement curve with the velocity curve just found [Figure 11.3(a)]. The required acceleration curve is shown in Figure 11.3(b).

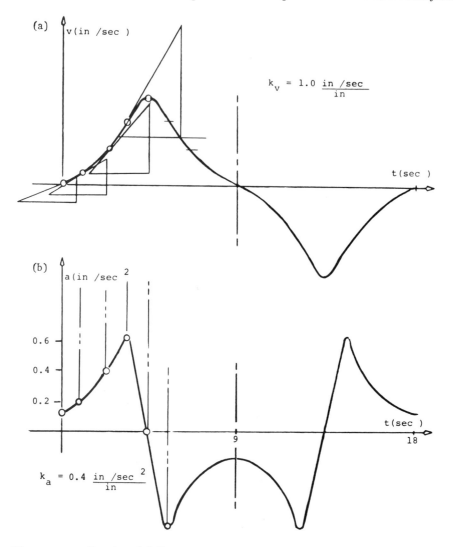

Figure 11.3 Graphical differentiation—tangent method: (a) Velocity curve; (b) acceleration curve.

Polar Method

An alternative graphical differentiation method commonly used is the *polar method*. This method has been found to produce greater accuracy than the tangent method mainly because it requires less graphical construction. The method is best described by considering the displacement-time curve given in Figure 11.4(a). Let it be required to develop the velocity and acceleration diagrams from the given curve.

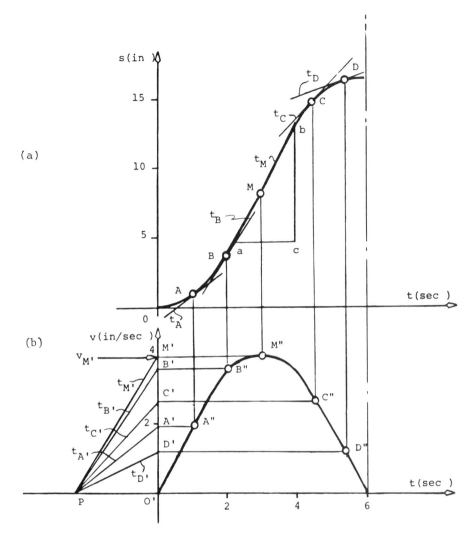

Figure 11.4 Graphical differentiation—polar method: (a) Displacement curve;
(b) velocity curve.

PROCEDURE

1. Lay out the axes for the velocity-time graph directly below those
    for the displacement curve, maintaining the same time scale in both
    cases [see Figure 11.4(b)].
2. Through the point of steepest slope on the displacement curve
    (point M), draw a tangent line to define the point of maximum
    velocity (t_M). Now, since by definition, velocity equals slope, the

maximum velocity at point M is given by

$$V_{max} = \frac{\Delta s}{\Delta t}$$

where Δs is the change of displacement over any interval Δt, or

$$V_{max} = \left(\frac{bc}{ac}\right) \frac{k_s}{k_t}$$

where

k_s = displacement scale, in./in.

k_t = time scale, sec/in.

3. Define a point P (called the pole) on the time axis of the velocity-time curve, at any convenient distance left of the origin, and from it draw line $t_{M'}$ parallel to tangent line t_M to intersect the velocity axis. This line locates at point M' on the velocity axis that represents the point of maximum velocity, or

$$v_{M'} = v_{max}$$

4. Plot the velocity of point M at the intersection of the horizontal line drawn from M' on the velocity curve and a vertical line drawn from M on the displacement curve. Label this point M''.
5. Use the value obtained for v_{max} to establish the scale for the velocity axis of the velocity-time graph. To do this, compare the actual measurement on the diagram with the computed maximum velocity and then by ratio determine the velocity value that is represented by 1 in. on the diagram. In this example we find that

$$V_{max} = \left(\frac{bc}{ac}\right) \frac{k_s}{k_t}$$

where

k_s = 5 in./in.

k_t = 2 sec/in.

Therefore,

$$v_{max} = \left(\frac{1.6}{1.0}\right)\frac{5}{2} = 4 \text{ in./sec}$$

Hence, 2 in. on the velocity curve represents 4 in./sec, or 1 in. represents 2 in./sec, or

$$k_v = 2(\text{in./sec})/\text{in.}$$

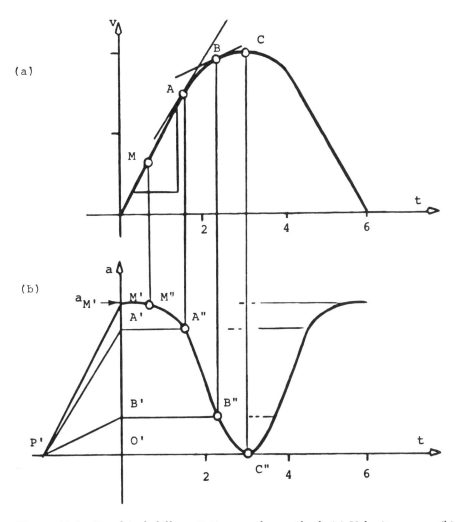

Figure 11.5 Graphical differentiation—polar method: (a) Velocity curve; (b) acceleration curve.

6. To establish the remaining velocity points at ordinates A, B, C, D, and so on, follow the procedure outlined in Steps 3 and 4. Figure 11.5(a) shows the completed velocity diagram.

7. After the velocity diagram has been completed, the acceleration diagram can be developed by following the procedure outlined above. Figure 11.5(b) shows the required curve.

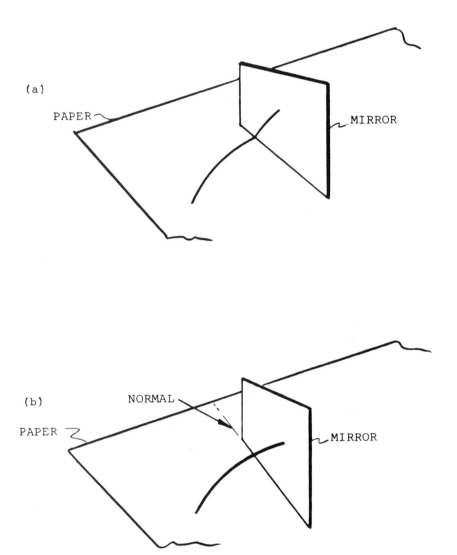

(a)

PAPER   MIRROR

(b)   NORMAL

PAPER   MIRROR

Figure 11.6   Use of a mirror in constructing tangents: (a) Incorrect mirror position; (b) correct mirror position.

Note on Accuracy

It should be emphasized that accuracy of the graphical differentiation methods depends primarily on

1. How accurately the tangents can be drawn
2. The number of intervals chosen
3. The ability to fit a smooth curve to a given set of points

For greater accuracy in constructing tangents, any straight edge having a reflective surface, such as the rectangular mirror shown in Figure 11.6, may be used. The straight edge, held vertical to the paper, is placed across the curve and rotated such that the visible curve and its reflection form a continuous curve. In this position, the straight edge is then normal to the curve. Therefore, a line drawn perpendicular to the normal and touching the curve is the required tangent.

Accuracy is increased with the number of intervals into which the curve is divided. The greater the number of intervals chosen, the greater the accuracy achieved. Finally, the ability to fit a smooth curve to a given set of points, particularly if double differentiation is required, greatly minimizes the risk of compounding errors from one curve to the other.

11.2 GRAPHICAL INTEGRATION

Just as it is possible to go from the displacement curve to the velocity curve and then to the acceleration curve using graphical differentiation, it is also possible to reverse the process, going from acceleration to velocity and then to displacement. The process by which this is achieved is called *graphical integration*.

The process of integration can be thought of as the procedure for obtaining the area under a given curve. As applied to the motion of a mechanism, the integration of the acceleration curve gives the velocity curve, and the integration of the acceleration curve gives the displacement curve. This is based on the motion laws, which state that

$$\Delta v = \bar{a}\,\Delta t \quad \text{or} \quad v_F - v_O = \bar{a}(t_F - t_O)$$

and

$$\Delta s = \bar{v}\,\Delta t \quad \text{or} \quad s_F - s_O = \bar{v}(t_F - t_O)$$

where the subscripts F and O denote final and original conditions, and $\bar{a}$ and $\bar{v}$ denote average acceleration and velocity. Hence, the following rules apply:

1. To derive the velocity-time curve from the acceleration-time curve, the change in velocity between any two times equals the area under the acceleration-time curve between the same two times.
2. To derive the displacement-time curve from the velocity-time curve, the change in displacement between any two times equals the area under the velocity-time curve between the same two times.

Midordinate Method

There are several graphical methods available to determine the area under the curve. One of the simplest is the **midordinate method**. The following example will illustrate the procedure.

EXAMPLE 11.2

The curve shown in Figure 11.7(a) represents a typical acceleration curve for a cam mechanism. It is desired to obtain the velocity-time curve.

SOLUTION

1. Divide the given curve into an equal number of sections, S_1, S_2, S_3, and so on, as shown in Figure 11.7(a), and construct mean ordinates $\bar{a}_1$, $\bar{a}_2$, $\bar{a}_3$, and so on, for each section to intersect the curve at 1, 2, 3, and so on.
2. Through points 1, 2, 3, and so on, draw horizontal lines extending between the boundaries of alternate sides of the curve. For example, in section S_1, triangles A and B are on alternate sides of the curve. From this it is easily seen that, provided that the slope of the curve within the section remains fairly constant, the alternate triangles are approximately equal. This means that in section S_1, the area of triangle A can be considered approximately equal to that of triangle B; similarly, in section S_2, the area of triangle C is equal to that of triangle D; and so on. Therefore, the area under the curve in section S_1 can be approximated by the rectangle $\bar{a}_1 \, \Delta t_1$; the area under section S_2 can be approximated by a rectangle $\bar{a}_2 \, \Delta t_2$; and so on.

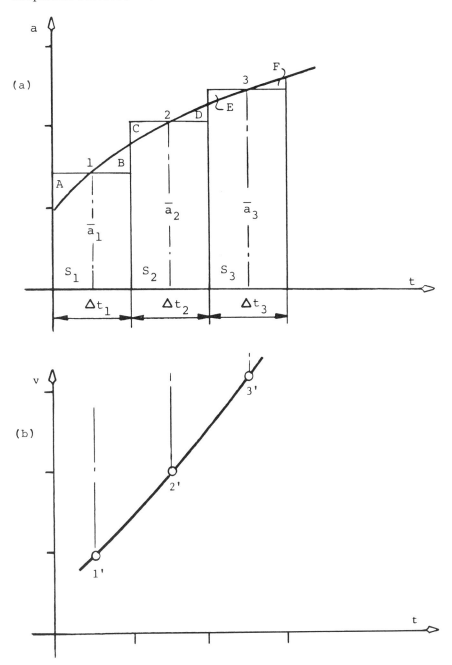

Figure 11.7 Graphical integration—midordinate method: (a) Acceleration curve; (b) velocity curve.

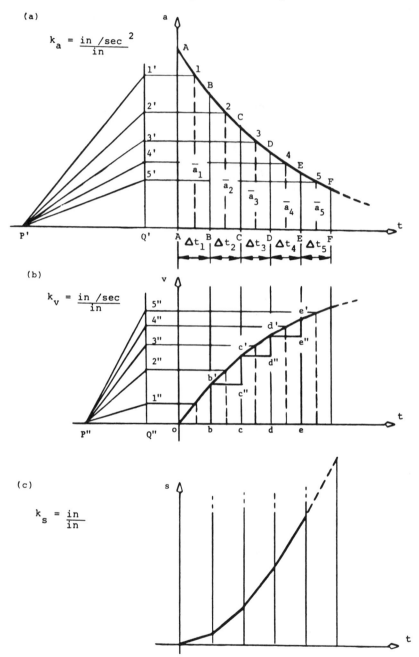

Figure 11.8   Graphical integration—polar method: (a) Acceleration curve; (b) velocity curve; (c) displacement curve.

3. Since the velocity equals the area under the curve, the velocities v_1, v_2, and v_3 may be found as follows:

$$v_1 = \bar{a}_1 \, \Delta t_1$$

$$v_2 = v_1 + \bar{a}_2 \, \Delta t_2$$

$$v_3 = v_2 + \bar{a}_3 \, \Delta t_3, \text{ and so on}$$

For example, in Figure 11.7(a),

$$v_1 = 1.4(1) = 1.4 \text{ in./sec}$$

$$v_2 = 1.4 + 2(1) = 3.4 \text{ in./sec}$$

$$v_3 = 3.4 + 2.4(1) = 5.8 \text{ in./sec}$$

The velocity curve is shown in Figure 11.7(b).
4. After the required velocity curve has been completed, repeat Steps 1 and 3 to develop the required displacement curve.

Polar Method

Another graphical integration method that is commonly used is the **polar method**, which is relatively fast and easy to apply. To illustrate the method, let us consider the acceleration curve shown in Figure 11.8(a). From this curve we will develop the velocity curve and, in turn, use the velocity curve to develop the displacement curve.

PROCEDURE

1. Erect coordinates *BB*, *CC*, *DD*, and so on, to divide the curve into an equal number of line intervals, as shown in Figure 11.8(a). In this case, five intervals have been chosen with ordinates $\bar{a}_1$ to $\bar{a}_5$ defined at the midpoints of the intervals. Also, since the horizontal axis is the time axis, each interval is defined at Δt_1, Δt_2, Δt_3, and so on. Note that all time scales for acceleration, velocity, and displacement curves must be the same, as with graphical differentiation.

2. From points 1, 2, 3, and so on, project horizontals to meet a line parallel to the a axis at points $1'$, $2'$, $3'$, and so on.
3. From a point P', located on the time axis, at any convenient distance left of the origin, draw connecting straight lines to points $1'$, $2'$, $3'$, and so on.
4. From the origin of the velocity axis [Figure 11.8(b)], draw line ob' parallel to $P'1$ to meet ordinate BB at b', line $b'c'$ parallel to $P'2'$ to meet ordinate CC at c', line $c'd'$ parallel to $P'3'$ to meet ordinate DD at d, and so on, until the complete velocity curve is obtained. From this velocity curve, the velocities at points 1, 2, 3, and so on, are given by

$$v_1 = P'Q'(bb')k_a k_t$$

$$v_2 = P'Q'(cc')k_a k_t$$

$$v_3 = P'Q'(dd')k_a k_t, \quad \text{and so on}$$

where b, c, d, and so on, are the intercepts of the first, second, third, and so on, ordinates with the time axis, and k_a [(in./sec^2)/in.] and k_t (sec/in.) are the acceleration and time scales, respectively.

The relationships above can be verified by applying the properties of similar triangles to triangles $P'Q'1'$ and obb', where bb' is the vertical leg of the velocity curve segment ob. From these triangles, we obtain

$$\frac{Q'1'}{bb'} = \frac{P'Q'}{ob} \tag{11.1}$$

However,

$$Q'1' = \bar{a}_1$$

and

$$ob = \Delta t_1$$

Therefore, Eq. (11.1) becomes

$$\frac{\bar{a}_1}{bb'} = \frac{P'Q'}{\Delta t_1}$$

from which follows

$$\bar{a}_1 \, \Delta t_1 \; = \; P'Q'(bb') \quad \text{(all parameters in inches)}$$

However,

$$\bar{a}_1 \, \Delta t(k_a k_t) \; = \; v_1 - v_0$$

where

$$v_0 \; = \; 0$$

Therefore,

$$v_1 \; = \; P'Q'(bb')k_a k_t$$

Similarly, from triangles $P'Q'2'$ and $b'c''c'$, where $c''c'$ is the vertical leg of the velocity curve segment $b'c''$, we obtain

$$\frac{Q'2'}{c''c'} \; = \; \frac{P'Q'}{b'c''}$$

from which follow

$$\bar{a}_2 \, \Delta t_2 \; = \; P'Q'(c''c')$$

$$v_2 - v_1 \; = \; P'Q'(c''c')k_a k_t$$

$$v_2 \; = \; v_1 + P'Q'(c''c')k_a k_t$$

$$v_2 \; = \; P'Q'(bb')k_a k_t + P'Q'(c''c')k_a k_t$$

$$= \; P'Q'(bb' + c''c')k_a k_t$$

$$= \; P'Q'(cc')k_a k_t$$

and from triangles $P'Q'3'$ and $c'd''d'$,

$$\frac{Q'3'}{d''d'} \; = \; \frac{P'Q'}{c'd''}$$

from which follow

$$\bar{a}_3 \, \Delta t_3 \; = \; P'Q'(d''d')$$

$$v_3 - v_2 \; = \; P'Q'(d''d')k_a k_t$$

$$v_3 = v_2 + P'Q'(d''d')k_a k_t$$

$$v_3 = P'Q'(cc')k_a k_t + P'Q'(d''d')k_a k_t$$

$$= P'Q'(cc' + d''d')k_a k_t$$

$$= P'Q'(dd')k_a k_t$$

To obtain the displacement diagram, repeat Steps 1 to 4, replacing the acceleration curve with the velocity curve. Figure 11.8(c) shows the required displacement diagram.

Note on Accuracy

As in the case of the graphical differentiation methods, the accuracy of the graphical integration method increases with the number or size of the intervals chosen. The smaller the interval chosen, the more likely the slope of the curve within that interval (or section) will remain constant, and hence the closer the approximation of equality between the alternate triangles. It is therefore desirable to make the intervals or sections as small as possible for greater accuracy. Also, if double differentiation is required, accuracy will depend on the ability to fit a smooth curve to a set of given points.

12

Miscellaneous Methods

12.1 COMPLETE GRAPHICAL ANALYSIS METHOD

In the graphical methods for acceleration analysis presented so far, one of the required calculations to be performed was that used to determine the normal acceleration components for points in the mechanism. In this section we demonstrate a complete graphical method, in which, by the proper choice of link, velocity, and acceleration scales, both the magnitude and direction of all normal acceleration components can be determined without the need for calculations.

To illustrate this method, let us consider link AB, shown in Figure 12.1(a), which rotates at an angular velocity of ω_{AB} rad/sec. First, we obtain the magnitude of the velocity of point B relative to A ($V_{B/A}$) by multiplying the length of the link by the angular velocity

$$V_{B/A} = AB \times \omega_{AB} \tag{12.1}$$

We then lay out the vector perpendicular to the link as shown. Here it is useful to recall that the magnitude of $\overline{V}_{B/A}$ is related to that of $\overline{A}^N_{B/A}$ by the expression

$$A^N_{B/A} = \frac{V^2_{B/A}}{BA} \tag{12.2}$$

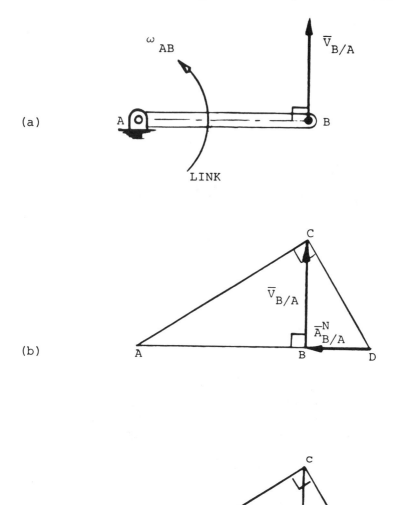

Figure 12.1   Normal acceleration construction.

Rearranging this equation, we obtain the relationship

$$\frac{A_{B/A}^N}{V_{B/A}} = \frac{V_{B/A}}{BA} \tag{12.3}$$

which can be represented as

$$\frac{BD}{BC} = \frac{BC}{BA} \tag{12.4}$$

by using two similar triangles, ACB and CDB, as shown in Figure 12.1(b), where

$$BD = A_{B/A}^N$$

and

$$BC = V_{B/A}$$

Here it can be seen that if the scales of link AB, velocity $V_{B/A}$, and acceleration $A_{B/A}^N$ in Eq. (12.3) are properly chosen, in accordance with the relationship expressed in Eq. (12.4), the length of line BD will accurately represent the required normal component.

Let the scales be defined as follows:

k_s = space scale, ft/in.

k_v = velocity scale, (ft/sec)/in.

k_a = acceleration scale, (ft/sec^2)/in.

where the inch units in the denominators represent actual measurements of the drawing in inches. For example, if the space scale is given as $k_s = 5$ in./in., a link length of 5 in. will be represented by a line 1 in. long on the drawing. Similarly, if the velocity scale is given as $k_v = 10$ (ft/sec) in., a velocity of 20 ft/sec will be represented by a line 2 in. long on the drawing; and for an acceleration scale of 50 (ft/sec^2)/in., an acceleration of 150 ft/sec^2 will be represented by a line 3 in. long.

The required relationship of the scales is obtained by considering triangle abc in Figure 12.1(c), where sides x, y, and z are proportional, respectively, to sides AB, BC, and BD of triangle ABC in Figure 12.1(b). In other words, if x represents the link length, y the velocity, and z the acceleration, we can write

$$xk_s = BA$$

or

$$x = \frac{BA}{k_s}$$

$$yk_v = BC \tag{12.5}$$

or

$$y = \frac{BC}{k_v}$$

$$zk_a = BD \tag{12.6}$$

or

$$z = \frac{BD}{k_a} \tag{12.7}$$

and from similarity,

$$\frac{z}{y} = \frac{y}{x} \tag{12.8}$$

Substitution of Eqs. (12.5) to (12.7) into Eq. (12.8) yields

$$\frac{BD/k_a}{BC/k_v} = \frac{BC/k_v}{BA/k_s} \tag{12.9}$$

or

$$\frac{A^N_{B/A}k_a}{V_{B/A}k_v} = \frac{V_{B/A}/k_v}{BA/k_s} \tag{12.10}$$

from which

$$A_{B/A}^N = \frac{V_{B/A}^2}{BA} \frac{k_a k_s}{k_v^2} \tag{12.11}$$

Thus, the scales must be related by the equation

$$\frac{k_a k_s}{k_v^2} = 1 \tag{12.12}$$

or

$$k_a k_s = k_v^2 \tag{12.13}$$

This means that any two scales may be chosen arbitrarily, but the third must be chosen from Eq. (12.12).

EXAMPLE 12.1

For the four-bar mechanism ABCD shown in Figure 12.2, make a complete graphical acceleration analysis given that crank AB is rotating with an angular velocity of 2 rad/sec (counterclockwise) and angular acceleration of 1 rad/sec² (clockwise).

SOLUTION

1. Select a space scale k_s = 2 in./in. or 1 in. (space scale) = 2 in. and velocity scale k_v = 2 (in./sec)/in. or 1 in. (velocity scale) = 2 in./sec, and from the relationship

$$\frac{k_s k_a}{k_v^2} = 1$$

obtain

$$k_a = \frac{(2/1)^2}{2/1} = 2 \text{ (in./sec}^2)/\text{in.}$$

or 1 in. (acceleration scale) = 2 in./sec².

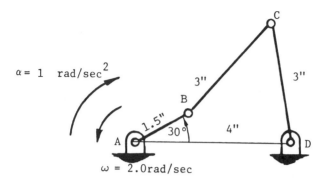

Figure 12.2   Four-bar mechanism.

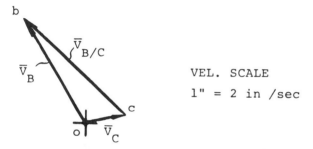

Figure 12.3   Velocity polygon.

2. Construct the velocity polygon (Figure 12.3) using the k_v scale. This requires the calculation of the velocity magnitude $V_{B/A}$ as a first step, then the layout of the vectors in accordance with the velocity polygon procedure already discussed.

3. Transfer vectors $\overline{V}_B$, $\overline{V}_C$, and $\overline{V}_{B/C}$ from the velocity polygon to points B and C on the linkage, maintaining the same orientation. $\overline{V}_B$ is drawn from B perpendicular to AB; $\overline{V}_C$ is drawn from C perpendicular to CD; and $\overline{V}_{B/C}$ is drawn from B perpendicular to BC.

4. Determine $\overline{A}_B^N$, $\overline{A}_C^N$, and $\overline{A}_{B/C}^N$ using the construction outlined above. For example, in Figure 12.4, $\overline{A}_B^N$ is obtained by construct-a right-angle triangle Abx, in which $\angle b$ is $90°$; Ax is the extended link AB, and the line Bb, perpendicular to AB, is the velocity magnitude of point B. The line segment Bx then represents the required value of $\overline{A}_B^N$.

5. Construct the acceleration polygon using the k_a scale (1 in. = 2 in./sec²), starting with the calculation of A_B^T, then following the acceleration polygon procedure already discussed. Figure 12.5

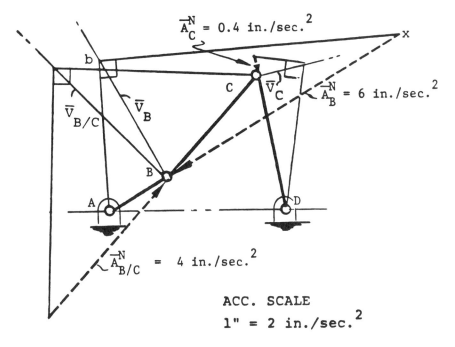

Figure 12.4   Normal acceleration construction.

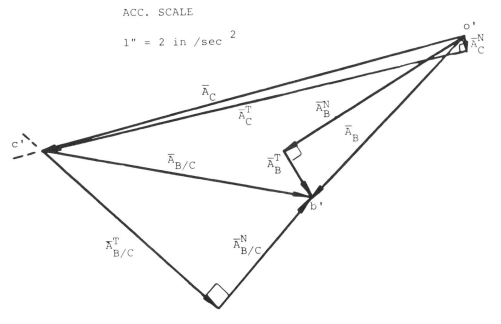

Figure 12.5   Acceleration polygon.

shows the acceleration polygon, from which the required acceleration $\overline{A}_C$ is determined to be

$$\overline{A}_C = 12.2 \text{ in./sec}^2 \quad \text{(as directed)}$$

In the example above, it is to be noted that only two calculations were necessary to complete the analysis after the scales were determined. These were to determine V_B and A_B^T, the velocity and acceleration magnitudes of the fist link. If this link were to rotate with a constant angular velocity, the calculation of A_B^T would not have been necessary, since the value of this acceleration would be zero. Thus, for a complete graphical analysis, the maximum number of calculations necessary is two.

12.2 EQUIVALENT LINKAGE METHOD

Determining the acceleration of points on many **higher-paired mechanisms**, such as those with rolling and sliding contacts, can become rather involved if a point-to-point analysis is attempted. This is because of the need to know the curvature of the path traced by a point on one link relative to the other and to apply the Coriolis law. If no easily recognized path is found, the analysis can be difficult.

To simplify this problem, the use of equivalent linkages has been found most effective. In application, an equivalent linkage replaces a higher-paired contact with appropriate lower pairs that will produce the correct values of velocities and accelerations for the instantaneous phase under consideration. An **equivalent linkage** may then be defined as one that produces identical motion as the part being analyzed for a given position or phase.

Figures 12.6 and 12.7 show several mechanisms with their equivalent linkages depicted as line diagrams on the right. Note that the rolling and sliding surfaces have been replaced by pin joints as part of a more simplified four-bar linkage. Note also that in each case, the floating link of the equivalent linkage is drawn along the common normal of the two contacting surfaces and connects the centers of curvature of the surfaces.

Although an equivalent linkage is generally valid only for a given instant or phase and does not ordinarily apply to a complete cycle, there are some instances in which the equivalent linkages of some higher-paired mechanisms will duplicate the input/output motion of those mechanisms throughout their motion cycle. Some examples are shown in Figures 12.6 and 12.7.

MECHANISM   EQUIVALENT LINKAGE

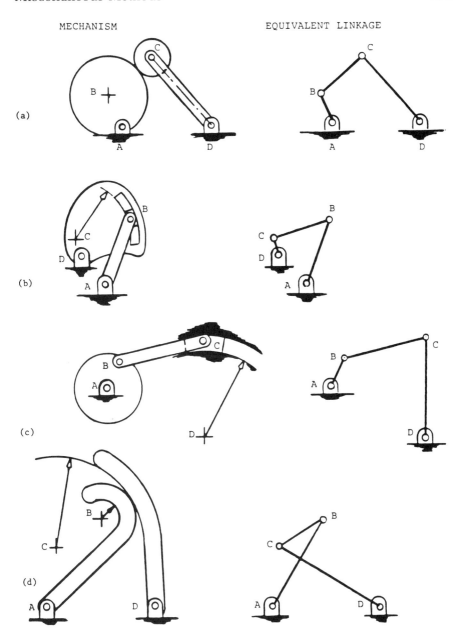

Figure 12.6   Kinematically equivalent four-bar linkages.

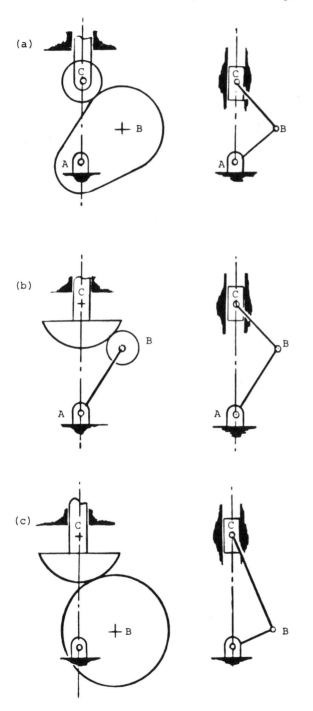

Figure 12.7   Kinematically equivalent slider-crank linkages.

EXAMPLE 12.2

Consider the cam mechanism shown in Figure 12.8(a). The cam (2) rotates counterclockwise at a constant angular velocity of 2 rad/sec. Find the acceleration of follower (4) using the equivalent linkage method.

SOLUTION

The equivalent mechanism for the cam mechanism given is the simple slider-crank ABC shown in Figure 12.8(b), for which the velocity and acceleration diagrams are readily obtained, as shown in Figures 12.8(c) and (d). Applying the velocity polygon construction procedure, we obtain

$$V_B = 0.9(2) = 1.8 \text{ in./sec}$$

$$V_C = 1.45 \text{ in./sec} \quad \text{(from the velocity polygon)}$$

$$V_{C/B} = 0.6 \text{ in./sec} \quad \text{(from the velocity polygon)}$$

Applying the acceleration construction procedure, we obtain

$$A_B^N = 0.9(2)^2 = 3.6 \text{ in./sec}^2$$

$$A_B^T = 0$$

$$A_{C/B}^N = \frac{(0.6)^2}{2.0} = 0.18 \text{ in./sec}^2$$

$$A_{C/B}^T = 3.8 \text{ in./sec}^2 \quad \text{(from the acceleration polygon)}$$

$$\overline{A}_C = 2.8 \text{ in./sec}^2 \quad \text{(directed as shown)}$$

Note that the resulting acceleration of point C ($\overline{A}_C$) is exactly the same as that obtained for point P4 (the same point) in Section 9.7, using an alternative method.

12.3 CENTER OF CURVATURE

The ability to locate the center of curvature of a path generated by a coupler point in plane motion is very important in the analysis or synthesis of a mechanism. With this ability, one is able not only to

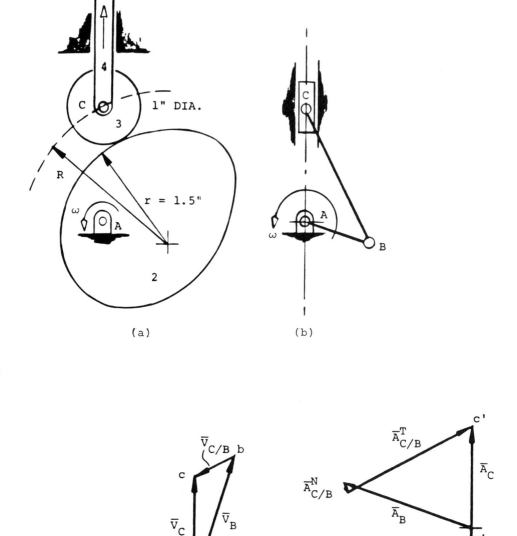

Figure 12.8   Acceleration analysis of a cam-follower mechanism: (a) Cam-follower mechanism; (b) equivalent linkage; (c) velocity polygon; (d) acceleration polygon.

determine the instantaneous velocity and acceleration of that point, but also to locate other key parameters related to the path geometry, such as the inflexion point, which is a useful indicator of the existence of straight-line motion for some finite distance during the mechanism cycle. Also, in rolling contact mechanisms such as cams, planetary gears, and roller bearings, one can, by locating the centers of curvature of the surfaces at the contact points, easily find the equivalent linkages that will simplify the analyses of those mechanisms.

Graphical Procedures

The following graphical methods will illustrate some of the various approaches that are available to determine the centers of curvature of paths described by points in a mechanism. We choose for this illustration a typical planetary gear mechanism where gear 2 rolls on gear 1, which is fixed. The centers of gears 1 and 2 are A_o and A, respectively, and the point of contact between the gears is defined as P. Let it be required to determine the center of curvature Bo of any point B on the revolving gear.

HARTMANN'S METHOD (Figure 12.9)

1. Assume gear 1 is held constant and find $\overline{V}_A = PA\omega_2$ ($\perp PA$).
2. Lay out pole velocity $\overline{V}_P$ by drawing a vector parallel to $\overline{V}_A$ to meet a line of proportionality connecting $\overline{V}_A$ to A_o.
3. Determine $\overline{V}_B$ from proportionality as follows:
   a. Rotate point B about P to point B' on the common normal.
   b. From B', draw vector $\overline{V}_{B'}$ parallel to $\overline{V}_A$ and terminating at a line of proportionality, drawn from P through the head of $\overline{V}_A$.
   c. Rotate vector $\overline{V}_B$, back to point B.
4. Determine V_P' by dropping a perpendicular from the head of $\overline{V}_P$ to meet a line drawn from P parallel to $\overline{V}_P$.
5. Draw a line of proportionality connecting heads of vectors $\overline{V}_B$ and $\overline{V}_P'$ and intersecting ray PB at a point B_o. This point will define the required center of curvature.

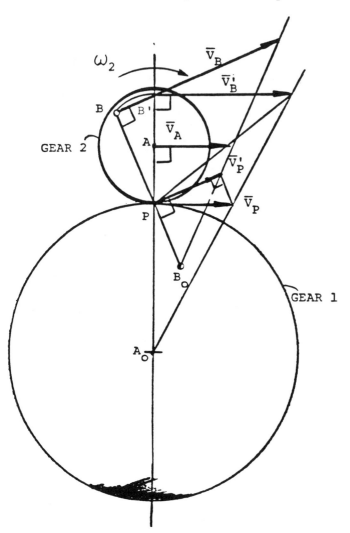

Figure 12.9   Hartmann's method.

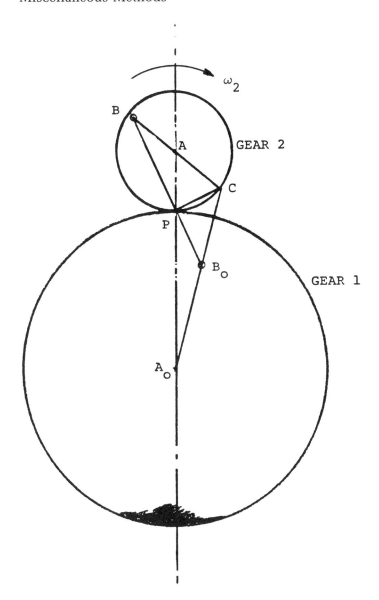

Figure 12.10   Bobillier's first method.

BOBILLIER'S FIRST METHOD (Figure 12.10)

1. Draw a line through points B and P.
2. Draw a line through points B and A.
3. Erect a perpendicular to PB (Step 1) from point B to meet line AB (Step 2) at point C.
4. Draw a line from point C to point A_o. This line will intersect line PB (Step 1) at the required center of curvature B_o.

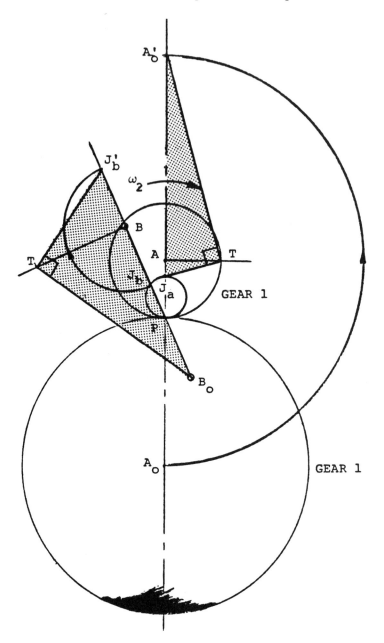

Figure 12.11   Mean proportional method.

THE MEAN PROPORTIONALITY METHOD
(Figure 12.11)

This method of center-of-curvature determination requires finding the inflexion point on the principal ray PA first, if this is not already known. Once this point is determined, the inflexion point on ray PB can be determined and then used to find the required center.

PROCEDURE

STEP 1   Finding the inflexion circle

a.   At point A, erect a perpendicular AT equal to PA on ray PA.
b.   With the center at A and radius AA_o, describe an arc to cut the ray at point A_o' on the opposite side of AT.
c.   Join point A_o' to point T with a straight line.
d.   At point T, construct a perpendicular to TA_o', cutting the ray at required point J_a.
e.   With PJ_a as the diameter, construct a circle through points P and J_a.

STEP 2   Finding the center of curvature B_o

a.   At point B, erect a perpendicular BT equal to PB on ray PB.
b.   With the center at B and radius BJ_b, describe an arc to cut the ray at point J_b' on the opposite side of BT.
c.   Join point J_b' to point T with a straight line.
d.   At point T, construct a perpendicular to TJ_b', cutting the ray at required point B_o.

In the above procedures, Steps c and d can be combined if a square-cornered straight edge is used to construct the right triangle. Also, observe that when the arc in Step b, is swung to one side of the mean proportional to create a tentative point such as A_o' or J_a', the actual point A_o or J_a is always found on the opposite side.

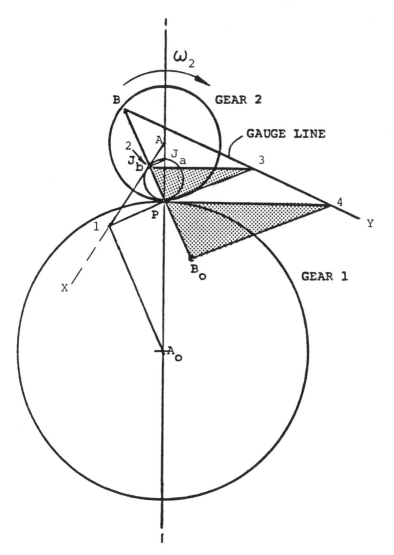

Figure 12.12   Bobillier's second method.

BOBILLIER'S SECOND METHOD (Figure 12.12)

Like the mean proportionality method, this method is based on the fact that the inflexion point on ray PB is known, which requires that the inflexion circle must be determined first.

PROCEDURE

STEP 1 Finding the inflexion circle

a. Draw an arbitrary line AX at any angle to AA_o less than 90°.
b. From A_o, draw a line to meet AX at point 1.
c. Join 1 to P with a straight line P1.
d. From P, draw a line parallel to A_o1 meeting AX at point 2.
e. From 2, draw a line parallel to P1 meeting AA_o at J_a the inflexion pole.
f. With PJ_a as the diameter, construct the inflexion circle through point P and J_a.

STEP 2 Finding the center of curvature

a. Connect point B to point P with a straight line. This line will cut the inflexion circle in Step 1f at J_b.
b. Draw an arbitrary line BY at any angle to AAo less than 90°.
c. From J_b, draw an arbitrary line to meet BY at 3.
d. Join 3 to P with straight line P3.
e. From P, draw a line parallel to J_b3 meeting BY at 4.
f. From 4, draw a line parallel to P3 meeting PB (extended) at B_o, the required center of curvature.

12.4   ACCELERATION ANALYSIS BY THE PARALLELOGRAM METHOD

Despite the wide use of the relative acceleration method in solving linkage problems, it is not uncommon for one to experience some confusion in correctly applying this method to the slider-crank mechanism. The confusion most often encountered arises from uncertainties or oversight as to the proper location of the relative radial acceleration vector in the acceleration polygon. As a result, many errors are made.

To avoid such confusion or oversight, the graphical method presented here makes use of simple parallelogram constructions that serve as guides in laying out the vectors. The method is not only simple to apply, but also saves time.

Slider-Crank Acceleration

Scope

In a typical slider-crank mechanism, as shown in Figure 12.13(a), crank AB has angular rotation ω and angular deceleration α about point A.

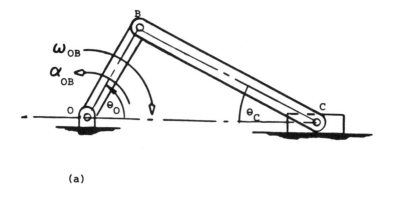

(a)

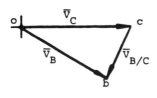

(b)

KEY POINT:

$\bar{A}^N_{B/C}$ (or $\bar{A}^N_{C/B}$ if used) is
always located on the 'reflected'
connecting rod.

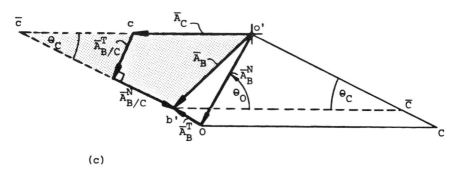

(c)

Figure 12.13 Graphical construction procedure: (a) Slider-crank mechanism;
(b) velocity polygon; (c) acceleration polygon.

Angle θ_O is the instantaneous angular position of the crank with respect to the line of action AC, and θ_C is the angle between the slider arm BC and line AC. It is required to determine slider acceleration $\overline{A}_C$ at any angle θ_O.

Construction Development

1. Define o' as the point of zero acceleration coincident with pivot B on the mechanism and draw vector $o'b'$ [see the acceleration diagram in Figure 12.13c] to represent acceleration $\overline{A}_B$ as determined by

$$\overline{A}_B = \overline{A}_B^N + \overline{A}_B^T \qquad (12.14)$$

where the magnitudes of $\overline{A}_B^N$ and $\overline{A}_B^T$ are given by

$$A_B^N = OB\omega^2$$

$$A_B^T = OB\alpha$$

2. With $o'b'$ as the diagonal and θ_C as the subtended angle, construct a parallelogram $oB'\overline{C}b'\overline{c}$ such that $o'\overline{C}$ is equal and parallel to $\overline{c}b'$, and $b'\overline{C}$ is equal and parallel to $o'\overline{c}$. (Note the geometric identity between triangular sections $o'b'\overline{c}$ and $o'b'\overline{C}$. Vector $o'b'$ may be considered the axis of asymmetry to the parallelogram.)

3. Determine $\overline{V}_{B/C}$ from the vectorial relationship

$$\overline{V}_{B/C} = \overline{V}_B - \overline{V}_C \qquad (12.15)$$

where

$$\overline{V}_B = OB\omega \quad \text{(directed perpendicular to } OB) \qquad (12.16)$$

and $\overline{V}_C$ is known in direction only (along the slider path). $\overline{V}_{B/C}$ can therefore be obtained from a velocity diagram as follows. From a point o [see the velocity diagram in Figure 12.13(b)], draw line ob scaled to represent $\overline{V}_B$ perpendicular to the instantaneous position of crank OB and pointing in the direction of the motion. Then draw a line through o parallel to the direction of the slider motion to represent velocity $\overline{V}_C$. Finally, draw a line from point b perpendicular to arm BC so that it intersects the $\overline{V}_C$ line at point c. The velocity magnitude $V_{B/C}$ is determined by measuring line bc and converting to the appropriate velocity according to the chosen scale.

4.   Compute the acceleration magnitude $A_{B/C}^N$ from

$$A_{B/C}^N = \frac{V_{B/C}^2}{BC} \tag{12.17}$$

Next, mark a segment of line $b'\bar{c}$ on the parallelogram to represent vector $\overline{A}_{B/C}^N$ scaled the same as $o'b'$ and heading toward b'. (This requirement is in keeping with the polygon convention, in which the acceleration vector relative to a point on a link must be directed toward the corresponding point on the acceleration polygon. Alternatively, if $\overline{A}_{C/B}^N$ is considered, this vector must point from b' to c' on the polygon.)

5.   From the tail of vector $\overline{A}_{B/C}^N$, draw a perpendicular line to intersect line $o'\bar{c}$ at point c'. Point c' defines the acceleration $\overline{A}_C$ along line $o'\bar{c}$, and the perpendicular drawn from $\overline{A}_{B/C}^N$ represents the tangential acceleration $\overline{A}_{B/C}^T$. The value of $\overline{A}_C$ may be checked with the vectorial relationship

$$\overline{A}_C = \overline{A}_B^N + \overline{A}_B^T - \overline{A}_{B/C}^N - \overline{A}_{B/C}^T \tag{12.18}$$

(Note that for a uniform velocity of crank OB, triangle $o'b'\bar{c}$ is identical to the configuration of the given mechanism OBC, in which case $\overline{C}$ and C are coincident points. Note also that $A_{B/C}^N$ always lies on the "reflected" slider arm.)

EXAMPLE 12.3

Consider a slider-crank mechanism where OB = 1.5 in., BC = 3 in., ω = 1 rad/sec, and α = 0. Determine the slider acceleration $\overline{A}_C$ for θ_O = 45° (Figure 12.14), 135° (Figure 12.15), 225° (Figure 12.16), and 315° (Figure 12.17).

SOLUTION

1.   From Eq. (12.18), acceleration A_B = 1.5 in./sec and line $o'b'$ is drawn 1.5 in. in length to represent this vector.
2.   Construct the parallelogram $o'\overline{C}b'\bar{c}$, using $o'b'$ as the diagonal.
3.   From Eq. (12.16), V_B = 1.5 in./sec. With line ob drawn 1.5 in. long to represent vector V_B, a velocity polygon is constructed from which $V_{B/C}$ = 1.12 in./sec.
4.   From Eq. (12.17), $A_{B/C}^N$ = 0.42 in. and a 0.42-in. segment of line $b'\bar{c}$ is denoted on the parallelogram to represent $\overline{A}_{B/C}^N$.

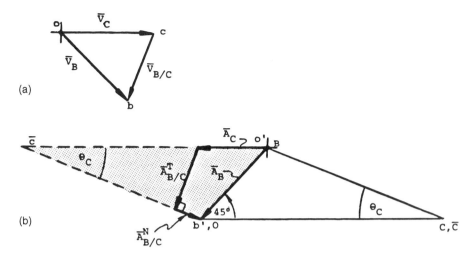

Figure 12.14   Crank angle at 45°: (a) Velocity polygon; (b) acceleration polygon.

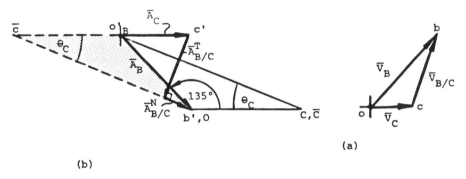

Figure 12.15   Crank angle at 135°: (a) Velocity polygon; (b) acceleration polygon.

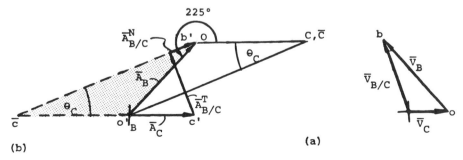

Figure 12.16   Crank angle at 225°: (a) Velocity polygon; (b) acceleration polygon.

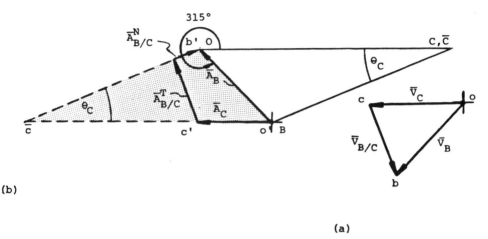

(b)

(a)

Figure 12.17   Crank angle at 315°: (a) Velocity polygon; (b) acceleration polygon.

5. A perpendicular line is drawn from the tail of vector $\overline{A}^{N}_{B/C}$ to intercept line $o'\overline{c}$ at point c'. The length of line $o'c'$ is measured to be 1.05 in., so that $\overline{A}_C = 1.05$ in./sec² (to the left). The use of Eq. (12.18) confirms this value as being accurate. Similarly, $\overline{A}_C$ for the crank positions $\theta_O = 135°$, 225°, and 315° are found to be 1.05 in./sec² (to the right), 1.05 in./sec² (to the right), and 1.05 in./sec² (to the left), respectively.

Four-Bar Acceleration

Scope

In the four-bar mechanism $ABCD$, shown in Figure 12.18a, crank AB has angular clockwise rotation ω and angular counterclockwise acceleration O about point A. Angle θ is the instantaneous angular position of AB with respect to the reference line AD. Let it be required to determine the linear acceleration $\overline{A}_C$ at any angle θ.

Construction Development

1. From the point B on the mechanism, draw vector Bb to represent the acceleration $\overline{A}_B$. This vector is determined from the vector equation

$$\overline{A}_B = \overline{A}^{N}_B + \overline{A}^{T}_B$$

where the magnitudes of $\overline{A}^{N}_B$ and $\overline{A}^{T}_B$ are given by

$$A^{N}_B = AB \times \omega^2$$

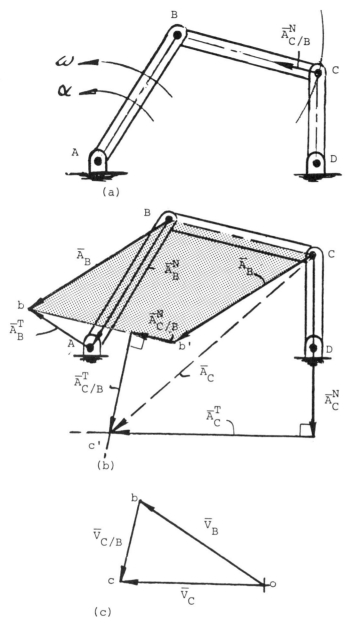

Figure 12.18   Four-bar mechanism and motion diagrams: (a) Four-bar mechanism; (b) acceleration diagram; (c) velocity diagram.

and

$$A_B^T = AB \times \alpha$$

2. Using the vector Bb and link BC as adjacent sides, construct a parallelogram $bBCb'$ on the mechanism where $Bb = Cb'$. Thus, the side Cb' also represents the acceleration $\overline{A}_B$.
3. Compute the magnitude of normal acceleration $\overline{A}_{C/B}^N$ from

$$A_{C/B}^N = \frac{V_{C/B}^2}{CB}$$

where $V_{C/B}$ is determined from a velocity diagram based on the vector relationship

$$\overline{V}_{C/B} = \overline{V}_C - \overline{V}_B$$

This diagram is constructed as follows:
a. From a point o, the polar origin (Figure 12.18c), draw a line ob, scaled to represent $\overline{V}_B$ perpendicular to the crank AB and pointing in the direction of the motion.

$$ob = AB \times \omega$$

b. Through point o, draw a line perpendicular to link CD to represent the line of action of velocity of point C, or $\overline{V}_C$.
c. From point b, draw a line perpendicular to link BC until it intersects the line of action of $\overline{V}_C$ at point c. By measuring the distances bc and oc, the magnitudes $\overline{V}_{C/B}$ and $\overline{V}_C$, respectively, are obtained.
4. Mark a segment of line bb' on the parallelogram to represent the vector $\overline{A}_{C/B}^N$, scaled the same as Cb', headed away from b'. This requirement is in keeping with the polygon convention in which the acceleration vector on a of point C relative to point B on a link must be directed toward point c' on the acceleration polygon. Alternatively, if $\overline{A}_{B/C}^N$ is considered, this vector must point to b' away from c' on the polygon. In either case, the relative acceleration vector always lies on side bb' of the parallelogram.
5. Locate the normal acceleration vector $\overline{A}_C^N$, directed from C toward D on link CD. The magnitude of this vector is given by

$$A_C^N = \frac{V_C^2}{CD}$$

where V_C is obtained from the velocity diagram in Figure 12.19.

6. Locate the terminus of the required vector AC by drawing perpendicular lines representing $\overline{A}_{C/B}^{N}$ and $\overline{A}_{C}^{N}$, respectively, from the termini of vectors $\overline{A}_{C/B}^{T}$ and $\overline{A}_{C}^{T}$ to meet at c'. This point will also define the terminus of $\overline{A}_{C/B}$.

7. Draw a vector from C to c' to represent the required acceleration $\overline{A}_{C}$.

The preceding construction is easily verified by checking the polygon for agreement with the known vectorial relationship

$$\overline{A}_{C} = \overline{A}_{B}^{N} + \overline{A}_{B}^{T} + \overline{A}_{C/B}^{N} + \overline{A}_{C/B}^{T}$$

EXAMPLE 12.4

Consider the four-bar mechanism $ABCD$ in Figure 12.19a where crank $AB = 1.5$ in., $BC = 2$ in., $CD = 2$ in., and $AD = 4$ in. Crank AB has a constant rotation of $\omega = 1$ rad/sec (which means that $\alpha = 0$). Determine the acceleration of point C, $\overline{A}_{C}$, for the crank position shown.

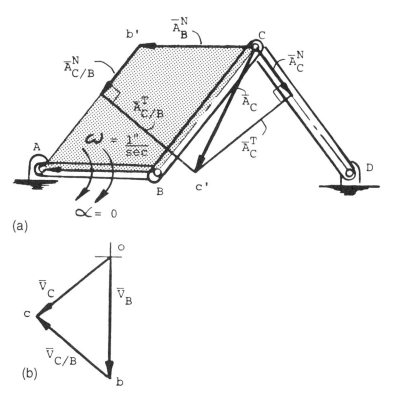

Figure 12.19   Example 12.4: (a) Four-bar mechanism and acceleration polygon; (b) velocity polygon.

SOLUTION

1. Compute the acceleration magnitude of point B, or $\overline{A}_B$, and draw line Bb to represent this magnitude.

$$A_B = \frac{V_B^2}{AB}$$

$$= \frac{1.5^2}{1.5} = 1.5 \text{ in./sec}^2$$

2. Construct a parallelogram $Bbb'C$ on the mechanism.
3. Compute the magnitude of normal acceleration $\overline{A}_{C/B}^N$ as follows:
   a. Determine the velocities $\overline{V}_{C/B}$ and $\overline{V}_C$ by constructing velocity polygon obc (see Figure 12.19b).

$$V_{C/B} = 1.2 \text{ in./sec (scaled)}$$

$$V_C = 1.2 \text{ in./sec (scaled)}$$

   b. Use the value found for $V_{C/B}$ to obtain

$$A_{C/B}^N = \frac{1.2^2}{2} = 0.72 \text{ in./sec}^2$$

4. Mark off line segment bb' on the parallelogram to represent vector $\overline{A}_{C/B}^N$.
5. Compute the magnitude of normal acceleration vector $\overline{A}_C^N$ and locate this vector, directed from C toward D on the link CD.

$$A_C^N = \frac{1.2^2}{2} = 0.72 \text{ in./sec}^2$$

6. Locate the terminus of required vector $\overline{A}_C$ by drawing perpendicular lines from the termini of $\overline{A}_{C/B}^N$ and $\overline{A}_B^N$ to meet at c'.
7. Draw a vector from C to c' to represent the required acceleration $\overline{A}$. From this,

$$\overline{A}_C = 1.7 \text{ in./sec}^2 \quad \text{(directed as shown)}$$

III

ANALYTICAL TECHNIQUES

Analytical techniques or mathematical methods for velocity and accelerations determination can be a powerful tool in the analysis and design of a mechanism. Compared to graphical techniques, analytical techniques offer two major advantages. They are faster and more accurate, provided that a valid expression has been derived for the mechanism. Graphical methods typically involve numerous diagrams that must be constructed for each shift in the position of the mechanism. The use of mathematics, on the other hand, allows general expressions for velocity and acceleration to be developed in terms of link geometry for all positions. Once a general expression is obtained for a mechanism, the velocity and acceleration values for points on that mechanism can readily be determined. This expression also reveals how various parameters such as lengths and angular positions of the links affect motion characteristics. This information is most important in the synthesis of a mechanism in which a specific output motion is desired.

The main disadvantage of analytical techniques is that the mathematical analysis required to obtain general velocity and acceleration expressions for a mechanism are typically complex, lengthy, and error-prone. Conventional mathematical methods require both complex numbers and calculus, and generally, these methods are too theoretical to provide the quick insight the designer needs to develop a practical linkage.

Fortunately, the mathematical analysis can be simplified using the simplified vector method presented in Chapters 14 through 17. Unlike the conventional methods, this method does not rely on calculus, but combines trigonometry, complex numbers, and principles of relative motion to obtain required motion relationships. Plane trigonometry and complex numbers are used to express motion relationships in concise vectorial forms, and relative motion principles help to simplify equation development.

Besides the simplified vector method, two alternative mathematical methods are demonstrated in this part. These are the modified vector method, which is basically a variation of the simplified vector method, and the calculus method, which is conventional. In both cases, it is clear that the application of calculus is essential.

13

Complex Algebra

13.1 INTRODUCTION

In mechanism analysis, a convenient way to describe the position of velocity or acceleration of a link is by the use of a complex number. A **complex number** consists of two parts, normally written in the form

$$a + ib$$

where a is the real part and b the imaginary part, denoted by the letter i, which has a value of $\sqrt{-1}$.

Two complex numbers that differ only in the sign of their imaginary parts, such as $a + ib$ and $a - ib$, are termed **conjugates**. Thus, $a - ib$ is the conjugate of $a + ib$; and conversely, $a + ib$ is the conjugate of $a - ib$.

13.2 COMPLEX VECTOR OPERATIONS

Computations involving complex numbers follow the rules for algebra, with the additional requirement that all powers of i be reduced to the lowest terms by applying the following properties:

$$i^2 = -1$$

$$i^3 = -i$$

$$i^4 = +1$$

$$i^5 = +i, \text{ etc.}$$

For example, if $\bar{r}_1 = a + ib$ and $\bar{r}_2 = c + id$, then

$$\bar{r}_1 + \bar{r}_2 = (a + ib) + (c + id)$$
$$= (a + c) + i(b + d)$$
$$\bar{r}_1 - \bar{r}_2 = (a + ib) - (c + id)$$
$$= (a - c) + i(b - d)$$
$$\bar{r}_1 \times \bar{r}_2 = (a + ib)(c + id)$$
$$= (ac - bd) + i(bc + ad)$$
$$\frac{\bar{r}_1}{\bar{r}_2} = \frac{a + ib}{c + id}$$
$$= \frac{(a + ib)(c - id)}{(c + id)(c - id)}$$
$$= \frac{ac + bd}{c^2 + d^2} + \frac{i(bc - ad)}{c^2 + d^2} = \frac{(ac + bd) + i(bc - ad)}{c^2 + d^2}$$

Note that in the division case, the quotient is conveniently found by multiplying both numerator and denominator by the conjugate of the denominator.

13.3 GEOMETRIC REPRESENTATION OF A COMPLEX VECTOR

A complex number is represented geometrically by a vector in the complex plane defined by two mutually perpendicular axes: the real and imaginary axes. For example, in Figure 13.1 the complex number $(a + ib)$ is shown to be represented by the vector $\bar{R}_1$, which extends from the origin o of the complex axes to a point P in the plane such that the horizontal component along the positive real axis represents the real part of the number a, and the vertical component along the positive imaginary axis represents the imaginary part b. Thus, the magnitude of $\bar{R}_1$ is given by

$$R_1 = |\bar{R}_1| = \sqrt{a^2 + b^2}$$

and the direction by

$$\theta_1 = \tan^{-1}\frac{b}{a} \quad (0° < \theta_1 < 90°)$$

Note that the angle θ_1 is conventionally defined as the angular displacement of the vector from the positive real axis and is positive when measured counterclockwise.

Similarly, the complex number $(-a + ib)$ is represented by a vector $\overline{R}_2$ as shown in Figure 13.1. Here the horizontal distance $-a$ along the negative real axis represents the real part of the number, and the vertical distance b along the positive imaginary axis represents the imaginary part. Therefore, the magnitude and direction of vector $\overline{R}$ are obtained as follows:

$$R_2 = |\overline{R}_2| = \sqrt{(-a)^2 + b^2}$$

$$\theta_2 = \tan^{-1}\frac{b}{-a} \quad (90° < \theta_2 < 180°)$$

Note that care should be taken in interpreting the value of θ_2 since in the range 0 to 360°, **there are two values of θ_2 for each value of tan**

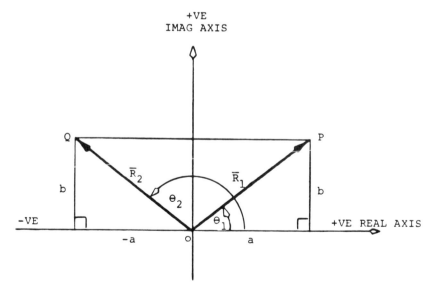

Figure 13.1 Graphical representation of complex numbers.

θ_2. To avoid ambiguity, it is usually advisable first to determine the quadrant in which the vector lies, using a simple sketch, before deciding which of the two angles applies.

In general, to represent a complex quantity, say, $\overline{R} = a + ib$, as a vector:

1. Find the absolute value of the components

$$|\overline{R}| = (a^2 + b^2)^{1/2}$$

2. Find the direction (θ) from

$$\theta = \tan^{-1} \frac{\text{imaginary}}{\text{real}}$$

or

$$\theta = \tan^{-1} \frac{b}{a}$$

noting that

If a is positive and b positive, θ is in first quadrant.
If a is negative and b positive, θ is in second quadrant.
If a is negative and b negative, θ is in third quadrant.
If a is positive and b negative, θ is in fourth quadrant.

13.4 COMPLEX FORMS

The vector $\overline{R}$ may be expressed in one of four complex forms: (1) rectangular, (2) trigonometric, (3) exponential, and (4) polar.

In the **rectangular form**,

$$\overline{R} = a + ib$$

where, as before, a and b are, respectively, the real and imaginary components of $\overline{R}$.

In the **trigonometric form**,

$$\overline{R} = R \cos \theta + iR \sin \theta$$

where $R \cos \theta = a$ and $R \sin \theta = b$, as shown in Figure 13.2. Alternatively,

$$\overline{R} = R(\cos \theta + i \sin \theta)$$

where $(\cos \theta + i \sin \theta)$ is the trigonometric form of the unit vector that defines the direction of $\overline{R}$.

In the **exponential form**,

$$\overline{R} = Re^{i\theta}$$

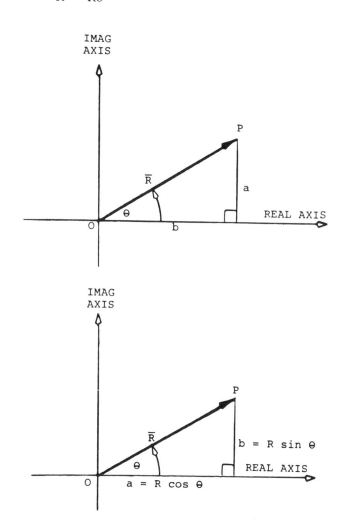

Figure 13.2   Graphical representation of the complex vector.

where $e^{i\theta}$ is the exponential equivalent of the unit vector ($\cos \theta + i \sin \theta$) given above.

In the **polar form**,

$$\overline{R} = R \angle \theta$$

where $\angle \theta$ defines the angular position of $\overline{R}$.

13.5 THE UNIT VECTOR

A **unit vector** is a vector that has a magnitude of unity. If $\overline{V}$ is a vector with magnitude $V \neq 0$, then $\overline{V}/V$ is a unit vector having the same direction of $\overline{V}$. Defining this unit vector as $\overline{v}$, we can write

$$\frac{\overline{V}}{V} = \overline{v}$$

or

$$\overline{V} = V\overline{v}$$

which means that any vector $\overline{V}$ can be represented by the magnitude of $\overline{V}$ multiplied by the unit vector $\overline{v}$ in the direction of $\overline{V}$.

The unit vector may be expressed in exponential form, such as $e^{i\theta}$ and $e^{i(\theta + 90°)}$, where θ and $(\theta + 90°)$ are the position angles of the vectors, measured from the real axis (see Figure B.1 in Appendix B) and i is equal to $\sqrt{-1}$. However, for the purpose of computation, it is more convenient to convert these unit vectors to their equivalent trigonometric forms, as follows:

$$e^{i\theta} = \cos \theta + i \sin \theta$$

and

$$e^{i(\theta + 90°)} = \cos (\theta + 90°) + i \sin (\theta + 90°)$$

Thus, from Figure 13.3,

$$\overline{V}_1 = V_1 e^{i\theta}$$
$$\overline{V}_1 = V_1(\cos \theta + i \sin \theta)$$

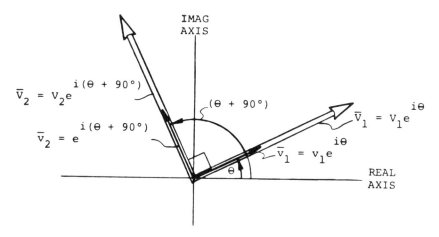

Figure 13.3   Unit vector.

and

$$\overline{V}_2 = V_2 e^{i(\theta + 90°)}$$

$$\overline{V}_2 = V_2[\cos(\theta + 90°) + i\sin(\theta + 90°)]$$

13.6 LINKAGE APPLICATION

In linkage analysis it is convenient to express vector quantities such as displacement velocity and acceleration as complex numbers that can be written in any of the complex forms discussed previously. Consider the case of a link AB that rotates counterclockwise with angular velocity ω and angular acceleration α. The displacement or change in position of the point B can be described by a position factor $\overline{R}$ directed from the origin of the complex axes to that point [Figure 13.4(a)]. That is, the displacement of link AB is given by

$$\overline{R} = Re^{i\theta}$$

where R is the link length and $e^{i\theta}$ the exponential form of the unit reactor that defines the instantaneous angular position of the link.

From previous velocity studies we know that the magnitude of velocity of point B can be obtained from the relationship

$$V_B = R\omega$$

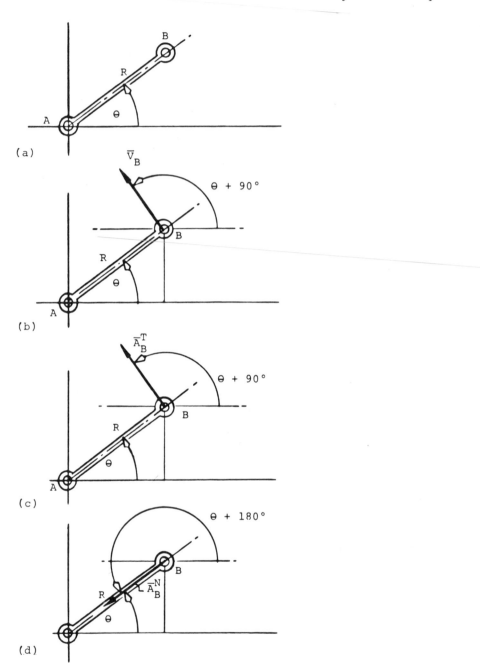

Figure 13.4   Graphical representations: (a) Displacement; (b) velocity; (c) tangential acceleration; (d) normal acceleration.

and that the vector describing this velocity acts perpendicularly to the link in the same sense as θ. This velocity vector $\overline{V}_B$, shown in Figure 13.4(b), can therefore be expressed in the complex form

$$\overline{V}_B = R\omega e^{i(\theta + 90°)}$$

where $e^{i(\theta + 90°)}$ is the unit vector used to indicate the direction of the velocity.

Similarly, the tangential acceleration shown in Figure 13.4(c) may be written as

$$\overline{A}_B^T = R\alpha e^{i(\theta + 90°)}$$

where $R\alpha$ is the magnitude of the acceleration of $e^{i(\theta + 90°)}$ the unit vector defining the direction. The normal acceleration, shown in Figure 13.4(d), may be written as

$$\overline{A}_B^N = R\omega^2 e^{i(\theta + 180°)} = -R\omega^2 e^{i\theta}$$

where $R\omega^2$ is the magnitude of the acceleration and $-e^{i\theta}$ the unit vector defining the direction.

EXAMPLE 13.1

Link AB of length R in Figure 13.5(a) rotates counterclockwise at an angular velocity ω.

a. Develop a complex expression for the velocity $\overline{V}_B$.
b. Evaluate the expression found in part a given that R = 1.5 in., ω = 1 rad/sec, and θ = 120°.

SOLUTION

Proceed as follows:

1. Using point B as the intersection of the real and imaginary axes of the complex plane [Figure 13.5(b)], draw line Bb, perpendicular to AB, to represent $\overline{V}_B$.
2. Drop a perpendicular from point b or terminus of $\overline{V}_B$ to intersect the imaginary axis at point o [Figure 13.5(c)].

From this construction we obtain the vector equation

$$\overline{V}_B = \overline{ob} + \overline{Bo} \quad \text{(vectorial summation)}$$

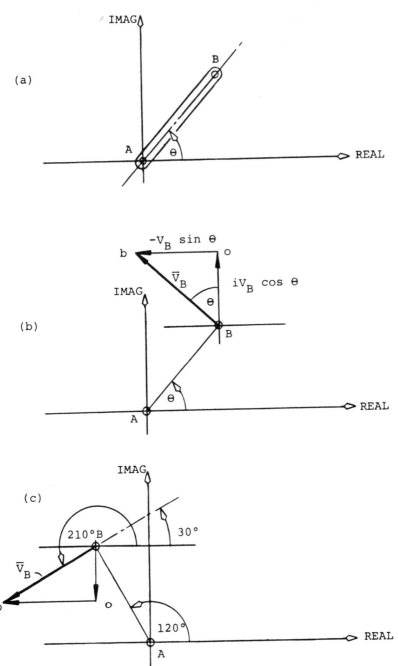

Figure 13.5   Velocity representation for Example 13.1.

where

$$ob = -V_B \sin \theta \quad \text{(real)}$$

$$Bo = iV_B \cos \theta \quad \text{(imaginary)}$$

Therefore,

$$\overline{V}_B = -V_B \sin \theta + iV_B \cos \theta$$

$$= -V_B(\sin \theta - i \cos \theta)$$

$$= -R\omega(\sin \theta - i \cos \theta)$$

$$\overline{V}_B = -R\omega(\sin \theta - i \cos \theta)$$

$$= -(1.5)(1)(\sin 120° - i \cos 120°)$$

$$= -1.5[0.866 - i(-0.5)]$$

$$= -1.5(0.866 + i\,0.5)$$

$$= -1.5\sqrt{(0.866)^2 + 0.5^2}$$

$$= -1.5(1) \quad \text{at} \quad \tan^{-1} \frac{0.5}{0.866}$$

$$= -1.5 \text{ in./sec at } 30°$$

$$= 1.5 \text{ in./sec at } 210°$$

14

Four-Bar Mechanism Analysis: Simplified Vector Method

14.1 INTRODUCTION

The four-bar mechanism is often considered the most basic of all kinematic mechanisms. Consisting of four rigid links (AB, BC, CD, and AD) and four turning pairs (A, B, C, and D) as shown in Figure 14.1, this mechanism can be arranged in three basic configurations and, accordingly, is classified as follows:

The **crank-rocker mechanism**, where drive crank AB is capable of complete rotation while the follower CD oscillates. Here the drive crank must be the shortest link.

The **drag link mechanism**, where both drive crank AB and follower CD are capable of complete rotation. Here the frame AD must be the shortest link.

The **double rocker mechanism**, where both drive crank AB and follower CD oscillate and neither is capable of complete rotation. Here the coupler BC must be the shortest link.

The four-bar mechanism is important in many ways:

1. It is the simplest possible plane linkage that can provide virtually any type of output motion.
2. Variations and combinations of this linkage make possible an almost limitless variety of mechanisms. Typical examples can be seen in some common applications shown in Figures 14.2 to 14.4.

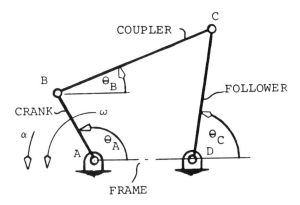

Figure 14.1   Four-bar mechanism.

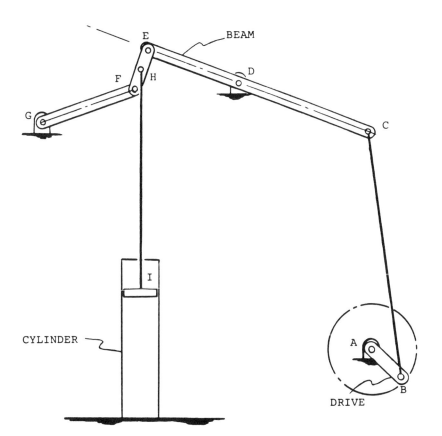

Figure 14.2   Beam pump mechanism.

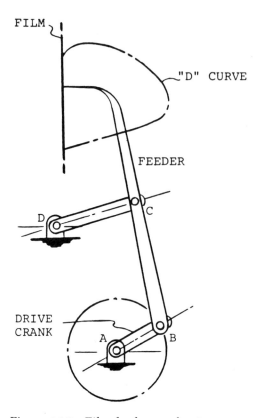

Figure 14.3   Film feeder mechanism.

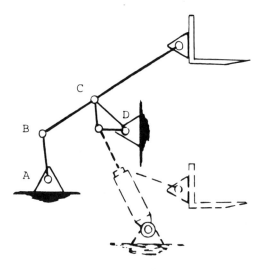

Figure 14.4   Forklift truck mechanism. (From A. S. Hall, Kinematics and Linkage Design, Prentice Hall, Englewood Cliffs, N.J. 1961.)

3. Analysis of many complex direct contact mechanisms can be sim-
    plified by replacing the mechanism with an equivalent four-bar
    linkage.

For these reasons, the motion characteristics of the four-bar linkage are
studied more than those of other kinematic mechanism.

The simplified mathematical method presented here quickly determines linear velocity and acceleration relationships of key points in
the four-bar linkage for any given position of the drive crank during its
motion cycle.

14.2 SCOPE AND ASSUMPTIONS

The mechanism $ABCD$ in Figure 14.1 represents a typical four-bar linkage mechanism in which links AB, BC, CD, and AD have known lengths,
and θ_A, θ_B, and θ_C are their respective angular positions measured from
base link AD. Crank AB rotates with an angular velocity ω and angular
acceleration α about pivot point A.

In this analysis, equations for computing linear velocities and
accelerations of the mechanism for a given instantaneous crank angle
θ_A will be determined. Note that all angular displacements, velocities,
and accelerations are considered positive for counterclockwise rotation
and negative for clockwise rotation.

14.3 GEOMETRIC RELATIONSHIPS

First, we determine the angular relationships of θ_B and θ_C by constructing a diagonal BD, as in Figure 14.5, to form two triangles, ABD and
BCD, where

$$ABD = \phi_B$$

$$ADB = \phi_D$$

$$CBD = \gamma_B$$

$$CDB = \gamma_D$$

Using basic trigonometric relationships, we can show that

$$\theta_B = \gamma_B - \phi_D \qquad\qquad (14.1)$$

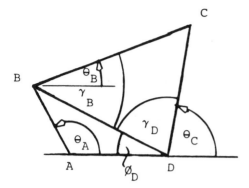

Figure 14.5   Geometric relationships.

and

$$\theta_C = 180° - \phi_D - \gamma_D \tag{14.2}$$

where

$$\gamma_B = \cos^{-1} \frac{BC^2 + BD^2 - CD^2}{2(BC)(BD)}$$

$$\phi_D = \sin^{-1}\left(\frac{AB}{BD} \sin \theta_A\right)$$

$$\gamma_D = \cos^{-1} \frac{BD^2 + CD^2 - BC^2}{2(BD)(CD)}$$

$$BD = [AB^2 + AD^2 - 2(AB)(AD) \cos \theta_A]^{1/2}$$

14.4   VELOCITY ANALYSIS

To determine the velocities $\overline{V}_B$, $\overline{V}_C$, and $\overline{V}_{C/B}$, we construct the velocity polygon Bcb for the mechanism, as in Figure 14.6, letting

Bb = magnitude of linear velocity of point B (V_B).
Bc = magnitude of linear velocity of point C (V_C).
bc = magnitude of linear velocity of C relative to B ($V_{C/B}$).

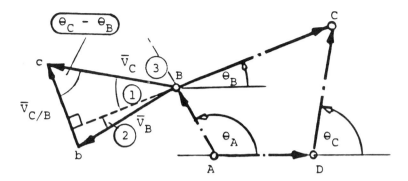

Figure 14.6   Velocity polygon.

Then from this polygon, we determine the angular relationships of the velocities as follows:

$$③ = \theta_C + 90° - \theta_A$$

$$① = 180° + \theta_B - \theta_A - ③$$

$$= 180° + \theta_B - \theta_A - (\theta_C + 90° - \theta_A)$$

$$= 180° + \theta_B - \theta_A - \theta_C - 90° + \theta_A$$

$$= 90° + \theta_B - \theta_C$$

$$② = 90° - ① - ③$$

$$= 90° - (90° + \theta_B - \theta_C) - (\theta_C + 90° - \theta_A)$$

$$= 90° - 90° - \theta_B + \theta_C - \theta_C - 90° + \theta_A$$

$$= \theta_A - \theta_B - 90°$$

$$\angle c = 90° - ① = 90° - (90° + \theta_B - \theta_C)$$

$$= 90° - 90° - \theta_B + \theta_C$$

$$= \theta_C - \theta_B$$

$$\angle b = 90° - ② = 90° - (\theta_A - \theta_B - 90°)$$

$$= 90° - \theta_A + \theta_B + 90°$$

$$= 180° - \theta_A + \theta_B$$

$$\underline{/bBc} = ① + ② = (90° + \theta_B - \theta_C) + (\theta_A - \theta_B - 90°)$$

$$= 90° + \theta_B - \theta_C + \theta_A - \theta_B - 90°$$

$$= \theta_A - \theta_C$$

V_B, V_C, and $V_{C/B}$ can now be found by applying the **Rule of Sines** from trigonometry.

$$\frac{V_B}{\sin(\theta_C - \theta_B)} = \frac{V_C}{\sin(180° - \theta_A + \theta_B)} = \frac{V_{C/B}}{\sin(\theta_A - \theta_C)} \qquad (14.3)$$

Noting that

$$\sin(180° - \theta_A + \theta_B) = \sin[180° - (\theta_A - \theta_B)] = \sin(\theta_A - \theta_B)$$

Eq. (14.3) can be written as

$$\frac{V_B}{\sin(\theta_C + \theta_B)} = \frac{V_C}{\sin(\theta_A - \theta_B)} = \frac{V_{C/B}}{\sin(\theta_A - \theta_C)}$$

or

$$\frac{V_B}{\sin(\theta_B - \theta_C)} = \frac{V_C}{\sin(\theta_B - \theta_A)} = \frac{V_{C/B}}{\sin(\theta_C - \theta_A)}$$

Also, since

$$V_B = \omega AB \qquad (14.4)$$

then, the scalar expressions for V_C and $V_{C/B}$ can be written as

$$V_C = \omega AB \, \frac{\sin(\theta_B - \theta_A)}{\sin(\theta_B - \theta_C)} \qquad (14.5)$$

$$V_{C/B} = \omega AB \, \frac{\sin(\theta_C - \theta_A)}{\sin(\theta_B - \theta_C)} \qquad (14.6)$$

Since $\overline{V}_B$, $\overline{V}_C$, and $\overline{V}_{C/B}$ are oriented, respectively, at angles $(\theta_A + 90°)$, $(\theta_C + 90°)$, and $(\theta_B + 90°)$ from the real or reference axis, we can write the required vectorial expressions as follows:

$$\overline{V}_B = V_B e^{i(\theta_A + 90°)} \qquad (14.7)$$

$$\overline{V}_C = V_C e^{i(\theta_C + 90°)} \qquad (14.8)$$

$$\overline{V}_{C/B} = V_{C/B} e^{i(\theta_B + 90°)} \qquad (14.9)$$

where $e^{i(\theta_A+90°)}$, $e^{i(\theta_C+90°)}$, and $e^{i(\theta_B+90°)}$ are unit vectors, used to define the direction of $\overline{V}_B$, $\overline{V}_C$, and $\overline{V}_{C/B}$, respectively.

(Note that $e^{i\theta} = \cos\theta + i\sin\theta$, where θ is the position angle.)

14.5 ACCELERATION ANALYSIS

To find the linear acceleration of point C $(\overline{A}_C)$, we apply the relative motion theory, which states that

$$\overline{A}_C = \overline{A}_B + \overline{A}_{C/B} \qquad (14.10)$$

Expanding this equation into its normal and tangential component form, we have

$$\overline{A}_C^N + \overline{A}_C^T = \overline{A}_B^N + \overline{A}_B^T + \overline{A}_{C/B}^N + \overline{A}_{C/B}^T \qquad (14.11)$$

where

$$\overline{A}_C^N = -\frac{V_C^2}{CD}\, e^{i\theta_C}$$

$$\overline{A}_C^T = A_C^T e^{i(\theta_C+90°)}$$

$$\overline{A}_B^N = -\omega^2 AB e^{i\theta_A}$$

$$\overline{A}_B^T = \alpha AB e^{i(\theta_A+90°)}$$

$$\overline{A}_{C/B}^N = -\frac{V_{C/B}^2}{BC}\, e^{i\theta_B}$$

$$\overline{A}_{C/B}^T = A_{C/B}^T e^{i(\theta_B+90°)}$$

or

$$-\frac{V_C^2}{CD}\, e^{i\theta_C} + A_C^T e^{i(\theta_C+90°)} = -\omega^2 AB e^{i\theta_A} + \alpha AB e^{i(\theta_A+90°)}$$

$$-\frac{V_{C/B}^2}{BC}\, e^{i\theta_B} + A_{C/B}^T e^{i(\theta_B+90°)} \qquad (14.12)$$

Examination of this complex equation indicates that the only unknown quantities are the magnitudes A_C^T and $A_{C/B}^T$. All other quantities (magnitudes and directions) are either known or can readily be determined from the problem data. Note that the directions of $\overline{A}_C^T$ and $\overline{A}_{C/B}^T$, although not precisely known, are assumed positive (or counterclockwise) for convenience only. If, in reality, any of the directions is reversed, the numerical solution of the equation will automatically produce a negative sign for the unknown quantity.

To solve Eq. (14.12), we equate the real and imaginary parts as follows:

Real:

$$-\frac{V_C^2}{CD} \cos \theta_C + A_C^T \cos(\theta_C + 90°)$$

$$= -\omega^2 AB \cos \theta_A + \alpha AB \cos(\theta_A + 90°)$$

$$-\frac{V_{C/B}^2}{BC} \cos \theta_B + A_{C/B}^T \cos(\theta_B + 90°) \qquad (14.13)$$

Imaginary:

$$-\frac{V_C^2}{CD} \sin \theta_C + A_C^T \sin(\theta_C + 90°)$$

$$= -\omega^2 AB \sin \theta_A + \alpha AB \sin(\theta_A + 90°)$$

$$-\frac{V_{C/B}^2}{BC} \sin \theta_B + A_{C/B}^T \sin(\theta_B + 90°) \qquad (14.14)$$

The solution of these simultaneous equations yields

$$A_C^T = \frac{C_1 B_2 - C_2 B_1}{A_1 B_2 - A_2 B_1} \qquad (14.15)$$

$$A_{C/B}^T = \frac{A_1 C_2 - A_2 C_1}{A_1 B_2 - A_2 B_1} \qquad (14.16)$$

where

$$A_1 = \cos(\theta_C + 90°) \qquad (14.17)$$

$$A_2 = \sin(\theta_C + 90°) \qquad (14.18)$$

$$B_1 = -\cos(\theta_B + 90°) \tag{14.19}$$

$$B_2 = -\sin(\theta_B + 90°) \tag{14.20}$$

$$C_1 = \frac{V_C^2}{CD} \cos\theta_C - \omega^2 AB \cos\theta_A$$

$$+ \alpha AB \cos(\theta_A + 90°) - \frac{V_{C/B}^2}{BC} \cos\theta_B \tag{14.21}$$

$$C_2 = \frac{V_C^2}{CD} \sin\theta_C - \omega^2 AB \sin\theta_A$$

$$+ \alpha AB \sin(\theta_A + 90°) - \frac{V_{C/B}^2}{BC} \sin\theta_B \tag{14.22}$$

Given that the values V_C, $V_{C/B}$, A_C^T, and $A_{C/B}^T$ have been determined, general equations for computing the linear accelerations are

$$\overline{A}_B = -\omega^2 AB e^{i\theta_A} + \alpha AB e^{i(\theta_A + 90°)} \tag{14.23}$$

$$\overline{A}_C = -\frac{V_C^2}{CD} e^{i\theta_C} + A_C^T e^{i(\theta_C + 90°)} \tag{14.24}$$

$$\overline{A}_{C/B} = -\frac{V_{C/B}^2}{BC} e^{i\theta_B} + A_{C/B}^T e^{i(\theta_B + 90°)} \tag{14.25}$$

EXAMPLE 14.1: Crank Rocker

Considering the crank rocker linkage in Figure 14.1, let $AB = 1.5$ in., $BC = 3$ in., $CD = 3$ in., $AD = 4$ in., $\theta_A = 30°$, $\omega = 2$ rad/sec (counterclockwise), and $\alpha = 1$ rad/sec². It is required to find $\overline{V}_C$, $\overline{V}_{C/B}$, $\overline{A}_C$, and $\overline{A}_{C/B}$.

SOLUTION

1. Determine θ_B and θ_C using Eqs. (14.1) and (14.2).

$$\theta_B = 62.14° - 15.51° = 46.63°$$

$$\theta_C = 180° - 15.51° - 62.14° = 102.35°$$

2. Determine V_B, V_C, and $V_{C/B}$ using Eqs. (14.4), (14.5), and (14.6).

$$V_B = 2(1.5) = 3.00 \text{ in./sec}$$

$$V_C = 2(1.5) \frac{\sin(46.6° - 30°)}{\sin(46.6° - 102.3°)} = -1.04 \text{ in./sec}$$

$$V_{C/B} = 2(1.5) \frac{\sin(102.3° - 30.0°)}{\sin(46.6° - 102.3°)} = -3.46 \text{ in./sec}$$

3. Determine $\overline{V}_B$, $\overline{V}_C$, and $\overline{V}_{C/B}$ using Eqs. (14.7), (14.8), and (14.9).

$$\overline{V}_B = 3.0e^{i(120°)} = 3.0(\cos 120° + i \sin 120°)$$
$$= 3.0 \text{ in./sec } \underline{/120°}$$

$$\overline{V}_C = -1.04e^{i(192.3°)} = -1.04(\cos 192.3° + i \sin 192.3°)$$
$$= 1.04 \text{ in./sec } \underline{/12.3°}$$

$$\overline{V}_{C/B} = -3.46e^{i(136.4°)} = -3.46(\cos 136.6° + i \sin 136.6°)$$
$$= 3.46 \text{ in./sec } \underline{/-43.4°}$$

4. Determine the constants A_1, A_2, B_1, B_2, C_1, and C_2 using Eqs. (14.17) through (14.22).

$$A_1 = \cos(102.35 + 90°)$$
$$= -0.98$$

$$A_2 = \sin(102.35 + 90°)$$
$$= -0.21$$

$$B_1 = -\cos(46.6° + 90°)$$
$$= 0.73$$

$$B_2 = -\sin(46.6° + 90°)$$
$$= -0.69$$

$$C_1 = \frac{(-1.04)^2}{3.0} \cos 102.35° - 2^2(1.5) \cos 30°$$

$$+ \; 1.0(1.5) \cos(30° + 90°) - \frac{(-3.46)^2}{3.0} \cos 46.6° = -8.76$$

$$C_2 = \frac{(-1.04)^2}{3.0} \sin 102.35° - 2^2(1.5) \sin 30°$$

$$+ \; 1.0(1.5) \sin(30° + 90°) - \frac{(-3.46)^2}{3.0} \sin 46.6° = -4.25$$

5. Determine A_C^T and $A_{C/B}^T$ using Eqs. (14.15) and (14.16).

$$A_C^T = \frac{(-8.76)(0.69) - (-4.25)(0.73)}{(-0.98)(-0.69) - (-0.21)(0.73)}$$

$$= 11.02 \text{ in./sec}^2$$

$$A_{C/B}^T = \frac{(-0.98)(-4.25) - (-0.21)(-8.76)}{(-0.98)(-0.69) - (-0.21)(0.73)}$$

$$= 2.76 \text{ in./sec}^2$$

6. Substitute the values found for A_C^T and $A_{C/B}^T$ in Eqs. (14.24) and solve for $\overline{A}_C$ and $\overline{A}_{C/B}$.

$$\overline{A}_C = -\frac{(-1.04)^2}{3.0} e^{i(102.3°)} + 11.02 e^{i(192.3°)}$$

$$= 0.36 \; \underline{/-77.6°} + 11.02 \; \underline{/-167.6°}$$

$$= 11.03 \text{ in./sec}^2 \; \underline{/-165.7°}$$

$$\overline{A}_{C/B} = -\frac{(-3.46)^2}{3.0} e^{i(46.63°)} + 2.76 e^{i(136.63°)}$$

$$= 3.99 \; \underline{/-133.4°} + 2.76 \underline{/136.6°}$$

$$= 4.85 \text{ in./sec}^2 \; \underline{/-168.04°}$$

See Figures 14.7 to 14.10 and Table 14.1 for velocity and acceleration profiles of the complete crank cycle ($\theta_A = 15$ to $360°$).

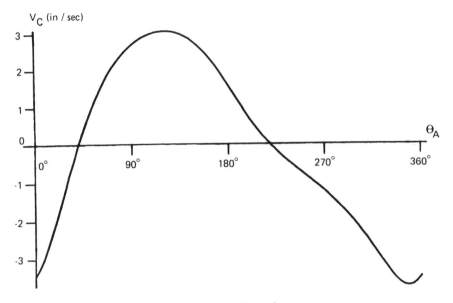

Figure 14.7   Velocity of point C vs crank angle.

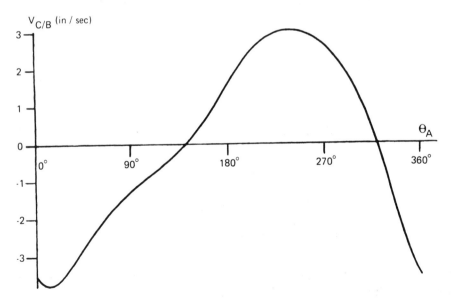

Figure 14.8   Velocity of C relative to B vs crank angle.

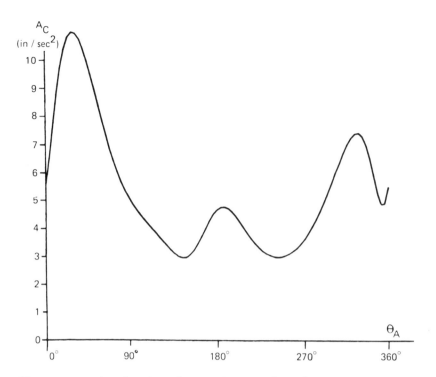

Figure 14.9   Acceleration of point C vs crank angle.

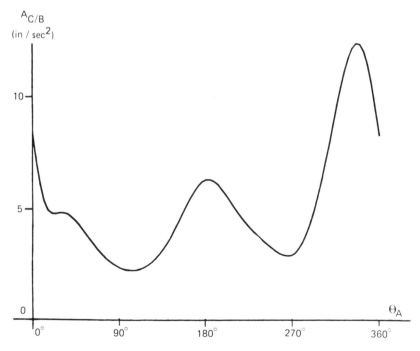

Figure 14.10   Acceleration of C relative to B vs crank angle.

Table 14.1 Velocity and Acceleration Profiles for Example Problem

$AB = 1.50$ $BC = 3.00$ $CD = 3.00$ $AD = 4.00$ $OMEGA(A) = 2.00$ $ALPHA(A) = 1.00$

θ(A)	θ(B)	θ(C)	V(B)	ARG	V(C)	ARG	V(CB)	ARG	A(B)	ARG	A(C)	ARG	A(CB)	ARG
0.	65.	115.	3.00	90.0	3.60	24.6	3.60	−24.6	6.18	166.0	5.55	−104.2	8.29	−56.0
15.	56.	107.	3.00	105.0	2.53	16.8	3.86	−34.1	6.18	−179.0	9.49	−150.2	5.05	−114.0
30.	47.	102.	3.00	120.0	1.04	12.3	3.46	−43.4	6.18	−164.0	11.03	−165.8	4.85	−168.0
45.	39.	102.	3.00	135.0	0.37	−168.5	2.81	−51.2	6.18	−149.0	9.84	−168.2	4.48	164.9
60.	33.	104.	3.00	150.0	1.46	−166.1	2.20	−57.5	6.18	−134.0	7.84	−160.9	3.63	148.8
75.	28.	109.	3.00	165.0	2.23	−161.4	1.68	−62.3	6.18	−119.0	6.15	−145.8	2.85	137.0
90.	24.	115.	3.00	180.0	2.74	−155.2	1.26	−66.0	6.18	−104.0	5.04	−125.4	2.37	127.0
105.	21.	122.	3.00	−165.0	3.03	−147.9	0.90	−68.6	6.18	−89.0	4.33	−102.7	2.23	118.3
120.	20.	130.	3.00	−150.0	3.15	−140.1	0.55	−70.5	6.18	−74.0	3.73	−77.9	2.48	111.9
135.	19.	138.	3.00	−135.0	3.07	−132.3	0.16	−71.4	6.18	−59.0	3.16	−46.9	3.17	108.8
150.	19.	145.	3.00	−120.0	2.80	−124.9	0.32	108.8	6.18	−44.0	3.03	−4.4	4.31	109.3

165.	20.	151.	3.00	-105.0	2.30	-118.5	0.93	110.4	6.18	-29.0	3.85	34.1	5.62	113.3
180.	24.	156.	3.00	-90.0	1.64	-113.6	1.64	113.6	6.18	-14.0	4.73	55.6	6.34	121.6
195.	29.	160.	3.00	-75.0	0.93	-110.4	2.30	118.5	6.18	1.0	4.69	66.1	5.99	135.7
210.	35.	161.	3.00	-60.0	0.32	-108.8	2.80	124.9	6.18	16.0	3.99	70.7	5.07	156.0
225.	42.	161.	3.00	-45.0	0.16	71.4	3.07	132.3	6.18	31.0	3.33	71.2	4.23	-179.6
240.	50.	160.	3.00	-30.0	0.55	70.5	3.15	140.1	6.18	46.0	3.02	68.6	3.59	-152.9
255.	58.	159.	3.00	-15.0	0.90	68.6	3.03	147.9	6.18	61.0	3.12	63.7	3.07	-121.8
270.	65.	156.	3.00	0.0	1.26	66.0	2.74	155.2	6.18	76.0	3.60	57.5	2.99	-81.6
285.	71.	152.	3.00	15.0	1.68	62.3	2.23	161.4	6.18	91.0	4.47	50.1	4.05	-42.8
300.	76.	147.	3.00	30.0	2.20	57.5	1.46	166.1	6.18	106.0	5.68	41.1	6.38	-20.3
315.	78.	141.	3.00	45.0	2.81	51.2	0.37	168.5	6.18	121.0	6.96	28.9	9.47	-11.8
330.	78.	133.	3.00	60.0	3.46	43.4	1.04	-12.3	6.18	136.0	7.39	10.7	12.07	-14.0
345.	73.	124.	3.00	75.0	3.86	34.1	2.53	-16.8	6.18	151.0	5.79	-25.0	11.97	-27.1
360.	65.	115.	3.00	90.0	3.60	24.6	3.60	-24.6	6.18	166.0	5.55	-104.2	8.29	-56.0

EXAMPLE 14.2: Drag Link

Considering the drag link mechanism in Figure 14.1, let AB = 2.0 in., BC = 1.5 in., CD = 2.5 in., AD = 1.0 in., θ_A = 240°, ω = 1 rad/sec (counterclockwise), and α = 0.5 rad/sec². It is required to find $\overline{V}_C$, $\overline{V}_{C/B}$, $\overline{A}_C$, and $\overline{A}_{C/B}$.

SOLUTION

1. Determine θ_B and θ_C using Eqs. (14.1) and (14.2).

$$\theta_B = 67.8° - (-40.9°) = 108.7°$$

$$\theta_C = 180° - (-40.9°) - 33.7° = 187.14°$$

2. Determine V_B, V_C, and $V_{C/B}$ using Eqs. (14.4), (14.5), and (14.6).

$$V_B = 1(2.0) = 2.0 \text{ in./sec}$$

$$V_C = 1(2.0)\,\frac{\sin(108.7° - 240°)}{\sin(108.7° - 187.14°)} = 1.53 \text{ in./sec}$$

$$V_{C/B} = 1(2.0)\,\frac{\sin(187.14° - 240°)}{\sin(108.7° - 187.14°)} = 1.63 \text{ in./sec}$$

3. Determine $\overline{V}_B$, $\overline{V}_C$, and $\overline{V}_{C/B}$ using Eqs. (14.7), (14.8), and (14.9).

$$\overline{V}_B = 2.0e^{i(330°)} = 2.0 \text{ in./sec } \underline{/-30°}$$

$$\overline{V}_C = 1.53e^{i(267.1°)} = 1.53 \text{ in./sec } \underline{/-82.9°}$$

$$\overline{V}_{C/B} = 1.63e^{i(198.7°)} = 1.63 \text{ in./sec } \underline{/-161.3°}$$

4. Determine the constants A_1, A_2, B_1, B_2, C_1, and C_2 using Eqs. (14.17) through (14.22).

$$A_1 = \cos(187.1° + 90°)$$

$$= 0.12$$

$$A_2 = \sin(187.1° + 90°)$$

$$= -0.99$$

$$B_1 = -\cos(108.7° + 90°)$$

$$= 0.95$$

$$B_2 = -\sin(108.7° + 90°)$$

$$= 0.32$$

$$C_1 = \frac{1.53^2}{2.5} \cos 187.1° + (1)^2(2.0) \cos 240°$$

$$+ 0.5(2.0) \cos(240° + 90°) - \frac{1.63^2}{1.5} \cos 108.7° = 1.49$$

$$C_2 = \frac{1.53^2}{2.5} \sin 187.1° + (1)^2(2.0) \sin 240°$$

$$+ 0.5(2.0) \sin(240° + 90°) - \frac{1.63^2}{1.5} \sin 108.7° = -0.56$$

5. Determine A_C^T and $A_{C/B}^T$ using Eqs. (14.15) and (14.16).

$$A_C^T = \frac{(1.49)(0.32) - (-0.56)(0.95)}{(0.12)(0.32) - (-0.99)(0.95)}$$

$$= 1.03 \text{ in./sec}^2$$

$$A_{C/B}^T = \frac{(0.12)(-0.56) - (-0.99)(1.49)}{(0.12)(0.32) - (-0.99)(0.95)}$$

$$= 1.45 \text{ in./sec}^2$$

6. Substitute the values found for A_C^T and $A_{C/B}^T$ in Eqs. (14.24) and (14.25) and solve for $\overline{A}_C$ and $\overline{A}_{C/B}$.

$$\overline{A}_C = -\frac{1.53^2}{2.5} e^{i(187.14°)} + 1.03e^{i(277.14°)}$$

$$= 0.94 \; \underline{/7.15°} + 1.03 \; \underline{/-82.9°}$$

$$= 1.39 \text{ in./sec}^2 \; \underline{/-40.4°}$$

$$\overline{A}_{C/B} = -\frac{163^2}{1.5} e^{i(108.7°)} + 1.45e^{i(198.7°)}$$

$$= 1.76 \; \underline{/-71.3°} + 1.45 \; \underline{/-161.3°}$$

$$= 2.28 \text{ in./sec}^2 \; \underline{/-110.7°}$$

Note that the four-bar mechanism may be presented in a crossed-phase configuration as shown in Figure 14.11, where, with crank *AB* in the first quadrant and rotating counterclockwise, link *BC* crosses the fixed link *AD*. In this configuration, the oscillation of the follower is below the fixed link and the equations for θ_B and θ_C do not apply directly. However, it is useful to note that this mechanism is the mirror image of the one in Figure 14.1 when the crank *AB* is in the fourth quadrant and its rotation is in the clockwise direction. Hence, the equations derived above are applicable to the velocity and acceleration analysis, provided that the reflected crank angle and reversed rotation sense are considered.

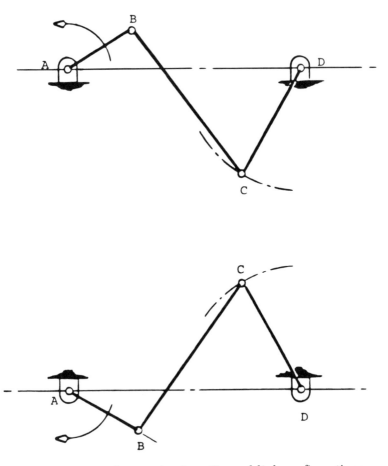

Figure 14.11   Four-bar mechanism: Crossed-link configurations.

Alternatively, if we elect to analyze the mechanism in the crossed phase as presented, Eqs. (14.1) and (14.2) must be modified as follows:

$$\theta_B = -\gamma_B - \phi_D$$

$$\theta_C = 180° + \gamma_D - \phi_D$$

All other relationships remain unchanged.

15

Slider-Crank Mechanism Analysis: Simplified Vector Method

15.1 INTRODUCTION

The slider-crank mechanism is probably the most common kinematic element to be found in most machines. A basic variation of the four-bar linkage, in which the follower crank is replaced by a sliding block, this mechanism is capable of converting rotary motion to linear motion, and vice versa.

The mechanism is generally found in three basic arrangements:

1. The **central or in-line type**, as in ABC of Figure 15.1, where the slider path of travel essentially passes through the crank pin or point A.
2. The **positive offset type**, as in ABCD of Figure 15.2, where the slider path is offset a distance CD (or eccentricity e) above a reference line AD, drawn parallel to the path and passing through point A.
3. The **negative-offset type**, as in ABCD of Figure 15.3, where the slider path is offset a distance CD below the reference line AD, drawn parallel to the path and passing through point A.

Among the countless applications of this mechanism are piston engines, pumps, compressors, saws, and other forms of reciprocating machinery. Some examples of these applications can be seen in Figures 15.4 to 15.7. Also, like the four-bar linkage, this mechanism is generally used in combination with other basic mechanisms to produce a wide variety

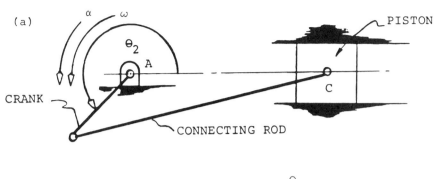

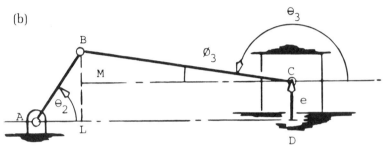

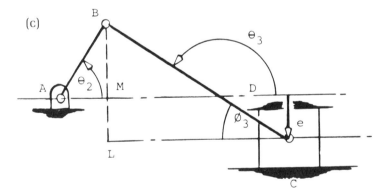

Figure 15.1   a) Central or in-line slider crank b) Slider crank with positive offset and c) Slider crank with negative offset.

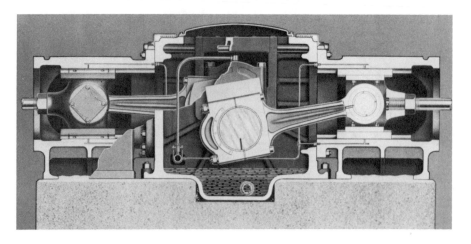

Figure 15.2   Slider-crank compressor. (Courtesy of Ingersoll Rand, Woodcliff Lake, N.J.)

Figure 15.3   Punch press.

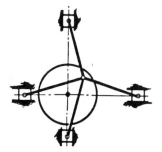

Figure 15.4   Radial engine.

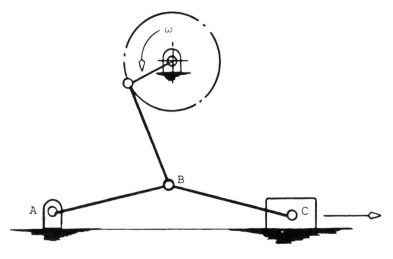

Figure 15.5   Toggle mechanism.

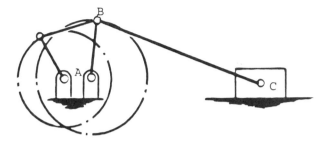

Figure 15.6   Drag-link quick-return mechanism.

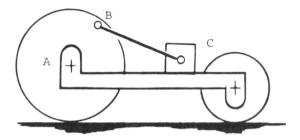

Figure 15.7   Steam locomotive.

of output motions as well as a model to simplify the analysis of more complex machines. For these reasons, the slider crank is one of the most frequently analyzed mechanisms. In this analysis, the linear velocity and acceleration relationships will be developed for any angular position of the drive crank.

15.2   SCOPE AND ASSUMPTIONS

The mechanism ABC in Figure 15.1 represents a typical slider-crank mechanism in which AB, the length of the drive crank 2, and BC, the length of the connecting rod 3, are known. AC is a variable distance depending on the angular position θ_2 of the crank during the motion cycle. Given that the crank rotates with an angular velocity ω_2 and angular acceleration α_2 (both in the counterclockwise direction), we will now derive general expressions to compute linear velocities and accelerations of points A, B, C, and point B relative to point C for any angular position of the crank.

In this analysis, all angular displacements, velocities, and accelerations are considered positive for counterclockwise rotation and negative for clockwise rotation. All velocity vectors are positive with respect to the positions of their respective links.

15.3 GEOMETRIC RELATIONSHIPS

To begin the analysis, we first seek to determine the angle of the connecting rod θ_3 in terms of the crank angle θ_2. Considering the typical slider-crank mechanism represented by the triangle ABC in Figure 15.8, we note that since the link lengths AB and BC are known, we can apply geometric and trigonometric relationships to obtain

$$\theta_3 = 180° + \phi_3 \tag{15.1}$$

where

$$\phi_3 = \sin^{-1} \frac{AB}{BC} \sin(360° - \theta_2)$$

$$= -\sin^{-1} \frac{AB}{BC} \sin \theta_2$$

or

$$\theta_3 = 180° - \phi_3 \tag{15.2}$$

where

$$\phi_3 = \sin^{-1}\left(\frac{AB}{BC} \sin \theta_2\right)$$

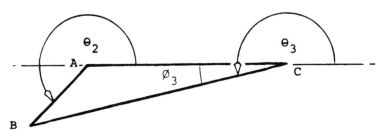

Figure 15.8   Geometric relationships.

For the positive-offset slider crank, represented by diagram $ABCD$ in Figure 15.2, we proceed by dropping a perpendicular from point B to meet AD at L, then extending the slider path to intersect the perpendicular BL. From this, we find the angle θ_3 as follows:

$$\theta_3 = 180° - \phi_3 \quad [\text{Eq. (15.2)}]$$

$$BC \sin \phi_3 + e = AB \sin \theta_2 \tag{15.3}$$

$$\phi_3 = \sin^{-1}\left(\frac{AB}{BC} \sin \theta_2 - \frac{e}{BC}\right) \tag{15.4}$$

$$\theta_3 = 180° - \sin^{-1}\left(\frac{AB}{BC} \sin \theta_2 - \frac{e}{BC}\right) \quad (e > 0) \tag{15.5}$$

Similarly, for the negative-offset slider crank, represented by diagram $ABCD$ in Figure 15.3, we obtain

$$\theta_3 = 180° - \sin^{-1}\left(\frac{AB}{BC} \sin \theta_2 + \frac{e}{BC}\right) \quad (e < 0) \tag{15.6}$$

Hence, the general equation may be written as

$$\theta_3 = 180° - \sin^{-1}\left(\frac{AB}{BC} \sin \theta_2 - \frac{e}{BC}\right) \tag{15.7}$$

where e has a positive value for positive offset and negative value for negative offset.

15.4 VELOCITY ANALYSIS

To determine the linear velocity relationships for $\overline{V}_B$, $\overline{V}_C$, and $\overline{V}_{B/C}$, let Bcb in Figure 15.9 represent the velocity polygon of the mechanism ABC, where

Bb = magnitude of linear velocity of point B (V_B).
Bc = magnitude of linear velocity of point C (V_C).
bc = magnitude of linear velocity of B relative to C ($V_{B/C}$).

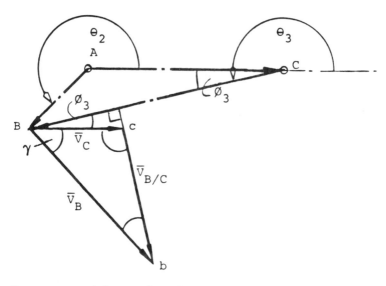

Figure 15.9   Slider-crank analysis.

Then the internal angles of the polygon can be obtained as follows:

$$\gamma = 360° - \theta_2 - 90°$$

$$= 270° - \theta_2$$

$$\angle b = 90° - \phi_3 - \gamma$$

$$= 90° - (\theta_3 - 180°) - (270° - \theta_2)$$

$$= 90° - \theta_3 + 180° - 270° + \theta_2$$

$$= \theta_2 - \theta_3$$

$$\underline{/bcB} = 180° - \gamma - \angle b$$

$$= 180° - (270° - \theta_2) - (\theta_2 - \theta_3)$$

$$= 180° - 270° + \theta_2 - \theta_2 + \theta_3$$

$$= \theta_3 - 90°$$

By applying the **Rule of Sines** from trigonometry, V_B, V_C, and $V_{B/C}$ can be expressed in the form

$$\frac{V_B}{\sin(\theta_3 - 90°)} = \frac{V_C}{\sin(\theta_2 - \theta_3)} = \frac{V_{B/C}}{\sin(270° - \theta_2)}$$

which reduces to

$$\frac{V_B}{\sin(90° - \theta_3)} = \frac{V_c}{\sin(\theta_3 - \theta_2)} = \frac{V_{B/C}}{\sin(90° - \theta_2)}$$

Thus, the scalar expressions for V_B, V_C, and $V_{B/C}$ are obtained as follows:

$$V_B = AB \times \omega_2 \tag{15.8}$$

$$V_C = V_B \frac{\sin(\theta_3 - \theta_2)}{\sin(90° - \theta_3)} \tag{15.9}$$

$$V_{B/C} = V_B \frac{\sin(90° - \theta_2)}{\sin(90° - \theta_3)} \tag{15.10}$$

To convert these equations to vectorial forms, we return to Figure 15.9 and note the following:

The directions of velocities $\overline{V}_B$ and $\overline{V}_{B/C}$ are assumed to be oriented at angles $(\theta_2 + 90°)$ and $(\theta_3 + 90°)$, respectively.
The direction of velocity $\overline{V}_C$ is known to act along a straight line only.
Therefore, we can write the vectorial expressions as follows:

$$\overline{V}_B = AB\omega_2 e^{i(\theta_2 + 90°)} \tag{15.11}$$

$$\overline{V}_C = AB\omega_2 \frac{\sin(\theta_3 - \theta_2)}{\sin(90° - \theta_3)} e^{i0°} \tag{15.12}$$

$$\overline{V}_{B/C} = AB\omega_2 \frac{\sin(90° - \theta_2)}{\sin(90° - \theta_3)} e^{i(\theta_3 + 90°)} \tag{15.13}$$

15.5 ACCELERATION ANALYSIS

To determine the linear acceleration relationships for point C $(\overline{A}_C)$, we apply the relative motion equation, which states that

$$\overline{A}_B = \overline{A}_C + \overline{A}_{B/C} \quad \text{(vectorial sum)} \tag{15.14}$$

Expanding this equation into its normal and tangential form, we obtain

$$\overline{A}_B^N + \overline{A}_B^T = \overline{A}_C^N + \overline{A}_C^T + \overline{A}_{B/C}^N + \overline{A}_{B/C}^T \tag{15.15}$$

where

$$\overline{A}_B^N = -\omega_2^2 ABe^{i\theta_2}$$

$$\overline{A}_B^T = \alpha_2 ABe^{i(\theta_2 + 90°)}$$

$$\overline{A}_C^N = -\frac{V_C^2}{\infty} e^{i90°} = 0 \quad \text{(straight line)}$$

$$\overline{A}_C^T = A_C^T e^{i0°} \quad \text{(real)}$$

$$\overline{A}_{B/C}^N = -\frac{V_{B/C}^2}{BC} e^{i\theta_3}$$

$$\overline{A}_{B/C}^T = A_{B/C}^T e^{i(\theta_3 + 90°)}$$

Note that point C has no normal acceleration since the slider path is a straight line of infinite radius. Therefore, the absolute acceleration of C is the tangential acceleration. Thus, Eq. (15.15) becomes

$$-\omega_2^2 ABe^{i\theta_2} + \alpha_2 ABe^{i(\theta_2 + 90°)} = A_C^T - \frac{V_{B/C}^2}{BC} e^{i\theta_3} + A_{B/C}^T e^{i(\theta_3 + 90°)}$$

(15.16)

This equation contains two unknown quantities: the magnitudes A_C^T and $A_{B/C}^T$. All other quantities (magnitudes and directions) are either known or can readily be determined from problem data. Note that the directions of $\overline{A}_C^T$ and $\overline{A}_{B/C}^T$, although not known precisely, are assumed to be positive for convenience. If the actual direction of either is reversed, the numerical solution of the equation will automatically produce a negative sign for the unknown quantity. To solve the unknowns, therefore, we equate the real and imaginary parts of the equation and rearrange as follows:

Real:

$$A_C^T + A_{B/C}^T \cos(\theta_3 + 90°)$$

$$= -\omega_2^2 AB \cos\theta_2 + \alpha_2 AB \cos(\theta_2 + 90°) + \frac{V_{B/C}^2}{BC} \cos\theta_3 \quad (15.17)$$

Imaginary:

$$A_{B/C}^T \sin(\theta_3 + 90°)$$

$$= -\omega_2^2 AB \sin \theta_2 + \alpha_2 AB \sin(\theta_2 + 90°) + \frac{V_{B/C}^2}{BC} \sin \theta_3 \quad (15.18)$$

The solution of these simultaneous equations yields

$$A_C^T = \frac{C_1 B_2 - C_2 B_1}{A_1 B_2 - A_2 B_1} \quad (15.19)$$

$$A_{B/C}^T = \frac{A_1 C_2 - A_2 C_1}{A_1 B_2 - A_2 B_1} \quad (15.20)$$

where

$$A_1 = 1 \quad (15.21)$$

$$A_2 = 0 \quad (15.22)$$

$$B_1 = \cos(\theta_3 + 90°) \quad (15.23)$$

$$B_2 = \sin(\theta_3 + 90°) \quad (15.24)$$

$$C_1 = -\omega_2^2 AB \cos \theta_2 + \alpha_2 AB \cos(\theta_2 + 90°) + \frac{V_{B/C}^2}{BC} \cos \theta_3$$

$$(15.25)$$

$$C_2 = -\omega_2^2 AB \sin \theta_2 + \alpha_2 AB \sin(\theta_2 + 90) + \frac{V_{B/C}^2}{BC} \sin \theta_3$$

$$(15.26)$$

Given that the values $V_{B/C}$, A_C^T, and $A_{B/C}^T$ have been found, the general equations for computing the linear accelerations can be summarized as follows:

$$\overline{A}_B = -\omega_2^2 AB e^{i\theta_2} + \alpha_2 AB e^{i(\theta_2 + 90°)} \quad (15.27)$$

$$\overline{A}_C = A_C^T e^{i0°} \quad \text{(real)} \quad (15.28)$$

$$\overline{A}_{B/C} = -\frac{V_{B/C}^2}{BC} e^{i\theta_3} + A_{B/C}^T e^{i(\theta_3 + 90°)} \quad (15.29)$$

EXAMPLE 15.1: Central Slider Crank

Considering the central slider-crank mechanism in Figure 15.1, let $AB = 1.5$ in., $BC = 3$ in., $\theta_2 = 150°$, $\omega_2 = 1.0$ rad/sec (counterclockwise), and $\alpha_2 = 0$ rad/sec². It is required to find $\overline{V}_C$, $\overline{V}_{B/C}$, $\overline{A}_C$, and $\overline{A}_{B/C}$.

SOLUTION[a]

1. Determine θ_3 using Eqs. (15.2) and (15.1).

$$\phi_3 = \sin^{-1}\left(\frac{1.5}{3.0}\sin 150°\right)$$

$$= 14.5°$$

$$\theta_3 = 180° - 14.5°$$

$$= 165.5°$$

2. Determine V_B, V_C, and $V_{B/C}$ using Eqs. (15.8), (15.9), and (15.10).

$$V_B = 1.5(1.0) = 1.50 \text{ in./sec}$$

$$V_C = (1.5)\frac{\sin(165.5° - 150°)}{\sin(90° - 165.5°)} = -0.41 \text{ in./sec}$$

$$V_{B/C} = (1.5)\frac{\sin(90° - 150°)}{\sin(90° - 165.5°)} = 1.34 \text{ in./sec}$$

3. Determine $\overline{V}_B$, $\overline{V}_C$, and $\overline{V}_{B/C}$ using Eqs. (15.11), (15.12), and (15.13).

$$\overline{V}_B = 1.5e^{i(240°)} = 1.5 \text{ in./sec } \underline{/-120°}$$

$$\overline{V}_C = -0.41e^{i0°} = 0.41 \text{ in./sec } \underline{/180°}$$

$$\overline{V}_{B/C} = 1.34\ e^{i(255.5°)} = 1.34 \text{ in./sec } \underline{/-104.5°}$$

---

[a]See Figure 15.10 for velocity and acceleration profiles of complete crank cycle from $\theta_2 = 0°$ to $\theta_2 = 360°$.

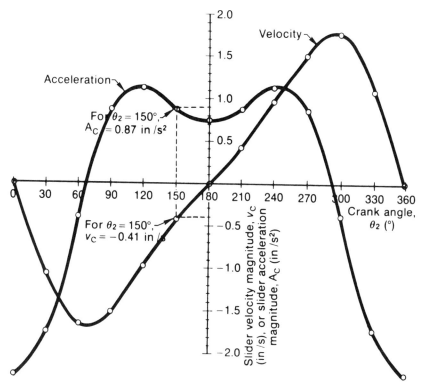

Figure 15.10   Slider motion characteristics.

4.   Determine the constants A_1, A_2, B_1, B_2, C_1, and C_2 using Eqs. (15.21) through (15.26).

$$A_1 = 1 \quad \text{[from Eq. (15.21)]}$$

$$A_2 = 0 \quad \text{[from Eq. (15.22)]}$$

$$B_1 = \cos(165.5° + 90°)$$
$$= -0.25$$

$$B_2 = \sin(165.5° + 90°)$$
$$= -0.97$$

$$C_1 = -(1)^2(1.5) \cos 150° + 0(1.5) \cos(150° + 90°)$$

$$+ \frac{1.34^2}{3} \cos 165.5°$$

$$= -(-1.3) + 0 + (-0.58)$$

$$= 0.72$$

$$C_2 = -(1)^2(1.5) \sin 150° + 0(1.5) \sin(150° + 90°)$$

$$+ \frac{1.34^2}{3} \sin 165.5°$$

$$= -0.75 + 0 + 0.15$$

$$= -0.60$$

5. Determine A_C^T and $A_{B/C}^T$ using Eqs. (15.19) and (15.20).

$$A_C^T = \frac{(0.72)(-0.97) - (-0.60)(-0.25)}{(1.00)(-0.97) - (0.00)(-0.25)}$$

$$= 0.87 \text{ in./sec}^2$$

$$A_{B/C}^T = \frac{(1.00)(-0.60) - (0.00)(0.72)}{(1.00)(-0.97) - (0.00)(-0.25)}$$

$$= 0.62 \text{ in./sec}^2$$

6. Substitute the values found for A_C^T and $A_{B/C}^T$ into Eqs. (15.28) and (15.29) and solve for $\overline{A}_C$ and $\overline{A}_{B/C}$.

$$\overline{A}_C = 0.87 \, e^{i0°}$$

$$= 0.87 \text{ in./sec}^2 \, \underline{/0°}$$

$$\overline{A}_{B/C} = -\frac{1.34^2}{3} \, e^{i(165.5°)} + 0.62 \, e^{i(255.5°)}$$

$$= 0.60 \, \underline{/-14.5°} + 0.62 \, \underline{/-104.5°}$$

$$= 0.86 \text{ in./sec}^2 \, \underline{/-60.4°}$$

EXAMPLE 15.2: Offset Slider Crank

Considering the offset slider-crank mechanism in Figure 15.2, let $AB = 1.5$ in., $BC = 3$ in., $e = 0.5$ in., $\theta_2 = 150°$, $\omega_2 = 1.0$ rad/sec (counterclockwise), and $\alpha_2 = 0$ rad/sec². It is required to find $\overline{V}_C$, $\overline{V}_{B/C}$, $\overline{A}_C$, and $\overline{A}_{B/C}$.

SOLUTION

1. Determine θ_3 using Eqs. (15.2) and (15.1).

$$\phi_3 = \sin^{-1}\left(\frac{1.5}{3.0}\sin 150° - \frac{0.5}{3.0}\right)$$

$$= 4.78°$$

$$\theta_3 = 180° - 4.78°$$

$$= 175.22°$$

2. Determine V_B, V_C, and $V_{B/C}$ using Eqs. (15.8), (15.9), and (15.10).

$$V_B = 1.5(1.0) = 1.50 \text{ in./sec}$$

$$V_C = (1.5)\frac{\sin(175.22° - 150°)}{\sin(90° - 175.22°)} = -0.64 \text{ in./sec}$$

$$V_{B/C} = (1.5)\frac{\sin(90° - 150°)}{\sin(90° - 175.22°)} = 1.30 \text{ in./sec}$$

3. Determine $\overline{V}_B$, $\overline{V}_C$, and $\overline{V}_{B/C}$ using Eqs. (15.11), (15.12), and (15.13).

$$\overline{V}_B = 1.5e^{i(240°)} = 1.5 \text{ in./sec } \underline{/-120°}$$

$$\overline{V}_C = -0.64\, e^{i0°} = 0.64 \text{ in./sec } \underline{/180°}$$

$$\overline{V}_{B/C} = 1.30\, e^{i(265.2°)} = 1.30 \text{ in./sec } \underline{/-94.78°}$$

4. Determine the constants A_1, A_2, B_1, B_2, C_1, and C_2 using Eqs. (15.21) through (15.26).

$$A_1 = 1 \quad \text{[from Eq. (15.21)]}$$

$$A_2 = 0 \quad \text{[from Eq. (15.22)]}$$

$$B_1 = \cos(175.2° + 90°)$$

$$= -0.08$$

$$B_2 = \sin(175.2° + 90°)$$

$$= -0.99$$

$$C_1 = -(1^2)(1.5) \cos 150° + 0(1.5) \cos(150° + 90°)$$

$$+ \frac{1.30^2}{3} \cos(175.2°)$$

$$= -(-1.3) + 0 + (-0.56)$$

$$= 0.73$$

$$C_2 = -(1^2)(1.5) \sin 150° + 0(1.5) \sin(150° + 90°)$$

$$+ \frac{130^2}{3} \sin(175.2°)$$

$$= -(0.75) + 0 + 0.084$$

$$= -0.70$$

5. Determine A_C^T and $A_{B/C}^T$ using Eqs. (15.19) and (15.20).

$$A_C^T = \frac{(0.73)(-0.99) - (-0.70)(-0.08)}{(1.00)(-0.99) - (0.00)(-0.08)}$$

$$= 0.79 \text{ in./sec}^2$$

$$A_{B/C}^T = \frac{(1.00)(-0.70) - (0.00)(0.73)}{(1.00)(-0.99) - (0.00)(-0.08)}$$

$$= 0.70 \text{ in./sec}^2$$

6. Substitute the values found for A_C^T and $A_{B/C}^T$ into Eqs. (15.28) and (15.29) and solve for $\overline{A}_C$ and $\overline{A}_{B/C}$.

$$\overline{A}_C = 0.79 \, e^{i0°}$$

$$= 0.79 \text{ in./sec}^2 \, \underline{/0°}$$

$$\overline{A}_{B/C} = -\frac{(1.30)^2}{3.0} \, e^{i(175.2°)} + 0.70 \, e^{i(265.2°)}$$

$$= 0.57 \, \underline{/-4.78°} + 0.70 \, \underline{/-94.8°}$$

$$= 0.90 \text{ in./sec}^2 \, \underline{/-56.0°}$$

16

Quick-Return Mechanism Analysis: Simplified Vector Method

16.1 INTRODUCTION

Quick-return mechanisms, typified by $ABCD$ in Figure 16.1, are a special class of sliding contact linkages that provide uniform velocity motion followed by a fast-return stroke to an initial position. These mechanisms, often used in equipment such as machine tools and production machinery to produce long, slow movements for cutting or feeding and fast-return strokes in which no work is done, are generally found in two basic arrangements:

1. The ***crank-shaper (or oscillating-beam) type*** (Figure 16.2), where the crank arm AB is shorter than the base AD, and as a result, the follower CD oscillates as the crank arm makes a complete revolution.
2. The ***Whitworth type*** (Figure 16.3), where the crank arm AB is longer than the base AD, and as a result, both crank arm and follower make complete revolutions.

Quick-return mechanisms, like other basic mechanisms, can be used as models to simplify the analysis of more complex machines or they can be arranged in combination with other mechanisms to produce specific output motions.

The Geneva mechanism shown in Figure 16.4 is a classical application of the quick-return mechanism.

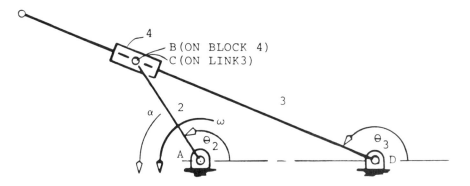

Figure 16.1   Quick-return mechanism.

Motion characteristics of quick-return mechanisms are usually difficult to analyze because of the complexity of the combined linkage and slider movements that give rise to the Coriolis acceleration. The simplified mathematical method presented here quickly determines linear velocity and acceleration relationships for a quick-return mechanism for any given drive crank position in its motion cycle.

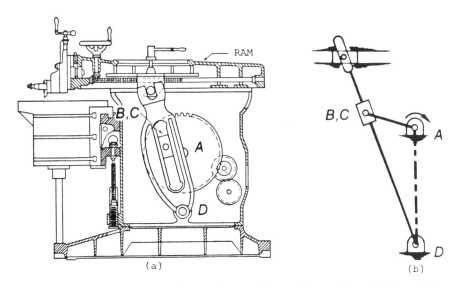

Figure 16.2   (a) Oscillating-beam shaper; (b) oscillating-beam linkage. (From *Applied Kinematics* by J. Harland Billings, © 1953 by D. Van Nostrand Company, Inc. Reprinted by permission of Wadsworth Publishing Company, Belmont, CA 94002.)

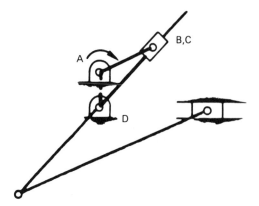

Figure 16.3   Whitworth linkage.

16.2   SCOPE AND ASSUMPTIONS

The mechanism $ABCD$ in Figure 16.1 represents a typical quick-return mechanism in which the drive crank AB (or link 2) and follower guide CD (or link 3) have known lengths. B and C are coincident points on the slider (link 4) and follower guide, respectively. AD (or link 1) is the distance between crank and follower pivot points. θ_2 is the angular position of link 2 measured from base link 1 or AD. The crank AB rotates with angular velocity ω_2 and angular acceleration α_2 (both counterclockwise).

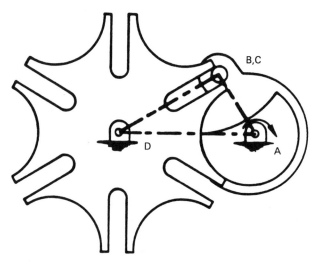

Figure 16.4   Geneva mechanism.

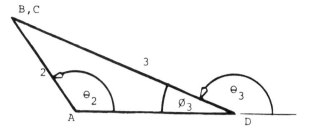

Figure 16.5   Geometric relationships.

In this analysis, equations for computing linear velocities and accelerations in terms of the variable angular position are to be determined. All angular displacements, velocities, and accelerations are considered positive for counterclockwise rotation and negative for clockwise rotation.

16.3   GEOMETRIC RELATIONSHIPS

To determine θ_3, consider triangle ACD in Figure 16.5. Since AB, AD, and θ_2 are known, basic geometric and trigonometric relationships yield

$$\theta_3 = 180° - \phi_3 \tag{16.1}$$

where

$$\phi_3 = \cos^{-1} \frac{CD^2 + AD^2 - AB^2}{2(CD)(AD)} \qquad (\theta_2 < 180°) \tag{16.2}$$

$$\phi_3 = -\cos^{-1} \frac{CD^2 + AD^2 - AB^2}{2(CD)(AD)} \qquad (\theta_2 > 180°) \tag{16.3}$$

$$CD = [AB^2 + AD^2 - 2(AB)(AD) \cos \theta_2]^{1/2} \tag{16.4}$$

16.4   VELOCITY ANALYSIS

To determine the relationships for velocities $\overline{V}_B$, $\overline{V}_C$, and $\overline{V}_{B/C}$, let Bcb in Figure 16.6 represent the velocity polygon of the mechanism $ABCD$, where

Bb = magnitude of linear velocity of point B (V_B).
Bc = magnitude of linear velocity of point C (V_C).
bc = magnitude of linear velocity of B relative to C ($V_{B/C}$).

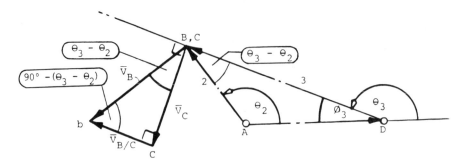

Figure 16.6   Velocity analysis.

Then V_B, V_C, and $V_{B/C}$ can be determined in terms of the angular positions of the links as follows:

$$\angle Bcb = 90°$$

$$\angle bBc = \theta_3 - \theta_2$$

$$\angle b = 90° - (\theta_3 - \theta_2)$$

By applying the **Rule of Sines** from trigonometry, the velocities can be expressed in the form

$$\frac{V_B}{\sin 90°} = \frac{V_C}{\sin(90° - \theta_3 + \theta_2)} = \frac{V_{B/C}}{\sin(\theta_3 - \theta_2)}$$

Since

$$V_B = \omega_2 AB \tag{16.5}$$

then

$$V_C = \omega_2 AB \sin(90 - \theta_3 + \theta_2) = \omega_2 AB \cos(\theta_3 - \theta_2) \tag{16.6}$$

and

$$V_{B/C} = \omega_2 AB \sin(\theta_3 - \theta_2) \tag{16.7}$$

Also, since $\overline{V}_B$, $\overline{V}_C$, and $\overline{V}_{B/C}$ are assumed to be oriented at angles $(\theta_2 + 90°)$, $(\theta_3 + 90°)$, and (θ_3), respectively, from the real or reference axis, the required vertical expressions can be written as follows:

$$\overline{V}_B = \omega_2 AB e^{i(\theta_2 + 90°)} \tag{16.8}$$

$$\overline{V}_C = \omega_2 AB \cos(\theta - \theta_2) e^{i(\theta_3 + 90°)} \tag{16.9}$$

$$\overline{V}_{B/C} = \omega_2 AB \sin(\theta - \theta_2) e^{i\theta_3} \tag{16.10}$$

where $e^{i(\theta_2 + 90°)}$, $e^{i(\theta_3 + 90°)}$, and $e^{i\theta_3}$ are unit vectors, used to define the directions of $\overline{V}_B$, $\overline{V}_C$, and $\overline{V}_{B/C}$, respectively.

Note that $e^{i\theta} = \cos\theta + i\sin\theta$, where θ is the position angle of the unit vector.

16.5   ACCELERATION ANALYSIS

To determine the linear accelerations, we apply the relative motion theory, which states that

$$\overline{A}_B = \overline{A}_C + \overline{A}_{B/C} \tag{16.11}$$

Expanding this equation into its normal and tangential component form, we obtain

$$\overline{A}_B^N + \overline{A}_B^T = \overline{A}_C^N + \overline{A}_C^T + \overline{A}_{B/C}^N + \overline{A}_{B/C}^T + \overline{A}_{B/C}^{Cor} \tag{16.12}$$

where

$$\overline{A}_B^N = -\omega_2^2 AB e^{i\theta_2}$$

$$\overline{A}_B^T = \alpha_2 AB e^{i(\theta_2 + 90°)}$$

$$\overline{A}_C^N = -\frac{V_C^2}{CD} e^{i\theta_3}$$

$$\overline{A}_C^T = A_C^T e^{i(\theta_3 + 90°)}$$

$$\overline{A}_{B/C}^N = -\frac{V_{B/C}^2}{\infty} e^{i(\theta_3 + 90°)} = 0$$

$$\overline{A}_{B/C}^T = A_{B/C}^T e^{i\theta_3}$$

$$\overline{A}_{B/C}^{Cor} = 2V_{B/C} \frac{V_C}{CD} e^{i(\theta_3 + 90°)}$$

Here it should be observed that an important difference between Eq. (16.12) and the corresponding equation for the four-bar linkage is that since points B and C are not rigidly connected, but slide relative to each other, it is necessary to determine the Coriolis acceleration $\overline{A}_{B/C}^{Cor}$, an acceleration component that results from the sliding motion.

Note that for the Coriolis acceleration, we consider the linear velocity of the slider relative to the rotating guide and angular velocity of the guide, namely, ω_3 or V_C/CD. Also, the Coriolis acceleration has the orientation of $\overline{V}_{B/C}$ when rotated 90° about its tail in the direction of ω_3.

Also, in Eq. (16.12), note that since the sliding path of B on C is a straight line, its radius of curvature is infinite. That is, $R = \infty$. Hence, $\overline{A}_{B/C}^N = 0$. Thus, Eq. (16.12) becomes

$$-\omega_2^2 ABe^{i\theta_2} + \alpha_2 ABe^{i(\theta_2 + 90°)} = -\frac{V_C^2}{CD}e^{i\theta_3} + A_C^T e^{i(\theta_3 + 90°)}$$

$$+ 2V_{B/C}\frac{V_C}{CD}e^{i(\theta_3 + 90°)} + A_{B/C}^T e^{i\theta_3}$$

$$(16.13)$$

This equation contains only two unknown quantities: the magnitudes A_C^T and $A_{B/C}^T$. All other quantities (magnitudes and directions) are either known or can readily be determined from the problem data. (Note that the directions of $\overline{A}_C^T$ and $\overline{A}_{B/C}^T$, although not precisely known, are assumed to be positive for convenience. If any of these directions is reversed, the numerical solution of the equation will automatically produce a negative sign for the unknown quantity.) To solve for the unknowns, therefore, we equate the real and imaginary parts of the equation and rearrange as follows:

Real:

$$A_C^T \cos(\theta_3 + 90°) + A_{B/C}^T \cos \theta_3$$

$$= -\omega_2^2 AB \cos \theta_2 + \alpha_2 AB \cos(\theta_2 + 90°)$$

$$- 2V_{B/C}\frac{V_C}{CD}\cos(\theta_3 + 90°) + \frac{V_C^2}{CD}\cos \theta_3 \qquad (16.14)$$

Imaginary:

$$A_C^T \sin(\theta_3 + 90°) + A_{B/C}^T \sin \theta_3$$

$$= -\omega_2^2 AB \sin \theta_2 + \alpha_2 AB \sin(\theta_2 + 90°)$$

$$- 2V_{B/C} \frac{V_C}{CD} \sin(\theta_3 + 90°) + \frac{V_C^2}{CD} \sin \theta_3 \qquad (16.15)$$

The solution of these simultaneous equations yields

$$A_C^T = \frac{C_1 B_2 - C_2 B_1}{A_1 B_2 - A_2 B_1} \qquad (16.16)$$

$$A_{B/C}^T = \frac{A_1 C_2 - A_2 C_1}{A_1 B_2 - A_2 B_1} \qquad (16.17)$$

where

$$A_1 = \cos(\theta_3 + 90°) \qquad (16.18)$$

$$A_2 = \sin(\theta_3 + 90°) \qquad (16.19)$$

$$B_1 = \cos \theta_3 \qquad (16.20)$$

$$B_2 = \sin \theta_3 \qquad (16.21)$$

$$C_1 = -\omega_2^2 AB \cos \theta_2 + \alpha_2 AB \cos(\theta_2 + 90°)$$

$$- 2V_{B/C} \frac{V_C}{CD} \cos(\theta_3 + 90°) + \frac{V_C^2}{CD} \cos \theta_3 \qquad (16.22)$$

$$C_2 = -\omega_2^2 AB \sin \theta_2 + \alpha_2 AB \sin(\theta_2 + 90°)$$

$$- 2V_{B/C} \frac{V_C}{CD} \sin(\theta_3 + 90°) + \frac{V_C^2}{CD} \sin \theta_3 \qquad (16.23)$$

Given that the values V_C, $V_{B/C}$, A_C^T, and $A_{B/C}^T$ are found, the general equations for computing the linear accelerations can be summarized as follows:

$$\overline{A}_B = -\omega_2^2 ABe^{i\theta_2} + \alpha_2 ABe^{i(\theta_2 + 90°)} \tag{16.24}$$

$$\overline{A}_C = -\frac{V_C^2}{CD} e^{i\theta_3} + A_C^T e^{i(\theta_3 + 90°)} \tag{16.25}$$

$$\overline{A}_{B/C} = \frac{2(V_{B/C})(V_C)}{CD} e^{i(\theta_3 + 90°)} + A_{B/C}^T e^{i\theta_3} \tag{16.26}$$

EXAMPLE 16.1: Shaper Mechanism

Considering the shaper mechanism in Figure 16.1, let $AB = 2$ in., $AD = 4$ in., $\theta_2 = 30°$, $\omega_2 = 62.83$ rad/sec (counterclockwise), and $\alpha_2 = 0$. It is required to find $\overline{V}_C$, $\overline{V}_{B/C}$, $\overline{A}_C$, and $\overline{A}_{B/C}$.

SOLUTION

1. Determine θ_3 using Eqs. (16.4), (16.2), and (16.1).

$$CD = [0.167^2 + 0.333^2 - 2(0.167)(0.333) \cos 30°]^{1/2} = 0.21 \text{ ft}$$

$$\phi_3 = \cos^{-1} \frac{0.21^2 + 0.33^2 - 0.167^2}{2(0.21)(0.33)} = 23.8°$$

$$\theta_3 = 180° - 23.8° = 156.2°$$

2. Determine V_B, V_C, and $V_{B/C}$ using Eqs. (16.5), (16.6), and (16.7).

$$V_B = (62.83)(0.167) = 10.47 \text{ ft/sec}$$

$$V_C = (62.83)(0.167) \cos(156.2° - 30°) = -6.18 \text{ ft/sec}$$

$$V_{B/C} = (62.83)(0.167) \sin(156.2° - 30°) = 8.45 \text{ ft/sec}$$

3. Determine $\overline{V}_B$, $\overline{V}_C$, and $\overline{V}_{B/C}$ using Eqs. (16.8), (16.9), and (16.10).

$$\overline{V}_B = 10.47 e^{i(120°)} = 10.47 \text{ ft/sec } \underline{/120°}$$

$$\overline{V}_C = -6.18 e^{i(246.2°)} = 6.18 \text{ ft/sec } \underline{/66.2°}$$

$$\overline{V}_{B/C} = 8.45 e^{i(156.2°)} = 8.45 \text{ ft/sec } \underline{/156.2°}$$

4. Determine constants A_1, A_2, B_1, B_2, C_1, and C_2 using Eqs. (16.18) through (16.23).

$$A_1 = \cos(156.2° + 90°)$$

$$= -0.4033$$

$$A_2 = \sin(156.2° + 90°)$$

$$= -0.9151$$

$$B_1 = \cos(156.2°)$$

$$= -0.9151$$

$$B_2 = \sin(156.2°)$$

$$= 0.4033$$

$$C_1 = -62.83^2(0.1666) \cos 30° + 0$$

$$- 2(8.45) \left(-\frac{6.18}{0.21} \right) \cos(156.2° + 90°)$$

$$+ \frac{(-6.18)^2}{0.21} \sin 156.2°$$

$$= -942.95$$

$$C_2 = -62.83^2(0.1666) \sin 30° + 0$$

$$- 2(8.45) \left(-\frac{6.18}{0.21} \right) \sin(156.2° + 90°)$$

$$+ \frac{(-6.18)^2}{0.21} \sin 156.2°$$

$$= -716.92$$

5. Determine A_C^T and $A_{B/C}^T$ using Eqs. (16.16) and (16.17).

$$A_C^T = \frac{(-942.95)(0.40) - (-716.92)(-0.92)}{(-0.40)(0.40) - (-0.92)(-0.92)}$$

$$= 1036.3 \text{ ft/sec}^2$$

$$A_{B/C}^T = \frac{(-0.40)(-716.92) - (-0.92)(-942.95)}{(-0.40)(0.40) - (-0.92)(-0.92)}$$

$$= 573.8 \text{ ft/sec}^2$$

6. Substitute the values found for A_C^T and $A_{B/C}^T$ into Eqs. (16.25) and (16.26) and solve for $\overline{A}_C$ and $\overline{A}_{B/C}$.

$$\overline{A}_C = -\frac{(-6.18)^2}{0.21} e^{i(156.2°)} + 1036.3 e^{i(246.2°)}$$

$$= 185.2 \underline{/-23.8°} + 1036.3 \underline{/-113.8°}$$

$$= 1052.7 \text{ ft/sec}^2 \underline{/-103.7°}$$

$$\overline{A}_{B/C} = \frac{2(8.45)(-6.18)}{0.207} e^{i(246.2°)} + 573.8 e^{i(156.2°)}$$

$$= 505.7 \underline{/66.2°} + 573.8 \underline{/156.2°}$$

$$= 764.8 \text{ ft/sec}^2 \underline{/114.8°}$$

See Figure 16.7 for velocity and acceleration profiles for complete crank cycle from 0 to 360°.

EXAMPLE 16.2: Whitworth Mechanism (Figure 16.8)

Considering the Whitworth mechanism in Figure 16.1, let $AB = 3$ in., $AD = 2$ in., $\theta_2 = 60°$, $\omega_2 = 30$ rad/sec (counterclockwise), and $\alpha_2 = 0$. It is required to find $\overline{V}_C$, $\overline{V}_{B/C}$, $\overline{A}_C$, and $\overline{A}_{B/C}$.

SOLUTION

1. Determine θ_3 using Eqs. (16.4), (16.2), and (16.1).

$$CD = [0.25^2 + 0.166^2 - 2(0.25)(0.166) \cos 60°]^{1/2}$$

$$= 0.22 \text{ ft}$$

$$\phi_3 = \cos^{-1} \frac{0.22^2 + 0.166^2 - 0.25^2}{2(0.22)(0.166)} = 79.1°$$

$$\theta_3 = 180° - 79.1°$$

$$= 100.9°$$

2. Determine V_B, V_C, and $V_{B/C}$ using Eqs. (16.5), (16.6), and (16.7).

$$V_B = 30(0.25) = 7.5 \text{ ft/sec}$$

$$V_C = 30(0.25) \cos(100.9° - 60°) = 5.67 \text{ ft/sec}$$

$$V_{B/C} = 30(0.25) \sin(100.9° - 60°) = 4.9 \text{ ft/sec}$$

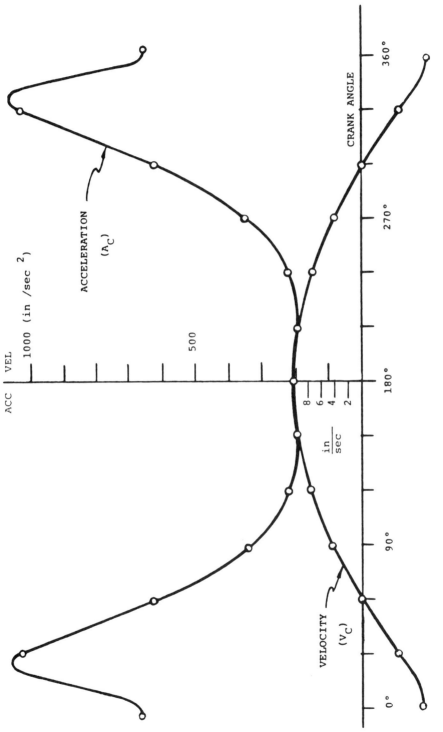

Figure 16.7   Shaper mechanism characteristics: Example 16.1.

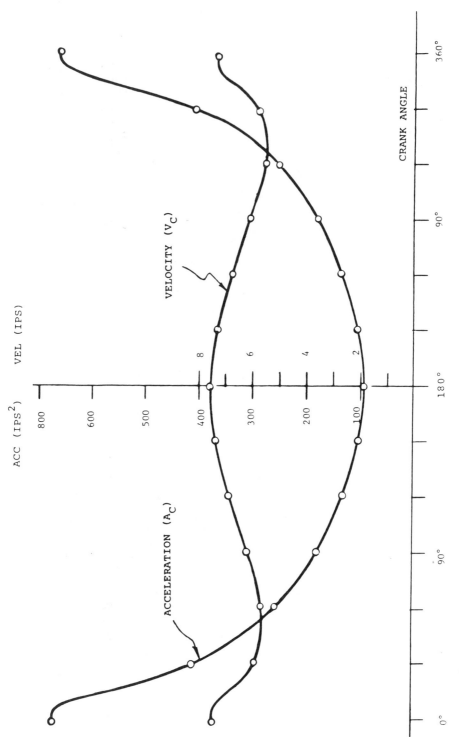

Figure 16.8   Whitworth mechanism characteristics: Example 16.2.

3. Determine $\overline{V}_B$, $\overline{V}_C$, and $\overline{V}_{B/C}$ using Eqs. (16.8), (16.9), and (16.10).

$$\overline{V}_B = 7.5e^{i150°} = 7.5 \text{ ft/sec } \underline{/150°}$$

$$\overline{V}_C = 5.67e^{i190.1°} = 5.67 \text{ ft/sec } \underline{/-169.1°}$$

$$\overline{V}_{B/C} = 4.9e^{i100.9°} = 4.9 \text{ ft/sec } \underline{/100.9°}$$

4. Determine constants A_1, A_2, B_1, B_2, C_1, and C_2 using Eqs. (16.18) through (16.23).

$$A_1 = \cos(100.9° + 90°)$$
$$= -0.98$$

$$A_2 = \sin(100.9° + 90°)$$
$$= -0.19$$

$$B_1 = \cos 100.9°$$
$$= -0.19$$

$$B_2 = \sin 100.9°$$
$$= 0.98$$

$$C_1 = -30^2(3) \cos(60°) + 0(3) \cos(60° + 90°)$$
$$- 2(4.9) \left(\frac{5.67}{0.22}\right) \cos(100.9° + 90°)$$
$$+ \left(\frac{5.67^2}{0.22}\right) \cos(100.9°)$$
$$= 107.9°$$

$$C_2 = -30^2(3) \sin(60°) + 0(3) \sin(60° + 90°)$$
$$- 2(4.9) \left(\frac{5.67}{0.22}\right) \sin(100.9° + 90°)$$
$$+ \left(\frac{5.67^2}{0.22}\right) \sin(100.9°)$$
$$= -3.96$$

5. Determine A_C^T and $A_{B/C}^T$ using Eqs. (16.16) and (16.17).

$$A_C^T = \frac{(107.94)(0.98) - (-3.96)(-0.19)}{(-0.98)(0.98) - (-0.19)(-0.19)}$$

$$= -105.3 \text{ ft/sec}^2$$

$$A_{B/C}^T = \frac{(-0.98)(-3.96) - (-0.19)(107.94)}{(-0.98)(0.98) - (-0.19)(-0.19)}$$

$$= -24.25 \text{ ft/sec}^2$$

6. Substitute the values found for A_C^T and $A_{B/C}^T$ into Eqs. (16.25) and (16.26) and solve for $\overline{A}_C$ and $\overline{A}_{B/C}$.

$$\overline{A}_C = -\frac{5.67^2}{0.22} e^{i(100.9°)} + (-105.3)e^{i(190.9°)}$$

$$= 145.9 \underline{/-79.1°} + 105.3 \underline{/10.9°}$$

$$= 179.9 \text{ ft/sec}^2 \underline{/-43.3°}$$

$$\overline{A}_{B/C} = \frac{2(4.9)(5.67)}{0.22} e^{i(190.9°)} + (-24.25)e^{i(100.9°)}$$

$$= 252.5 \underline{/-169.1°} + 24.25 \underline{/-79.1°}$$

$$= 253.7 \text{ ft/sec}^2 \underline{/-163.6°}$$

See Figure 16.8 for velocity and acceleration profiles for complete crank cycle from 0 to 360°.

17

Sliding Coupler Mechanism Analysis: Simplified Vector Method

17.1 INTRODUCTION

The sliding coupler mechanism, represented by ABC in Figure 17.1, is an important class of the slider crank chain in which the connecting rod or coupler BC (link 3) slides through a cylinder or block that is free to rotate, via trunnions, about a fixed axis. The mechanism is generally found in two basic arrangements: an *oscillating cylinder* (or rocking block) type, where the crank arm AB is shorter than the base AC, and the *rotating cylinder* (or rotating block type), where the crank arm is longer than the base. In the first arrangement, the cylinder (or block) oscillates as the crank arm rotates, whereas in the second arrangement, the cylinder (or block) rotates with the crank.

Typical applications for the sliding coupler include steam engines, some pumps and compressors, hydraulic actuators (such as front-end loaders), variable-speed, indexing drives, and various compound mechanisms (see Figures 17.2 to 17.9).

Motion characteristics of the sliding coupler mechanism are usually difficult to analyze because of the complexity of the coupler and slider motions that give rise to the Coriolis acceleration. The simplified method presented here quickly determines linear velocity and acceleration relationships for the sliding coupler mechanism for any angular position of the crank cycle.

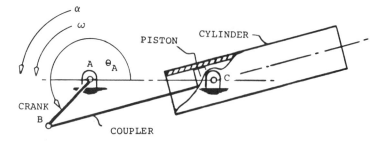

Figure 17.1   Sliding coupler mechanism.

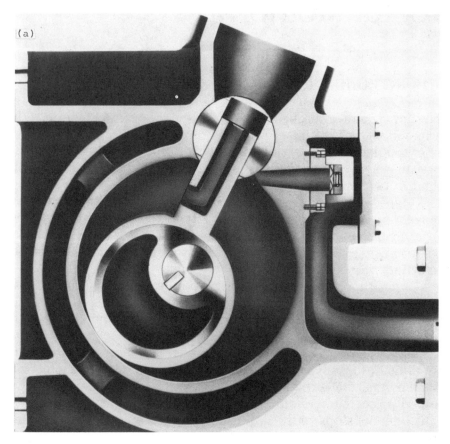

Figure 17.2   (a) Compressor. (Courtesy of Kinney Vacuum, Unit of General Signal).

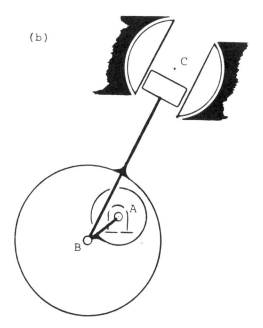

Figure 17.2 (Continued) (b) Mechanism.

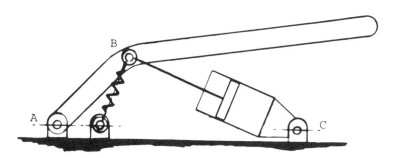

Figure 17.3 Application: Foot pump.

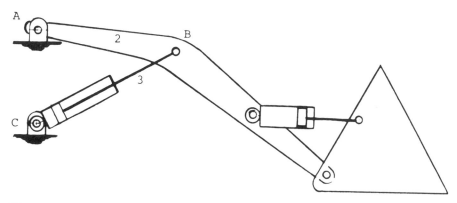

Figure 17.4 Loader actuator.

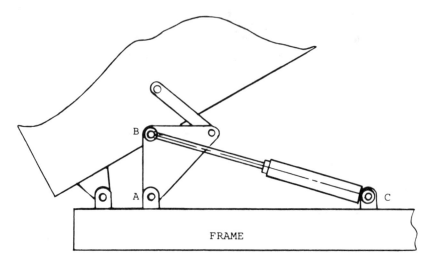

Figure 17.5   Dump truck.

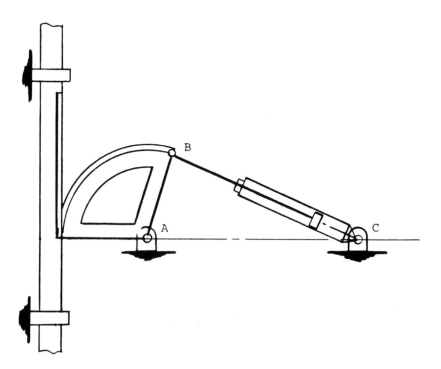

Figure 17.6   Rack and pinion mechanism.

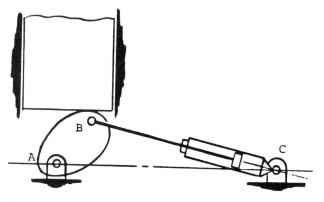

Figure 17.7   Cam-follower mechanism.

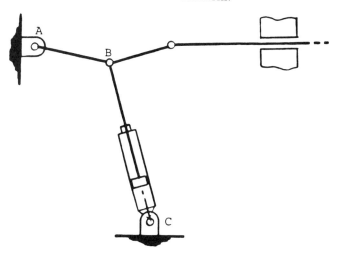

Figure 17.8   Toggle mechanism.

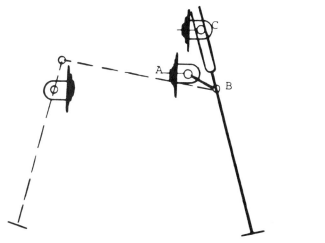

Figure 17.9   Walking mechanism. (From A. S. Hall, Kinematics and Linkage Design, Prentice Hall, Englewood Cliffs, N.J., 1961).

17.2 SCOPE AND ASSUMPTIONS

The sliding coupler mechanism ABC in Figure 17.1 consists of a crank AB of fixed length, coupler BC of variable length, and slider C that is constrained to move within a cylinder pivoted at a fixed point C. Given that the crank rotates with an angular velocity ω_A and angular acceleration α_A (both in a counterclockwise direction), we will now derive general expressions to compute linear velocities and accelerations of points A, B, and C relative to B for any angular position θ_A of the crank arm. All angular displacements, velocities, and accelerations are considered positive for counterclockwise rotation and negative for positive rotation.

17.3 GEOMETRIC RELATIONSHIPS

First, we determine the angel θ_C of the coupler BC in terms of crank angle θ_A. Considering triangle ABC in Figure 17.10, we note that

$$BC = [AB^2 + AC^2 - 2(AB)(AC) \cos \theta_A]^{1/2} \qquad (17.1)$$

$$\theta_C = 180° + \phi_C \qquad (17.2)$$

where it can be shown that

$$\phi_C = -\cos^{-1} \frac{BC^2 + AC^2 - AB^2}{2(BC)(AC)} \quad (\theta_A \le 180°)$$

$$\phi_C = \cos^{-1} \frac{BC^2 + AC^2 - AB^2}{2(BC)(AC)} \quad (\theta_A > 180°) \qquad (17.3)$$

17.4 VELOCITY ANALYSIS

To determine the relationships for velocities $\overline{V}_B$, $\overline{V}_C$, and $\overline{V}_{C/B}$, let Bcb in Figure 17.11 represent the velocity polygon of the mechanism ABC, where

Bb = magnitude of linear velocity of point B $\overline{V}_B$ (V_B).
Bc = magnitude of linear velocity of point C $\overline{V}_C$ (V_C).
bc = magnitude of linear velocity of C relative to B $\overline{V}_{C/B}$ ($V_{C/B}$).

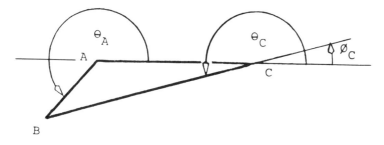

Figure 17.10   Geometric relationships.

Applying the **Rule of Sines** from trigonometry, the velocity magnitudes V_B, V_C, and $V_{C/B}$ can be determined in terms of angular positions of the links as follows:

$$\frac{V_B}{\sin 90°} = \frac{V_C}{\sin(\theta_A - \theta_C)} = \frac{V_{C/B}}{\sin(90° - \theta_A + \theta_C)} \tag{17.4}$$

where

$$\sin(\theta_A - \theta_C) = \sin(\theta_A - \phi_C - 180°)$$

$$= -\sin(180° - \theta_A + \phi_C)$$

$$= -\sin(\theta_A - \phi_C) \tag{17.5}$$

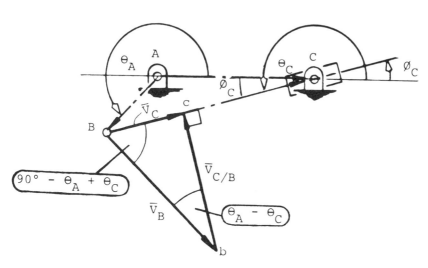

Figure 17.11   Velocity analysis.

$$\sin(90° - \theta_A + \theta_C) = \sin(90° - \theta_A + \phi_C + 180°)$$
$$= \sin(270° - \theta_A + \phi_C)$$
$$= -\sin(90° - \theta_A + \phi_C) \qquad (17.6)$$

Applying Eqs. (17.5) and (17.6) to (17.4) yields

$$\frac{V_B}{\sin 90°} = \frac{V_C}{-\sin(\theta_A - \phi_C)} = \frac{V_{C/B}}{-\sin(90° - \theta_A + \phi_C)} \qquad (17.7)$$

where

$$V_B = AB\omega \qquad (17.8)$$

$$V_C = -\frac{V_B \sin(\theta_A - \phi_C)}{\sin 90°} = -AB\omega \sin(\theta_A - \phi_C) \qquad (17.9)$$

$$V_{C/B} = -\frac{V_B \sin(90° - \theta_A + \phi_C)}{\sin 90°}$$
$$= -AB\omega \cos(\theta_A - \phi_C) \qquad (17.10)$$

An inspection of the velocity polygon shows that $\overline{V}_B$ and $\overline{V}_{C/B}$ are oriented, respectively, at angle $(\theta_A + 90°)$ and $(\phi_C + 90°)$, whereas the direction of $\overline{V}_C$ is constrained to that along the connecting rod BC, at angle ϕ_C. Therefore, the velocity vectors can be expressed in exponential form as follows:

$$\overline{V}_B = AB\omega e^{i(\theta_A + 90°)} \qquad (17.11)$$

$$\overline{V}_C = -AB\omega \sin(\theta_A - \phi_C)e^{i\phi_C} \qquad (17.12)$$

$$\overline{V}_{C/B} = -AB\omega \cos(\theta_A - \phi_C)e^{i(\phi_c + 90°)} \qquad (17.13)$$

17.5 ACCELERATION ANALYSIS

To determine the acceleration relationships, we apply the relative motion equation, which states that

$$\overline{A}_C = \overline{A}_B + \overline{A}_{C/B} \qquad (17.14)$$

Expanding this equation into its normal and tangential form, we obtain

$$\overline{A}_C^N + \overline{A}_C^T + \overline{A}_C^{Cor} = \overline{A}_B^N + \overline{A}_B^T + \overline{A}_{C/B}^N + \overline{A}_{C/B}^T \qquad (17.15)$$

where

$$\overline{A}_C^N = -\frac{V_C^2}{\infty}\, e^{i(\phi_C + 90°)} = 0$$

$$\overline{A}_C^T = A_C^T e^{i\phi_C}$$

$$\overline{A}_C^{Cor} = 2V_C\, \frac{V_{C/B}}{BC}\, e^{i(\phi_C + 90°)}$$

$$\overline{A}_B^N = -\omega^2 ABe^{i\theta_A}$$

$$\overline{A}_B^T = \alpha ABe^{i(\theta_A + 90°)}$$

$$\overline{A}_{C/B}^N = -\frac{V_{C/B}^2}{BC}\, e^{i\phi_C}$$

$$\overline{A}_{C/B}^T = A_{C/B}^T e^{i(\phi_C + 90°)}$$

As in the case of the quick-return mechanism, since the slider C is constrained to move relative to a rotating guide, the cylinder, it is necessary to determine Coriolis acceleration $\overline{A}_C^{Cor}$ resulting from this motion. In this case, for Coriolis acceleration, we consider the linear velocity of the slider or V_C and the angular velocity of the guide (or connecting rod), namely, ω_{CB} or $V_{C/B}/CB$. Also, Coriolis acceleration has the orientation of the $V_{C/B}$ vector when rotated 90° about its tail in the direction of ω_{CB}.

Expansion of Eq. (17.15) yields

$$A_C^T e^{i\phi_C} = -\omega^2 ABe^{i\theta_A} + \alpha ABe^{i(\theta_A + 90°)} - \frac{V_{C/B}^2}{BC}\, e^{i\phi_C}$$

$$+ A_{C/B}^T e^{i(\phi_C + 90°)} - 2V_C\, \frac{V_{C/B}}{BC}\, e^{i(\phi_C + 90°)} \qquad (17.16)$$

Equation (17.15) contains two unknown quantities: the magnitudes A_C^T and $A_{C/B}^T$. All other magnitudes and unit vectors are either known or can be determined from the geometry and operating characteristics of the mechanism. Although the directions for the vectors $\overline{A}_C^T$ and $\overline{A}_{C/B}^T$ are not precisely known, they can be assumed for convenience to

be positive. If any of these directions are reversed, the numerical solution of the equation will automatically produce a negative sign, indicating the correct direction. To solve for the unknowns, therefore, we
equate the real and imaginary parts of the equation and rearrange as
follows:

Real:

$$A_C^T \cos \theta_C - A_{C/B}^T \cos(\phi_C + 90°) = -\omega^2 AB \cos \theta_A$$

$$+ \alpha AB \cos(\theta_A + 90°) - \frac{V_{C/B}^2}{BC} \cos \phi_C - 2V_C \frac{V_{C/B}}{BC} \cos(\phi_C + 90°)$$

$$(17.17)$$

Imaginary:

$$A_C^T \sin \phi_C - A_{C/B}^T \sin(\phi_C + 90°) = -\omega^2 AB \sin \theta_A$$

$$+ \alpha AB \sin(\theta_A + 90°) - \frac{V_{C/B}^2}{BC} \sin \phi_C - 2V_C \frac{V_{C/B}}{BC} \sin(\phi_C + 90°)$$

$$(17.18)$$

Solution of the simultaneous equations (17.17) and (17.18) is obtained
by

$$A_C^T = \frac{C_1 B_2 - C_2 B_1}{A_1 B_2 - A_2 B_1} \qquad (17.19)$$

$$A_{C/B}^T = \frac{A_1 C_2 - A_2 C_1}{A_1 B_2 - A_2 B_1} \qquad (17.20)$$

where

$$A_1 = \cos \phi_C \qquad (17.21)$$

$$A_2 = \sin \phi_C \qquad (17.22)$$

$$B_1 = -\cos(\phi_C + 90°) \qquad (17.23)$$

$$B_2 = -\sin(\phi_C + 90°) \qquad (17.24)$$

$$C_1 = -\omega^2 AB \cos \theta_A + \alpha AB \cos(\theta_A + 90°)$$

$$- \frac{V^2_{C/B}}{BC} \cos \phi_C - 2V_C \frac{V_{C/B}}{BC} \cos(\phi_C + 90°) \qquad (17.25)$$

$$C_2 = -\omega^2 AB \sin \theta_A + \alpha AB \sin(\theta_A + 90°)$$

$$- \frac{V^2_{C/B}}{BC} \sin \phi_C - 2V_C \frac{V_{C/B}}{BC} \sin(\phi_C + 90°) \qquad (17.26)$$

Given that the values V_C, $V_{C/B}$, A^T_C, and $A^T_{C/B}$ have been determined, the general equations for computing the linear accelerations can be summarized as follows:

$$\overline{A}_B = -\omega^2 AB e^{i\theta_A} + \alpha AB e^{i(\theta_A + 90°)} \qquad (17.27)$$

$$\overline{A}_C = 2V_C \frac{V_{C/B}}{BC} e^{i(\phi_C + 90°)} + A^T_C e^{i\phi_C} \qquad (17.28)$$

$$\overline{A}_{C/B} = -\frac{V^2_{C/B}}{BC} e^{i\phi_C} + A^T_{C/B} e^{i(\phi_C + 90°)} \qquad (17.29)$$

EXAMPLE 17.1: Oscillating Cylinder

Considering the oscillating cylinder mechanism in Figure 17.1, let $AB = 8$ in., $AC = 10$ in., $\theta_A = 120°$, $\omega_A = 18$ rad/sec (counterclockwise), and $\alpha_A = 0$. It is required to find $\overline{V}_C$, $\overline{V}_{C/B}$, $\overline{A}_C$, and $\overline{A}_{C/B}$.

SOLUTION[a]

1. Determine ϕ_C using Eqs. (17.1) and (17.2).

$$BC = [0.67^2 + 0.83^2 - 2(0.67)(0.83) \cos 120°]^{1/2}$$

$$= 1.30 \text{ ft}$$

$$\phi_C = -\cos^{-1} \frac{1.3^2 + 0.83^2 - 0.67^2}{2(1.3)(0.83)}$$

$$= -26.33°$$

---

[a]See Figure 17.12 for velocity and acceleration profiles for complete crank cycle from 0 to 360°.

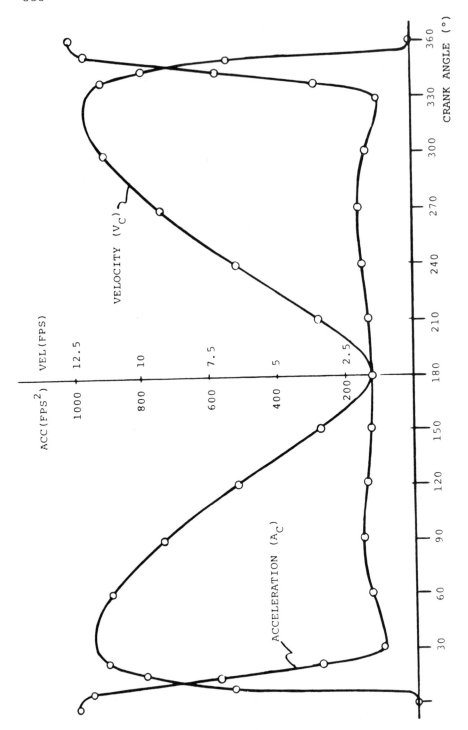

Figure 17.12 Oscillating cylinder characteristics: Example 17.1.

2. Determine V_B, V_C, and $V_{C/B}$ using Eqs. (17.8), (17.9), and (17.10).

$$V_B = 0.67(18) = 12 \text{ ft/sec}$$

$$V_C = -0.67(18) \sin[120° - (-26.33°)] = -6.65 \text{ ft/sec}$$

$$V_{C/B} = -(0.67)(18) \cos[120° - (-26.33°)] = 9.98 \text{ ft/sec}$$

3. Determine $\overline{V}_B$, $\overline{V}_C$, $\overline{V}_{C/B}$ using Eqs. (17.11), (17.12), and (17.13).

$$\overline{V}_B = 12.0e^{i(120° + 90°)} = 12.0 \text{ ft/sec} \underline{/-150°}$$

$$\overline{V}_C = -6.65e^{i(-26.3°)} = 6.65 \text{ ft/sec} \underline{/153.6°}$$

$$\overline{V}_{C/B} = 9.98e^{i(-26.3° + 90°)} = 9.98 \text{ ft/sec} \underline{/63.7°}$$

4. Determine constants A_1, A_2, B_1, B_2, C_1, and C_2 using Eqs. (17.21) through (17.26).

$$A_1 = \cos(-26.3°) = 0.89$$

$$A_2 = \sin(-26.3°) = -0.44$$

$$B_1 = -\cos(-26.3° + 90°) = -0.44$$

$$B_2 = -\sin(-26.3° + 90°) = -0.89$$

$$C_1 = -(18^2)(0.67) \cos 120° + 0(0.67) \cos(120° + 90°)$$
$$- \frac{9.98^2}{1.3} \cos(-26.3°) - 2(-6.65)\left(\frac{9.98}{1.3}\right)$$
$$\times \cos(-26.3'' + 90°) = 84.61$$

$$C_2 = -(18^2)(0.67) \sin 120° + 0(0.67) \sin(120° + 90°)$$
$$- \frac{9.98^2}{1.3} \sin(-26.3°) - 2(-6.65)\left(\frac{9.98}{1.3}\right)$$
$$\times \sin(-26.3° + 90°) = -61.58$$

5. Determine A_C^T and $A_{C/B}^T$ using Eqs. (17.19) and (17.20).

$$A_C^T = \frac{(84.61)(-0.89) - (-61.58)(0.44)}{(0.89)(-0.89) - (-0.44)(-0.44)}$$
$$= 103.14 \text{ ft/sec}^2$$

$$A_{C/B}^T = \frac{(0.89)(-61.58) - (-0.44)(84.61)}{(0.89)(-0.89) - (-0.44)(-0.44)}$$

$$= 17.66 \text{ ft/sec}^2$$

6. Substitute the values found for A_C^T and $A_{C/B}^T$ into Eqs. (17.28) and (17.29) and solve for $\overline{A}_C$ and $\overline{A}_{C/B}$.

$$\overline{A}_C = \frac{2(-6.65)(9.98)}{1.3} e^{i(-26.3° + 90°)} + 103.14 e^{i(-26.3°)}$$

$$= 102.1 \underline{/-116.3°} + 103.14 \underline{/-26.3°}$$

$$= 145.12 \text{ ft/sec}^2 \underline{/-71.04°}$$

$$\overline{A}_{C/B} = -\frac{9.98^2}{1.3} e^{i(-26.3°)} + 17.66 e^{i|-26.3° + 90°}$$

$$= 76.6 \underline{/153.6°} + 17.66 \underline{/63.7°}$$

$$= 78.64 \text{ ft/sec}^2 \underline{/140.7°}$$

EXAMPLE 17.2: Rotating Cylinder

Considering the rotating cylinder version of the mechanism in Figure 17.1, let $AB = 3$ in., $AC = 2$ in., $\theta_A = 210°$, $\omega_A = 30$ rad/sec (counterclockwise), and $\alpha_A = 0$. It is required to find $\overline{V}_C$, $\overline{V}_{C/B}$, $\overline{A}_C$, and $\overline{A}_{C/B}$.

SOLUTION

1. Determine ϕ_C using Eqs. (17.1) and (17.3).

$$BC = [3.0^2 + 2.0^2 - 2(3.0)(2.0) \cos 210°]^{1/2}$$

$$= 4.8 \text{ in.}$$

$$\phi_C = \cos^{-1} \frac{4.8^2 + 2.0^2 - 3.0^2}{2(4.8)(2.0)}$$

$$= 18.06°$$

2. Determine magnitudes V_B, V_C, and $V_{C/B}$ using Eqs. (17.8), (17.9), and (17.10).

$$V_B = 3.0(30) = 90.0 \text{ in./sec}$$

$$V_C = -3.0(30) \sin(210° - 18.06°) = 18.6 \text{ in./sec}$$

$$V_{C/B} = -3.0(30) \cos(210° - 18.06°) = 88.06 \text{ in./sec}$$

3. Determine complete vectors $\overline{V}_B$, $\overline{V}_C$, and $\overline{V}_{C/B}$ using Eqs. (17.11), (17.12), and (17.13).

$$\overline{V}_B = 90e^{i(210° + 90°)} = 90 \text{ in./sec} \underline{/-60°}$$

$$\overline{V}_C = 18.6e^{i(18.06°)} = 18.6 \text{ in./sec} \underline{/18.06°}$$

$$\overline{V}_{C/B} = 88.06e^{i(18.06° + 90°)} = 88.06 \text{ in./sec} \underline{/108.6°}$$

4. Determine constants A_1, A_2, B_1, B_2, C_1, and C_2 using Eqs. (17.21) through (17.26).

$$A_1 = \cos(18.06°) = 0.95$$

$$A_2 = \sin(18.06°) = 0.31$$

$$B_1 = -\cos(18.06° + 90°) = 0.31$$

$$B_2 = -\sin(18.06° + 90°) = -0.95$$

$$C_1 = -(30.0)^2(3.0) \cos 210° + 0(3.0) \cos (210° + 90°)$$

$$-\frac{88.06^2}{4.8} \cos(18.06°) - 2(18.6) \left(\frac{88.06}{4.8}\right)$$

$$\times \cos(18.06° + 90°)$$

$$= 1024.3$$

$$C_2 = -(30.0)^2(3.0) \sin 210° + 0(3.0) \sin(210° + 90°)$$

$$-\frac{88.06^2}{4.8} \sin(18.06°) - 2(18.6) \left(\frac{88.06}{4.8}\right)$$

$$\times \sin(18.06° + 90°)$$

$$= 208.64$$

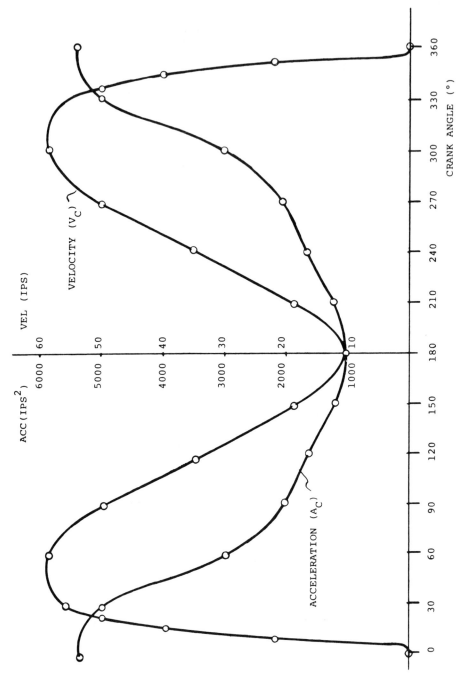

Figure 17.13   Rotating cylinder characteristics: Example 17.2.

5. Determine A_C^T and $A_{C/B}^T$ using Eqs. (17.19) and (17.20).

$$A_C^T = \frac{(1024.3)(-0.95) - (208.6)(0.31)}{(0.95)(-0.95) - (0.31)(0.31)}$$

$$= 1038.5 \text{ in./sec}^2$$

$$A_{C/B}^T = \frac{(0.95)(208.6) - (0.31)(1024.3)}{(0.95)(-0.95) - (0.31)(0.31)}$$

$$= 119.32 \text{ in./sec}^2$$

6. Substitute the values found for A_C^T and $A_{C/B}^T$ into Eqs. (17.28) and (17.29) and solve for $\overline{A}_C$ and $\overline{A}_{C/B}$.

$$\overline{A}_C = \frac{2(18.6)(88.06)}{4.8} e^{i(18.06° + 90°)} + 1038.5 e^{i(18.06°)}$$

$$= 677.57 \,\underline{/108.07°} + 1038.5 \,\underline{/18.06°}$$

$$= 1240 \text{ in./sec}^2 \,\underline{/51.18°}$$

$$\overline{A}_{C/B} = -\frac{88.06^2}{4.8} e^{i(18.06°)} + 119.32 e^{i(18.06° + 90°)}$$

$$= 1603.15 \,\underline{/-161.9°} + 119.32 \,\underline{/108.06°}$$

$$= 1607.58 \text{ in./sec}^2 \,\underline{/166.18°}$$

See Figure 17.13 for velocity and acceleration profiles for complete crank cycle for 0 to 360°.

18

Slider-Crank Mechanism Analysis: Modified Vector Method

18.1 INTRODUCTION

In the slider-crank analysis by the simplified vector method (Chapter 15), the motion relationships were obtained completely by applying relative motion principles. There the velocity relationships were derived with the aid of the velocity polygon and the acceleration relationships were derived by expressing the relative acceleration equation in terms of the velocity relationships. In this analysis, the procedure is basically the same except that the acceleration equations are not obtained by relative motion, but instead by differentiation of the velocity relationships.

18.2 SCOPE AND ASSUMPTIONS

Let ABC in Figure 18.1 represent a typical slider-crank mechanism where the crank AB rotates counterclockwise at an angular velocity ω about joint A, and θ_A is its instantaneous angular position away from **top dead center**. ϕ_C is the angle between the slider arm BC and line AC.

It is required to develop linear velocity and acceleration relationships for points B and C and also for point B relative to point C at any position of the crank. It is assumed that the angular velocity ω

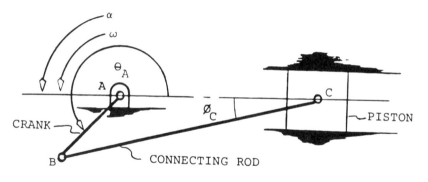

Figure 18.1   Slider-crank analysis: Mechanism.

for counterclockwise rotation is positive and for clockwise rotation negative.

18.3   GEOMETRIC CONSIDERATIONS

As in previous analysis, we first seek to determine the angle of the connecting rod θ_A. There it was shown that

$$\theta_C = 180° + \phi_C$$

where

$$\phi_C = -\sin^{-1} \frac{AB}{BC} \sin \theta_A$$

or

$$\theta_C = 180 - \phi_C$$

where

$$\phi_C = \sin^{-1} \frac{AB}{BC} \sin \theta_A$$

18.4 VELOCITY ANALYSIS

Let Bcb be the velocity polygon for the mechanism ABC (Figure 18.2).
In Section 15.4 it was shown that the scalar velocity equation for a
typical slider slider crank may be expressed as

$$\frac{V_B}{\sin(90° - \theta_C)} = \frac{V_C}{\sin(\theta_C - \theta_A)} = \frac{V_{B/C}}{\sin(90° - \theta_A)} \tag{18.1}$$

where

        V_B = magnitude of linear velocity of point B

        V_C = magnitude of linear velocity of point C

      $V_{B/C}$ = magnitude of linear velocity of B relative to C

Then we would obtain

$$\sin(90° - \theta_C) = \sin(90° - 180° + \phi_C)$$

$$= -\sin(90° - \phi_C)$$

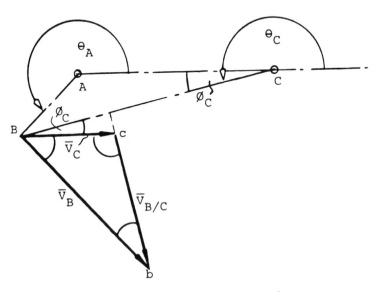

Figure 18.2   Slider-crank analysis: Velocity polygon.

$$\sin(\theta_C - \theta_A) = \sin(180° - \phi_C - \theta_A)$$

$$= \sin(\phi_C + \theta_A)$$

Hence, Eq. (18.1) can be rewritten as

$$\frac{V_B}{-\sin(90 - \phi_C)} = \frac{V_C}{\sin(\phi_C + \theta_A)} = \frac{V_{B/C}}{\sin(90 - \theta_A)}$$

from which

$$V_B = AB\omega$$

$$V_C = -V_B \frac{\sin(\phi_C + \theta_A)}{\sin(90° - \phi_C)}$$

$$V_{B/C} = -V_B \frac{\sin(90° - \theta_A)}{\sin(90° - \phi_C)}$$

In vector form (magnitudes and directions considered), these equations can be expressed in polar form as follows:

$$\overline{V}_B = -V_B(\sin \theta_A - i \cos \theta_A)$$

$$\overline{V}_C = V_C \cos 0°$$

$$\overline{V}_{B/C} = \overline{V}_B - \overline{V}_C$$

where $-(\sin \theta_A - i \cos \theta_A)$ and $\cos 0°$ are unit vectors used to orient the vectors $\overline{V}_B$ and $\overline{V}_C$.

Finally, we obtain the following:

$$\overline{V}_B = -AB\omega(\sin \theta_A - i \cos \theta_A) \tag{18.2}$$

$$\overline{V}_C = -AB\omega \frac{\sin(\phi_C + \theta_A)}{\sin(90° - \phi_C)} \tag{18.3}$$

$$\overline{V}_{B/C} = -AB\omega \left[\sin \theta_A - \frac{\sin(\phi_C + \theta_A)}{\sin(90° - \phi_C)} - i \cos \theta_A \right] \tag{18.4}$$

18.5   ACCELERATION ANALYSIS

With the velocity equations for the mechanism determined, the corresponding acceleration expressions are obtained by differentiating these equations with respect to time, noting that $\theta_A = \omega \times t$.

Acceleration of B ($\overline{A}_B$)

$$\overline{A}_B = \frac{d\overline{V}_B}{dt}$$

Using Eq. (18.2), we have

$$\overline{V}_B = -AB \times \omega(\sin \theta_A - i \cos \theta_A)$$

$$\overline{A}_B = -AB \times \omega^2(\cos \theta_A + i \sin \theta_A)$$

$$\qquad - AB \times \dot{\omega}(\sin \theta_A - i \cos \theta_A) \qquad\qquad (18.5)$$

where the first and second terms on the right-hand side represent the normal and tangential accelerations, respectively, of point B, and $\dot{\omega}$ (or α) is the angular acceleration of the same point.

Acceleration of C ($\overline{A}_C$)

$$\overline{A}_C = \frac{d\overline{V}_C}{dt}$$

Using Eq. (18.3), we have

$$\overline{V}_C = -AB \times \omega \frac{\sin(\theta_A + \phi_C)}{\sin(90° - \phi_C)}$$

and if we substitute for ϕ_C, it follows that

$$\overline{V}_C = -AB \times \omega \left[\sin \theta_A + \frac{AB \cos \theta_A \sin \theta_A}{(BC^2 - AB^2 \sin^2 \theta_A)^{1/2}} \right]$$

from which

$$\overline{A}_C = -AB \times \omega^2 \left[\cos \theta_A + \frac{AB(\cos^2 \theta_A - \sin^2 \theta_A)}{(BC^2 - AB^2 \sin^2 \theta_A)^{1/2}} \right.$$
$$\left. + \frac{AB^3 \cos^2 \theta_A \sin^2 \theta_A}{(BC^2 - AB^2 \sin^2 \theta_A)^{3/2}} \right]$$
$$- AB \times \dot{\omega} \left[\sin \theta_A + \frac{AB \cos \theta_A \sin \theta_A}{(BC^2 - AB^2 \sin^2 \theta_A)^{1/2}} \right] \quad (18.6)$$

Acceleration of B Relative to C $(\overline{A}_{B/C})$

$$\overline{A}_{B/C} = \overline{A}_B - \overline{A}_C$$

Therefore, using Eqs. (18.5) and (18.6), we have

$$\overline{A}_{B/C} = AB \times \omega^2 \left[\frac{AB(\cos^2 \theta_A - \sin^2 \theta_A)}{(BC^2 - AB^2 \sin^2 \theta_A)^{1/2}} \right.$$
$$\left. + \frac{AB^3 \cos^2 \theta_A \sin^2 \theta_A}{(BC^2 - AB^2 \sin^2 \theta_A)^{3/2}} - i \sin \theta_A \right]$$
$$+ AB \times \dot{\omega} \left[\frac{AB \cos \theta_A \sin \theta_A}{(BC^2 - AB^2 \sin^2 \theta_A)^{1/2}} + i \cos \theta_A \right]$$

$$(18.7)$$

Note that when there is no angular acceleration (or $\dot{\omega} = 0$), the second terms on the right-hand side of both Eqs. (18.6) and (18.7) vanish.

EXAMPLE 18.1

Let

$$AB = 1.5 \text{ in.}$$
$$BC = 3.0 \text{ in.}$$
$$\omega_{AB} = 1 \text{ rad/sec} \quad \text{(clockwise)}$$
$$\theta_A = 30°$$

Find $\overline{V}_C$, $\overline{V}_{B/C}$, $\overline{A}_C$, and $\overline{A}_{B/C}$.

SOLUTION

Using Eq. (18.3) we have

$$\overline{V}_C = -AB \times \omega \frac{\sin(\theta_A + \phi_C)}{\sin(90° - \phi_Z)}$$

where

$$\omega = -\omega_{AB} = -1 \text{ rad/sec}$$

$$\phi_C = \sin^{-1}\left(\frac{AB}{BC} \sin \theta_A\right)$$

$$= \sin^{-1}(0.5 \sin 30°) = 14.5°$$

$$\overline{V}_C = -1.5(-1) \frac{\sin(30° + 14.5°)}{\sin(90° - 14.5°)}$$

$$= 1.5 \left(\frac{0.70}{0.96}\right)$$

$$= 1.09 \text{ in./sec } \underline{/0°}$$

Using Eq. (18.4), we have

$$\overline{V}_{B/C} = -AB \times \omega \left[(\sin \theta_A - i \cos \theta_A) - \frac{\sin(\theta_A + \phi_C)}{\sin(90° - \phi_C)} \right]$$

$$= -AB \times \omega \left[(\sin 30° - i \cos 30°) - \frac{\sin 44.5°}{\sin 75.5°} \right]$$

$$= -1.5(-1)(0.5 - 0.866i - 0.729)$$

$$= 1.5(-0.229 - 0.866i)$$

$$= 1.34 \tan^{-1} \frac{866}{229} \text{ in./sec } \underline{/-104.5°}$$

Using Eq. (18.6), we have

$$
\begin{aligned}
\overline{A}_C &= -1.5(1) \left[\cos 30° + \frac{1.5(\cos^2 30° - \sin^2 30°)}{(3^2 - 1.5^2 \sin^2 30°)^{1/2}} \right. \\
&\quad \left. + \frac{1.5^3 \cos^2 30° \sin^2 30°}{(3^2 - 1.5^2 \sin^2 30°)^{3/2}} \right] \\
&= -1.5(0.866 + 0.258 + 0.0258) \\
&= 1.72 \text{ in./sec}^2 \, \underline{/180°}
\end{aligned}
$$

Using Eq. (18.7) we have

$$
\begin{aligned}
\overline{A}_{B/C} &= 1.5(1) \left[\frac{1.5(\cos^2 30° - \sin^2 30°)}{(3^2 - 1.5^2 \sin^2 30°)^{1/2}} \right. \\
&\quad \left. + \frac{1.5^3 \cos^2 30° \sin^2 30°}{(3^2 - 1.5^2 \sin^2 30°)^{3/2}} - i \sin 30° \right] \\
&= 1.5(0.289 - 0.5i) \\
&= 0.86 \tan^{-1} - \frac{500}{239} \text{ in./sec}^2 \, \underline{/-60.4°}
\end{aligned}
$$

EXAMPLE 18.2

Let

$$AB = 1.5 \text{ in.}$$

$$BC = 3.0 \text{ in.}$$

$$\omega_{AB} = 1 \text{ rad/sec} \quad \text{(counterclockwise)}$$

$$\theta_A = 120°$$

Find $\overline{V}_C$, $\overline{V}_{B/C}$, $\overline{A}_C$, and $\overline{A}_{B/C}$.

SOLUTION

Using Eq. (18.3), we have

$$
\overline{V}_C = -AB \times \omega \, \frac{\sin(\theta_A + \phi_C)}{\sin(90° - \phi_C)}
$$

where

$$\omega = +\omega_{AB} = 1 \text{ rad/sec}$$

$$\phi_C = \sin^{-1}\left(\frac{AB}{BC} \sin \theta_A\right)$$

$$= \sin^{-1}(0.5 \sin 120°) = 25.6°$$

$$\overline{V}_C = -AB \times \omega \frac{\sin(120° + 25.6°)}{\sin(90° - 25.6°)}$$

$$= -1.5(1)\left(\frac{0.565}{0.90}\right)$$

$$= 0.94 \text{ in./sec } \underline{/180°}$$

Using Eq. (18.4) we have

$$\overline{V}_{B/C} = -AB \times \omega \left[(\sin \theta_A - i \cos \theta_A) - \frac{\sin(\theta_A + \phi_C)}{\sin(90° - \phi_C)}\right]$$

$$= -AB \times \omega \left[(\sin 120° - i \cos 120°) - \frac{\sin 145.6°}{\sin 64.4°}\right]$$

$$= -1.5(1)\left(0.866 + 0.51 - \frac{0.565}{0.90}\right)$$

$$= 1.5(-0.239 - 0.5i)$$

$$= 0.830 \tan^{-1}\frac{500}{239} \text{ in./sec } \underline{/-115.6°}$$

Using Eq. (18.5), we have

$$\overline{A}_C = -1.5(1)\left[\cos 120° + \frac{1.5(\cos^2 120° - \sin^2 120°)}{(3^2 - 1.5^2 \sin^2 120°)^{1/2}}\right.$$

$$\left. + \frac{1.5^3 \cos^2 120° \sin^2 120°}{(3^2 - 1.5^2 \sin^2 120°)^{3/2}}\right]$$

$$= -1.5(-0.5 - 0.277 + 0.032)$$

$$= 1.12 \text{ in./sec}^2 \underline{/0°}$$

Using Eq. (18.7) we have

$$\overline{A}_{B/C} = 1.5(1) \left[\frac{1.5(\cos^2 120° - \sin^2 120°)}{(3^2 - 1.5^2 \sin^2 120°)^{1/2}} \right.$$

$$\left. + \frac{1.5^3 \cos^2 120° \sin^2 120°}{(3^2 - 1.5^2 \sin^2 120°)^{3/2}} - i \sin 120° \right]$$

$$= 1.5(-0.245 - 0.866i)$$

$$= 1.35 \tan^{-1} - \frac{866}{245} \text{ in./sec}^2 \underline{/-105.8°}$$

19
Slider-Crank Mechanism Analysis: Calculus Method

19.1 INTRODUCTION

An analytical approach, commonly used in analyzing the motion characteristics of a mechanism, consists of

1. Writing a mathematical expression to describe the displacement or
   position of the mechanism
2. Differentiating the displacement expression with respect to time to
   obtain the velocity expression
3. Differentiating the velocity expression with respect to time to obtain
   the acceleration expression

This method is illustrated with the familiar slider-crank mechanism.

19.2 SCOPE AND ASSUMPTIONS

Let the mechanism ABC in Figure 19.1 represent a typical slider-crank
where AB, in turn, represents the crank, BC the connecting rod, and C
the slider. Given that the angular displacement of the crank AB at any
instant in θ, we will now develop general expressions to compute the
linear displacement, velocity, and acceleration of the slider C in terms
of θ.

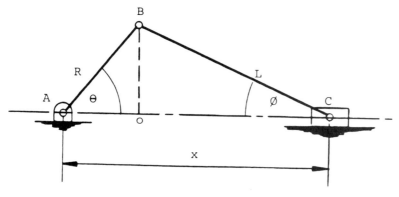

Figure 19.1   Slider-crank model.

19.3   DISPLACEMENT, VELOCITY, AND ACCELERATION ANALYSIS

Let

$$AB = R \quad \text{and} \quad BC = L$$

Then

$$x = R \cos \theta + L \cos \phi \tag{19.1}$$

$$\frac{dx}{dt} = -R \sin \theta \, \frac{d\theta}{dt} - L \sin \phi \, \frac{d\phi}{dt} \tag{19.2}$$

where

$$L \sin \phi = R \sin \theta \tag{19.3}$$

and

$$L \cos \phi \, \frac{d\phi}{dt} = R \cos \theta \, \frac{d\theta}{dt} \tag{19.4}$$

or

$$\frac{d\phi}{dt} = \frac{R \cos \theta}{L \cos \phi} \frac{d\theta}{dt} \tag{19.5}$$

Therefore, by substituting Eqs. (19.3) and (19.5) in (19.2), we obtain

$$\frac{dx}{dt} = -R \sin \theta \frac{d\theta}{dt} - R \sin \theta \frac{R \cos \theta}{L \cos \phi} \frac{d\theta}{dt} \tag{19.6}$$

$$= -R \frac{d\theta}{dt} \left(\sin \theta + \frac{R \sin \theta \cos \theta}{L \cos \phi} \right) \tag{19.7}$$

or

$$v = -R \frac{d\theta}{dt} \left(\sin \theta + \frac{R \sin 2\theta}{2L \cos \phi} \right) \tag{19.8}$$

Let

$$A = -R \frac{d\theta}{dt} \quad \text{and} \quad B = \sin \theta + \frac{R \sin 2\theta}{2L \cos \phi} \tag{19.9}$$

Then

$$\frac{d^2x}{dt^2} = B \frac{d}{dt} (A) + A \frac{d}{dt} (B)$$

$$= \left(\sin \theta + \frac{R \sin 2\theta}{2L \cos \phi} \right) \frac{d}{dt} \left(-R \frac{d\theta}{dt} \right)$$

$$+ \left(-R \frac{d\theta}{dt} \right) \frac{d}{dt} \left(\sin \theta + \frac{R \sin 2\theta}{2L \cos \phi} \right)$$

$$= -R \frac{d^2\theta}{dt^2} \left(\sin \theta + \frac{R \sin 2\theta}{2L \cos \phi} \right)$$

$$- R \frac{d\theta}{dt} \frac{d}{dt} \left(\sin \theta + \frac{R \sin 2\theta}{2L \cos \phi} \right) \tag{19.10}$$

Let

$$u = R \sin 2\theta \quad \text{and} \quad v = 2L \cos \phi \tag{19.11}$$

Then

$$\frac{d}{dt} \frac{R \sin 2\theta}{2L \cos \phi} = \frac{v \, du - u \, dv}{v^2}$$

$$= \frac{2L \cos \phi (2R \cos 2\theta) \, d\theta/dt - R \sin 2\theta (-2L \sin \phi) \, d\theta/dt}{4L^2 \cos^2 \phi}$$

$$= \frac{(4L \cos \phi \, R \cos 2\theta) \, d\theta/dt}{4L^2 \cos^2 \phi}$$

$$+ \frac{2R \sin 2\theta (R \sin \theta)}{4L^2 \cos^2 \phi} \frac{R \cos \theta}{L \cos \phi} \frac{d\theta}{dt}$$

$$= \left(\frac{R \cos 2\theta}{L \cos \phi} + \frac{R^3 \sin^2 2\theta}{4L^3 \cos^3 \phi} \right) \frac{d\theta}{dt} \qquad (19.12)$$

$$\frac{d}{dt} \left(\sin \theta + \frac{R \sin 2\theta}{2L \cos \phi} \right) = \left(\cos \theta + \frac{R \cos 2\theta}{L \cos \phi} + \frac{R^3 \sin^2 2\theta}{4L^3 \cos^3 \phi} \right) \frac{d\theta}{dt}$$

$$(19.13)$$

$$\frac{d^2x}{dt^2} = -R \frac{d^2\theta}{dt^2} \left(\sin \theta + \frac{R \sin 2\theta}{2L \cos \phi} \right)$$

$$- R \left(\frac{d\theta}{dt} \right)^2 \left(\cos \theta + \frac{R \cos 2\theta}{L \cos \phi} + \frac{R^3 \sin^2 2\theta}{4L^3 \cos^3 \phi} \right) \qquad (19.14)$$

If $L \gg R$, we can set $\phi = 0$. This yields

$$a = \frac{d^2x}{dt^2} = -R \frac{d^2\theta}{dt^2} \left(\sin \theta + \frac{R}{2L} \sin 2\theta \right)$$

$$- R \left(\frac{d\theta}{dt} \right)^2 \cos \theta + \frac{R}{L} \cos 2\theta + \frac{R^3 \sin^2 2\theta}{4L^3} \qquad (19.15)$$

Note that the first term of this equation becomes zero for constant angular velocity of the crank (i.e., no acceleration). Also, by making use of the relationship

$$L \cos \phi = (L^2 - R^2 \sin^2 \theta)^{1/2} \qquad (19.16)$$

we can rewrite the general expressions for velocity and acceleration, Eqs. (19.8) and (19.15), as follows:

$$v = -R\omega \left[\sin \theta + \frac{R \sin 2\theta}{2(L^2 - R^2 \sin^2 \theta)^{1/2}} \right] \tag{19.17}$$

$$a = -R\alpha \left[\sin \theta + \frac{R \sin 2\theta}{2(L^2 - R^2 \sin^2 \theta)^{1/2}} \right]$$

$$- R\omega^2 \left[\cos \theta + \frac{R \cos 2\theta}{(L^2 - R^2 \sin^2 \theta)^{1/2}} + \frac{R^3 \sin^2 2\theta}{4(L^2 - R^2 \sin^2 \theta)^{3/2}} \right]$$

$$\tag{19.18}$$

IV
Gears

Gears belong to a special class of mechanisms whose principal function is to transmit motion and power from one shaft to another. Specifically, gears are usually in the form of cylinders whose active surfaces are provided with teeth that interlock or "mesh" such that the rotation of one accurately controls that of the other and the relationship between their angular speeds as well as torques is constant. When two gears are in mesh, the smaller of the pair is generally called the *pinion*, whereas the larger is referred to as the *gear*. When the pinion is on the drive shaft, the pair acts as a speed reducer. Conversely, when the gear drives the pinion, the pair acts as an increaser. Gears are more frequently used as **speed reducers** than as **speed increasers**.

20

Gear Fundamentals

20.1 TYPES OF GEARS

There are many types of gears, each designed according to its special application. The most common types are as follows:

Spur gear (Figure 20.1). This is the simplest and most common form of all gear types. In this type, the teeth are cut straight and parallel to the axis of the shaft, and motion is transmitted from one parallel shaft to another. Spur gears are very widely used in machine construction because of their simplicity and economy. However, their use is limited to low to moderate speeds because of their low contact ratios.

Helical gear (Figure 20.2). In this type of gear, the teeth are cut across the face in the form of a helix, where if the gear had sufficient axial length, the teeth would appear as threads on a screw. The helix may be either right- or left-handed. Helical gears may be used to transmit motion between parallel as well as nonparallel shafts. Compared to spur gears, they are quieter and more suited for high-speed and heavy load operations. They also have greater strength due to the fact that several teeth are in mesh at the same time. A main disadvantage of this gear type, however, is the side thrust produced due to the helix angle. The larger the helix angle, the larger the side thrust. A common example where helical gears are used is in automobile transmission.

Figure 20.1   Spur. (Courtesy Mobil Oil Corp.)

Herringbone gear (Figure 20.3). This is a variation of the helical gear, where the teeth are cut in the form of right- and left-hand helices. When a herringbone gear is fabricated with a central groove between the right- and left-hand helices, it is termed a ***double-helical gear***. The right- and left-hand helices make it possible to

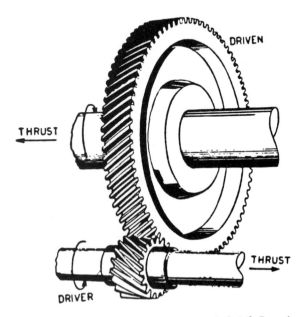

Figure 20.2   Helical. (Courtesy Mobil Oil Corp.)

Figure 20.3   Herringbone. (Courtesy Mobil Oil Corp.)

absorb the axial thrust within the gear. For proper operation, herringbone gears must be aligned very precisely in an axial direction if each half-tooth is to take its share of the load. Herringbone (or double-helical gears) are used extensively for power transmission from high-speed turbines to low-speed propeller shafts. Their main disadvantage is that they are costly.

Bevel gears (Figure 20.4). These are basically truncated cones with teeth on their outer surfaces. They are used where rotary motion is to be transmitted between nonparallel intersecting shafts. A bevel gear having straight teeth is called a **plain or straight bevel gear** (see Figure 20.4), whereas one having teeth curved in the

Figure 20.4   Plain bevel. (Courtesy Mobil Oil Corp.)

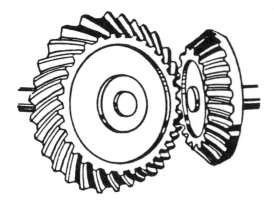

Figure 20.5   Spiral bevel. (Courtesy Mobil Oil Corp.)

form of a spiral around the conical surface is termed a **spiral bevel gear** (see Figure 20.5). Spiral bevel gears tend to run more quietly than plain bevel gears and can sustain somewhat larger loads, since several more teeth are in contact at any time during engagement. A disadvantage of this gear type is that it produces a large axial thrust force. Hence, care must be exercised at installation to ensure that adequate thrust bearings are provided.

When identical plain bevel gears are used for right-angle drives, they are referred to as **miter gears**.

Hypoid gear (Figure 20.6). This gear is similar in appearance to the spiral bevel gear, except that the motion is transmitted between

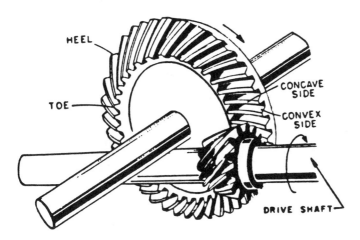

Figure 20.6   Hypoid. (Courtesy Mobil Oil Corp.)

Figure 20.7   Rack and pinion. (Courtesy Mobil Oil Corp.)

two nonintersecting shafts at right angles. Hypoid gears were originally developed for rear axle drives on automobiles since they
permit the drive shaft to enter the gear box below the centerline
of the rear wheels, thus lowering the center of gravity of the car.

Rack gear (Figure 20.7). This is a straight gear of infinite radius, normally used with a spur gear called the **pinion**, to convert rotary
motion to translation motion or vise versa. Racks are used extensively in machine tools.

Worm gear (Figure 20.8). This gear is similar to a helical screw that,
when meshed with a helical gear referred to as **worm gear or wheel**,
is used mainly to obtain large-speed reductions between nonintersecting shafts at right angles. The number of threads on a worm

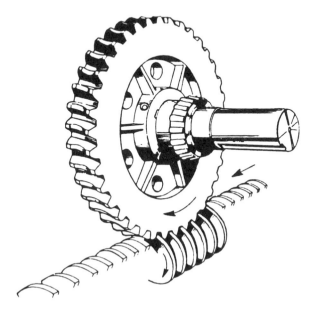

Figure 20.8   Worm. (Courtesy Mobil Oil Corp.)

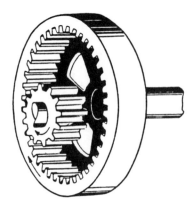

Figure 20.9   Internal. (Courtesy Mobil Oil Corp.)

is indicative of the number of teeth the worm gear will advance
with each revolution of the worm. Worms may be made with either
right- or left-hand threads similar to those on screws or bolts.

Internal or annular gear (Figure 20.9). This is a ring-type gear in which
the teeth are cut on the inner surface and are pointed toward the
gear axis. In this type of gear, the positions of the addendum and
dedendum are reversed from those of the external or spur gear.
This results in a different tooth action. Compared to external gear
drive, internal gear drive is more compact, it runs more smoothly,
and the teeth are stronger. When meshed with a smaller external
spur gear, termed a **pinion**, this combination makes planetary
motions possible. Also, when it is necessary to maintain the same
sense of rotation of two parallel shafts, the internal gear is espe-
cially desirable.

A summary of the above gear types is given in Table 20.1.

20.2 GEAR TERMINOLOGY

Basic Spur: Gear Terminology

In order to analyze or design a gear system, it is important to understand
the following terms and definitions (see Figure 20.10):

Pitch circle. The theoretical circle that passes through the pitch point.
When two gears turn together, their pitch circles turn in rolling
contact like the rolling cylinders that they represent.

Pitch point (P). The point of tangency of the pitch circles.

Pitch diameter (D_p). The diameter of the pitch circle.

Table 20.1   Summary of Gear Types

Types	Shaft relationships	Advantages	Disadvantages
External spur	Parallel shafting, opposite shaft directions	No axial thrust, moderate cost	Small contact ratio, moderate speeds
Internal spur	Parallel shafting, same shaft directions	Large contact ratio, compact, no axial thrust, short centers	Moderate speeds, costly
Rack and pinion	Rotary to linear motion, linear to rotary motion	Compact	Small contact ratio, low speeds
Helical	Parallel shafting nonparallel shafting	Large contact, ratio, large load capacity, high speeds, quiet, smooth operation	Axial thrust
Herringbone	Parallel shafting	Large contact ratio, large load capacity, no axial thrust	Costly
Plain bevel	Right-angle drive, intersecting shafts	Least costly of bevel type	Low to moderate speeds
Spiral bevel	Right-angle drive, intersecting shafts	Large contact ratio, large load capacity	Costly
Hypoid bevel	Right-angle drive, nonintersecting shafts	Large load capacity, rigid support	Costly
Worm and wheel	Right-angle drive, nonintersecting shafts	High-speed reduction	Axial thrust on worm

Addendum (A). The radial distance from the pitch circle to the outside circle of the gear.

Addendum radius (R_A). The maximum radius of the gear.

Dedendum (B). The radial distance from the pitch circle to the root circle that defines the bottom of the gear tooth.

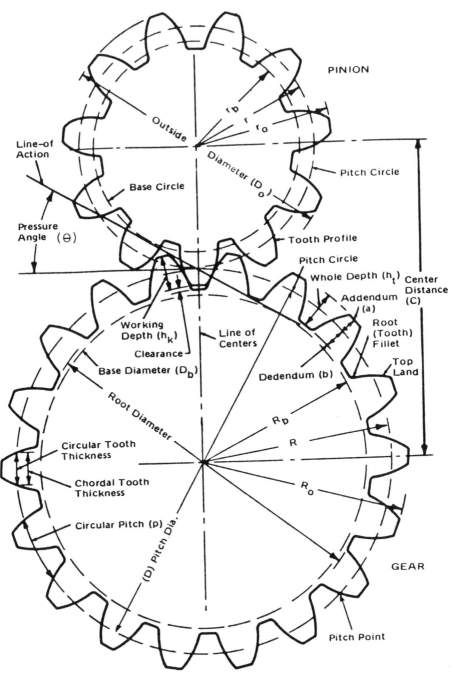

Figure 20.10(a)   Basic gear terminology. (Courtesy of Stock Drive Products—
Division of Designatronics.)

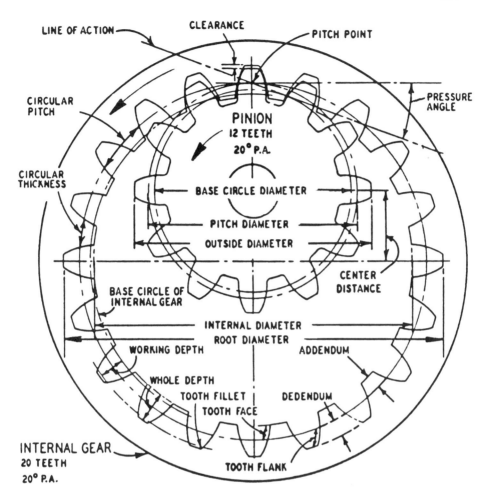

Figure 20.10(b) Internal gear and pinion terminology. (Courtesy Fellows Corp.)

Dedendum Radius (R_D). The radius of the root circle.

Tooth space. Distance between faces of adjacent teeth, measured along the pitch circle.

Clearance. The space measured on the line of centers between the addendum circle of its mating gear.

Backlash. The difference between the tooth width and tooth space measured along the pitch circle (see Figure 20.11). Some backlash is necessary tocompensate for inaccuracies in the forming of the tooth to provide space for a lubricant and allow for thermal expansion. Without backlash, gears could not mesh without jamming.

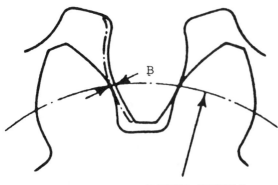

PITCH CIRCLE

Figure 20.11   Backlash between gears. (Courtesy of Stock Drive Products—Division of Designatronics.)

Working depth (H_K). The sum of the addendums of two mating teeth.
Face width. The thickness of the gear measured from the front face of the gear to the back face. This is also termed the *face of the gear*.
Tooth face. The contacting surface of the tooth from the pitch circle to the addendum circle.
Flank of tooth. The surface of the tooth from the pitch circle to the dedendum circle.
Fillet (F). The rounded corner between the flank and dedendum circle.
Circular pitch (P_C). The distance from a point on one tooth to the corresponding point on an adjacent tooth measured on the pitch circle. Alternatively, it is the circumference of the pitch circle divided by the number of teeth

$$P_C = \frac{\pi D_P}{T}$$

(Note: Mating gears have the same circular pitch.)
Diametral pitch (P_D). The ratio of the number of teeth on a gear to pitch diameter in inches

$$P_D = \frac{T}{D_P}$$

The diametral pitch is a more convenient number to use in defining the size of a gear than a physical dimension. It denotes the relative size of the teeth. The chart in Figure 20.12 shows the relative sizes of spur teeth of different diametral pitches. Note that

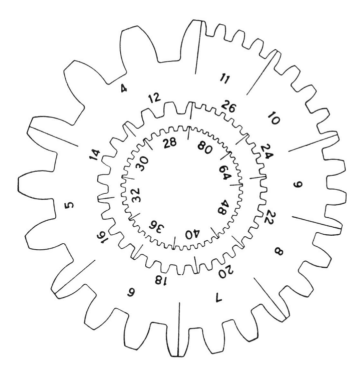

Figure 20.12 Tooth size and diametral pitch. (Courtesy of Barber-Colman Co., Rockford, Ill.)

the smaller the diametral pitch, the larger the teeth. For example, for two gears A and B having the same pitch diameter, the teeth of gear A whose P_D is 2 will be larger than those of gear B whose P_D is 6. Also, since circular pitch is defined as

$$P_C = \frac{\pi D_P}{T}$$

and diametral pitch is given by

$$P_D = \frac{T}{D_P}$$

then the relationship between the circular pitch and diametral pitch is obtained as

$$P_D \, P_C = \pi$$

(Note: Mating gears have the same diametral pitch.)

Root circle. The circle coinciding with the bottom of the tooth space.

Root diameter. The diameter of the root circle.

Top land. The top surface of a tooth.

Bottom land. The surface of the bottom of the tooth space.

Circular tooth thickness (or tooth thickness). The width of tooth measured along the pitch circle.

Chordal tooth thickness. The length of the chord subtending the circular tooth thickness arc.

Pressure angle (θ). The angle between the tooth profile and radial line at its pitch point.

Chordal addendum. The height from the top of a tooth to the chord subtending the circular thickness arc.

Center distance (C). Sum (for external gears) of difference (for internal gears) of the pitch radii of two mating gears. For example, the center distance between two external gears, 1 and 2, is given by

$$C = \frac{D_1 + D_2}{2}$$

$$= \frac{1}{2}\left(\frac{T_1}{P_D} + \frac{T_2}{P_D}\right)$$

$$= \frac{T_1 + T_2}{2P_D}$$

where

$$T_1 = \text{tooth number for gear 1}$$

$$T_2 = \text{tooth number for gear 2}$$

Module (M). The pitch diameter divided by the number of teeth, or the reciprocal of the diametral pitch

$$M = \frac{1}{P_D}$$

The module system is generally used in countries that have adopted the metric system.

Helical Gear Terminology

The terminology applied to spur gears is also applicable to helical gears, with the addition of the following (see Figure 20.13):

Helix angle (ψ). The angle between a tangent to the helix and an element of the pitch cylinder.
Normal circular pitch (P_{CN}). The circular pitch in the normal plane.

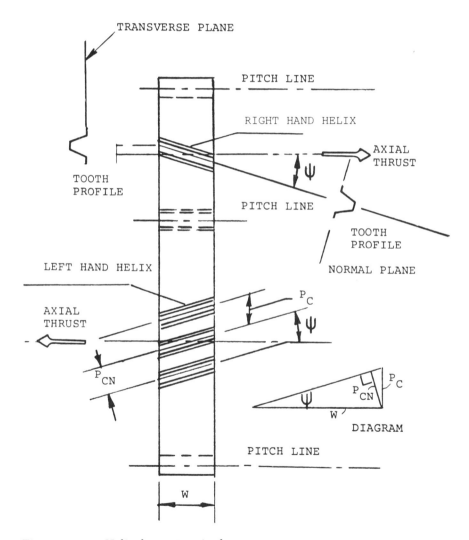

Figure 20.13   Helical gear terminology.

Transverse circular pitch (P_C). The circular pitch in the transverse plane. The normal and transverse circular pitches are related by

$$P_{CN} = P_c \cos \psi$$

where

$$P_C = \frac{\pi D}{T}$$

and ψ is the helix angle.

Normal diametral pitch (P_{DN}). The diametral pitch corresponding to the normal circular pitch

$$P_{DN} = \frac{\pi}{P_{CN}}$$

Face width (W). The length of the teeth in the axial plane

$$W = \frac{P_C \cos \psi}{\sin \psi} = P_C \tan \psi$$

Bevel Gear Terminology

The terminology applicable to spur gears is also applicable to helical gears, with the addition of the following (see Figure 20.14):

Pitch cone. Cone formed by pitch elements of the teeth.
Cone center. The apex of the pitch cone.
Face cone. Cone formed by elements of the top surface of the teeth. This cone may or may not have the same apex as the pitch cone.
Root cone. Cone formed by elements of the roots of the teeth.
Back cone. Cone formed by elements of the outside edges of the teeth.
Face angle. Angle between an element of the face cone and its axis.
Pitch angle. Angle between an element of the pitch cone and its axis.
Back angle. Angle between an element of the back cone and its axis.
Back cone distance. Length of the back cone element.
Dedendum angle. Angle between the root cone element and pitch cone element.
Addendum angle. Angle between the pitch cone element and face cone element.

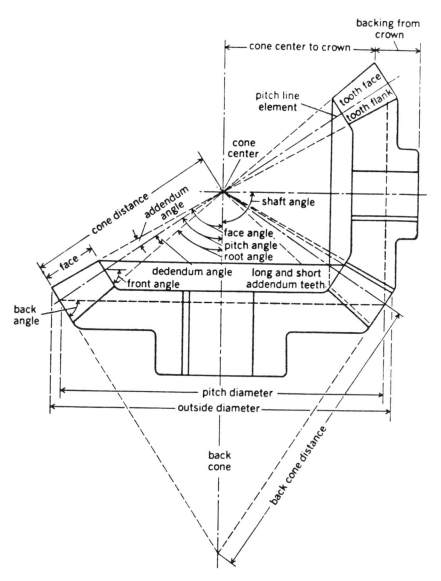

Figure 20.14   Bevel gear terminology.

Worm Gear Terminology

The terminology applied to spur and helical gear types is also applicable to worm gears, with the addition of the following (see Figure 20.15):

Worm lead (L). The axial distance a threaded shaft between advances in one revolution

$$L = NP_C$$

where N is the number of threads of the worm. For example, when N = 1 (single-threaded worm), lead equals pitch; when N = 2 (double-threaded worm), lead equals twice pitch; and when N = 3 (triple-threaded worm), lead equals three times pitch

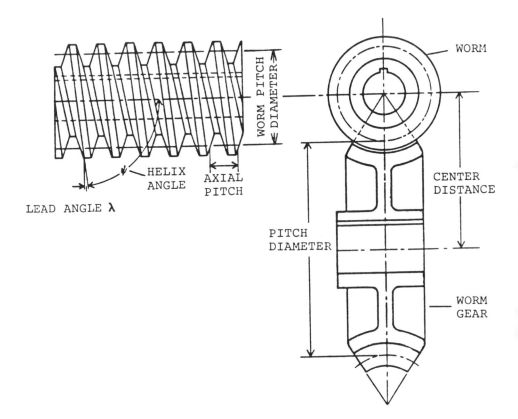

Figure 20.15   Worm gear terminology.

Lead angle (λ). The angle of a worm between a tangent to the pitch helix and the plane of rotation, or the complement of its helix angle. That is,

$$\tan \lambda = \frac{L}{\pi D_P}$$

Note that for shafts 90° apart (which is the normal case), the lead angle of a worm is equal to the helix angle of the worm gear.

20.3   LAW OF GEARING

The law of gearing requires that for mating gears, the velocity ratio of the pitch surfaces remain constant at all times. To produce this condition, the common normal at the point of contact must intersect the line of centers at a fixed point (referred to as the pitch point) (see Figure 20.16). The common normal is the line of action of the driving force or line along which the driving force acts.

Speed Ratio

The speed or velocity ratio of two meshing spur gears can be established by referring to Figure 20.17, where the pitch circles of gears A and B behave like two rolling cylinders in contact at point P. Since by definition of pure rolling there can be no slipping, the linear velocity of P on gear A must be the same as the linear velocity of P on gear B. That is,

$$V_P^A = V_P^B \tag{20.1}$$

or

$$R_A \omega_A = R_B \omega_B \tag{20.2}$$

Therefore,

$$\text{The speed ratio} = \frac{\omega_A}{\omega_B} = \frac{R_B}{R_A} \quad \text{(a constant)} \tag{20.3}$$

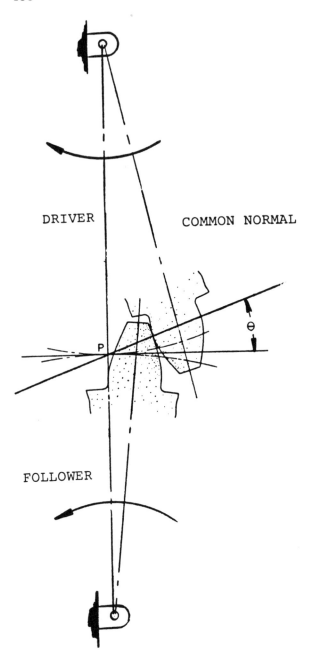

DRIVER   COMMON NORMAL

θ

P

FOLLOWER

Figure 20.16   Law of gearing.

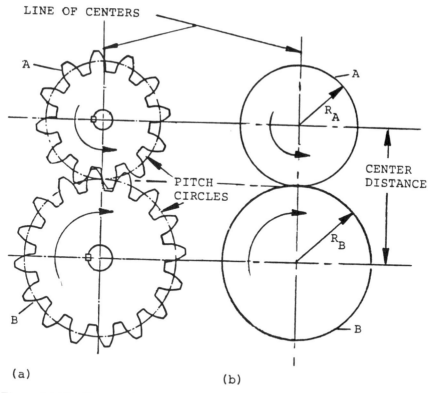

LINE OF CENTERS

A

PITCH
CIRCLES

A

R_A

CENTER
DISTANCE

R_B

B

B

(a)

(b)

Figure 20.17   Two meshing spur gears: (a) meshing pair; (b) theoretical rolling cylinders.

or

$$\frac{\omega_A}{\omega_B} = \frac{D_B}{D_A} \qquad (20.4)$$

The speed ratio may also be written, if we rate that $D = T/P_D$, as

$$\frac{\omega_A}{\omega_B} = \frac{T_B/P_D}{T_A/P_D} \qquad (20.5)$$

or

$$\frac{\omega_A}{\omega_B} = \frac{T_B}{T_A} \qquad (20.6)$$

Thus, the speed ratio is not only inversely proportional to the gear ratio, but also inversely proportional to the ratio of the tooth numbers.

EXAMPLE 20.1

A pair of mating spur gears of diametral pitch 4 have 24 and 32 teeth. Determine the center distance and pitch diameters if the gears mate (a) externally and (b) internally.

SOLUTION

1. For the external case,

$$C = \frac{T_1 + T_2}{2P_D}$$

$$= \frac{24 + 32}{2(4)}$$

$$= 7$$

Finding the pitch diameter for small gear:

$$\frac{T_1}{D_P} = \frac{T_1 + T_2}{2C}$$

$$D_P = \frac{2(7)24}{24 + 32}$$

$$= 6$$

Finding the pitch diameter of large gear, we get

$$D_P = \frac{2(7)32}{24 + 32}$$

$$= 8 \text{ in.}$$

2. For the internal case,

$$C = \frac{T_1 - T_2}{2P_D}$$

$$= \frac{24 - 32}{2(4)}$$

$$= 1 \text{ in. (abs.)}$$

Finding the pitch diameter of small gear, we obtain

$$D_P = \frac{2(1)24}{24 - 32}$$

$$= 6 \text{ in. (abs.)}$$

Finding the pitch diameter of large gear yields

$$D_P = \frac{2(1)32}{24 - 32}$$

$$= 8 \text{ in. (abs.)}$$

EXAMPLE 20.2

A pair of helical gears have a normal diameter pitch of 6 and center distance 10. The pinion has 12 teeth. If the speed reduction of this drive is 6 to 1, determine the pitch diameters, number of teeth of the large gear, and helix angle.

SOLUTION

Finding the number of teeth of large gear. From Eq. (20.6), we obtain

$$SR = \frac{T_P}{T_G}$$

$$\frac{1}{6} = \frac{12}{T_G}$$

$$T_G = 12(6) = 72$$

Finding the pitch diameter for gear:

$$C = \frac{T_P + T_G}{2P_D}$$

$$P_D = \frac{T_P + T_G}{2C}$$

$$\frac{72}{D_P} = \frac{12 + 72}{2(10)} = \frac{84}{20}$$

$$D_P = \frac{20}{84}(72)$$

$$= 17.1 \text{ in.}$$

Finding the pitch diameter for pinion from the above yields

$$D_P = \frac{20}{84}(12)$$

$$= 34.28 \text{ in.}$$

Finding the helix angle:

$$C = \frac{T_P + T_G}{2P_D \cos \Psi}$$

$$\cos \Psi = \frac{84}{2(6)(12)}$$

$$= \frac{7}{12}$$

$$\Psi = \cos^{-1}\left(\frac{7}{12}\right)$$

$$= 54.31°$$

Conjugate Curves

When two mating tooth profiles satisfy the law of gearing, they are said to be **conjugate curves**. Although it is possible within limits to select a profile for the teeth of a given gear and then construct a conjugate profile for the teeth of the mating gear, in practice, only two types of tooth profiles have been in common use. They are the *cycloidal* and *involute*.

An involute profile (see Figure 20.18). The path formed by a point on a string held taut as it is unwound from a cylinder. This is the most widely used tooth profile for several reasons, most important of which include its ease of manufacture and the fact that the line of centers of a pair of involute gears can be varied without changing the velocity ratio. The involute may also be described as the path of a point on a straight line as it rolls around a base circle.

A *cycloidal profile* (see Figure 20.19). The path formed by a point on the circumference of a circle as it rolls on a pitch circle. If the circle is rolled on the outside of a pitch circle, the curve generated is called an *epi-cycloid*. If the circle is rolled on the inside of the pitch circle, the curve generated is termed a *hypo-cycloid*. As shown in Figure 20.19, the cycloidal tooth profile is a combination

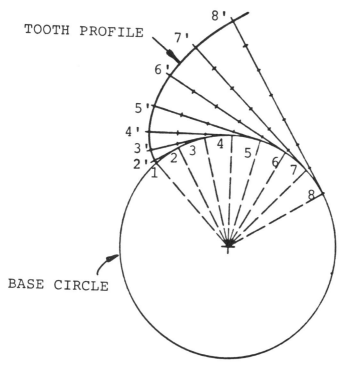

Figure 20.18   Involute profile.

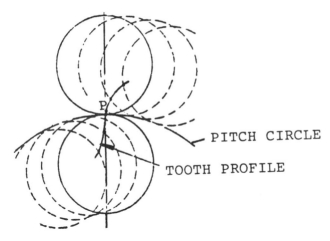

Figure 20.19   Cycloidal profile.

of both epi-cycloid (used to generate the tooth face) and hypo-cycloid (used to generate the tooth flank). The cycloidal profile was the earliest profile to be employed in gear design and it is still used in clocks and watches.

20.4   INVOLUTE TOOTH ACTION

It has been seen from Figure 20.20(a) that the involute is a curve traced by the end of a string as it is unwound from a cylinder and the circle that generates this curve is called the **base circle**. Since the instantaneous motion of this point relative to the release point on the cylinder must be normal to the string (considered the radius), the string conversely must always be normal to the involute curve for all positions as the point moves along the curve. This is an important property of the involute curve since if two cylinders D and F are wrapped with a string and turned, as shown in Figure 20.20, the common normal of the involute curves traced by a point in the string will lie along the string (or tangent to the surfaces of the cylinder). The common normal, therefore, for two mating involute gears may be defined as a straight line tangent to their base circles. This line is also referred to as the **pressure line** or **line of action** since it is along this line that the force (neglecting friction) between contacting tooth surfaces acts. Correspondingly, the angle that the pressure line makes with a perpendicular to the line of centers is termed the **pressure angle** θ.

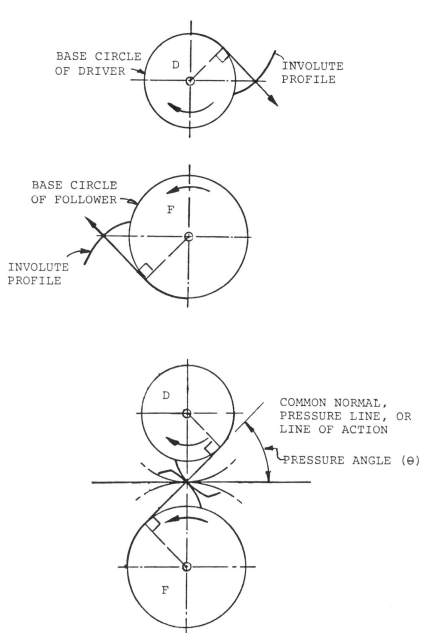

Figure 20.20   Tooth action.

Consider now the pair of involute gears 1 and 2 (driver and follower, respectively) shown in Figure 20.20(b). Here we see a pair of
teeth shown in solid lines at the beginning of contact, and the same
pair of teeth shown dotted at the end of contact. Note that contact begins
at point *A* where the addendum circle of the driver gear 1 cuts the line
of action, and ends at point *B* where the addendum circle of the driven
gear 2 cuts the line of action. Since the points *A* and *B* define the limits
of contact, the line *AB* is referred to as the ***path of contact***. This is the
locus of all contact points between the mating gears, as seen in Figure
20.21(a). Correspondingly, the arc of the pitch circle *MN* through which
the gear turns from the position where contact starts to the position
where contact ceases is called the ***arc of contact***. The arc of contact
has the same length on the pitch circle of either gear. The ***angle of
action*** [see Figure 20.21(b)] is the angle through which a gear turns
during contact between any pair of teeth. This is the angle MON subtended by the arc MN. Note that whereas the arcs of contact of mating
gears are equal in length, the angles of action will differ if the gears
differ in radii.

Approach and Recess Action

During the action cycle of two meshing gears, there is a certain amount
of sliding contact between the tooth profiles. This sliding is maximum
at the inception of contact and progressively decreases to zero as the
profiles approach the pitch point. At this point, since there is no sliding,
the gears are said to be in rolling contact. As the profiles move away

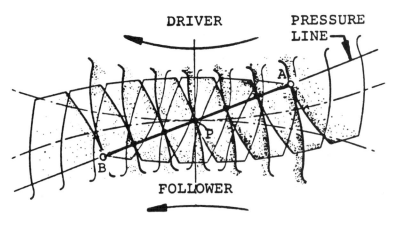

Figure 20.21(a)   Successive points of contact.

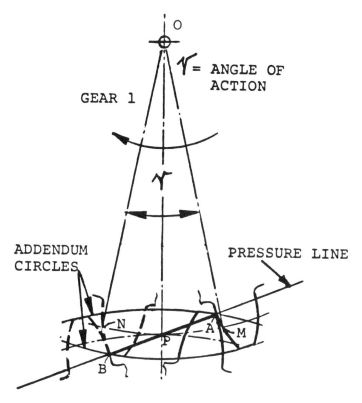

Figure 20.21(b) Angle of action.

from the pitch point, sliding again begins and progressively increases to a second maximum at which point contact ceases. The sliding action in which the pair of teeth move from the initial point of contact to the pitch point is termed **approach action**, whereas that in which the pair move from the pitch point to the end of contact is referred to as **recess action**.

Graphical Determination of Approach and Recess Angles
(see Figure 20.22)

1. From the initial point of contact A, outline the contacting face of a tooth to intersect the pitch circle at a point M.
2. From the final point of contact, outline the contacting face of the tooth in Step 1 to intersect the pitch circle at N. Points M and N,

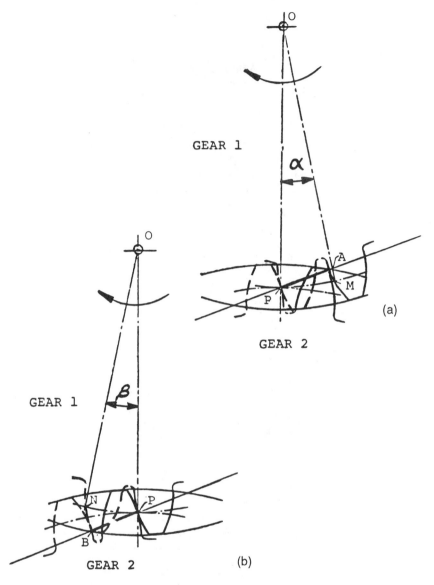

Figure 20.22   (a) Approach; (b) recess action.

as before, will define the limits of the arc of action (MN) and, hence, the arcs of approach (PM) and recess (PN).

In Figure 20.22(a), the approach action is defined by angle AOP (or alpha), whereas the recess action [see Figure 20.22(b)] is defined by angle BOP (or beta). From this it is seen that

The path of contact = path of approach + path of recess

$$AB = AP + PB$$

The arc of contact = arc of approach + arc or recess

$$\text{Arc } MN = \text{arc } MP + \text{arc } PN$$

The angle of action = angle of approach + angle of recess

$$\angle MON = \angle MOP + \angle PON$$

Length of Path of Contact

Graphical Approach

The length of the path of contact is a function of pitch radius, addendum, and pressure angle. A graphical layout using these parameters is the simplest means of determining paths of contact, and reasonable accuracy can be achieved if large scales are employed. Layouts of paths of contact for typical gear arrangements are shown in Figure 20.23(a–c). Note that contact is only possible within the shaded area defined by the overlapping addendum circles of the gears. Specifically, the path of contact for the three cases shown may be obtained graphically as follows.

The path of contact for two external spur gears [Figure 20.23(a)] begins where the line of action cuts the addendum circle of one gear and ends where the line of action cuts the addendum circle of the other gear.

The path of contact for a rack and pinion [Figure 20.23(b)] begins at the point where the line of action cuts the addendum line of the rack and ends where the line of action cuts the addendum circle of the pinion.

The path of contact for an internal gear and pinion [Figure 20.23(c)] begins where the line of action cuts the addendum circle of the gear and ends where the line of action cuts the addendum circle of the pinion.

Mathematical Approach

If extreme accuracy is desired, the path lengths can be determined from trigonometry. Consider Figure 20.24 where the path of contact BA is

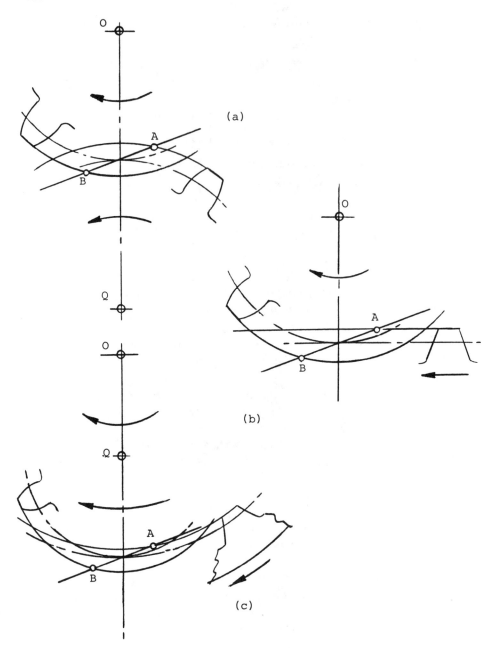

Figure 20.23   Paths of contact. (a) External spur gears, (b) rack and pinion, (c) internal gear and pinion.

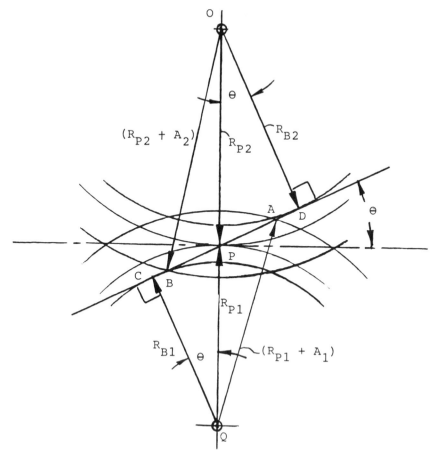

Figure 20.24   Length of path of contact.

to be determined. From the figure we see that

$$BA = BP + PA \tag{20.7}$$

where

$$BP = BD - PD \tag{20.8}$$

and

$$PA = AC - PC \tag{20.9}$$

Also,

$$OB = \text{pitch radius} + \text{addendum} = R_{P2} + A_2$$

and

$$OD = R_{B2} = R_{P2} \cos \theta$$

Therefore,

$$BD = \sqrt{OB^2 - OD^2} = \sqrt{(R_{P2} + A_2)^2 - R_{P2}^2 \cos^2 \theta}$$

Now

$$PD = R_{P2} \sin \theta$$

Therefore, Eq. 20.8 becomes

$$BP = \sqrt{(R_{P2} + A_2)^2 - R_{P2}^2 \cos^2 \theta} - R_{P2} \sin \theta \qquad (20.10)$$

The distance PA is found in a similar manner, using the triangle QCA, where

$$QA = P_{P1} + A_1$$
$$QC = R_{P1} \cos \theta$$
$$AC = \sqrt{QF^2 - QC^2} = \sqrt{(R_{P1} + A_1)^2 - R_{P1}^2 \cos^2 \theta}$$

and

$$PC = R_{P1} \sin \theta$$

Therefore, Eq. 20.9 becomes

$$PA = \sqrt{(R_{P1} + A_1)^2 - R_{P1}^2 \cos^2 \theta} - R_{P1} \sin \theta \qquad (20.11)$$

Substituting the values found for BP and PA in Eq. 20.7, we then obtain the required path of contact, as follows:

$$BA = \sqrt{(R_{P1} + A_1)^2 - R_{P1}^2 \cos^2 \theta}$$
$$+ \sqrt{(R_{P2} + A_2)^2 - R_{P2}^2 \cos^2 \theta} - (R_{P1} + R_{P2}) \sin \theta$$

Determining the Arc of Action

One method commonly used to determine the arc of action is to first determine its rectified length along the line of action. This length is determined graphically as follows.

PROCEDURE (see Figure 20.25)

1. Draw a line TT tangent to the pitch circle at P.
2. From points A and B (initial and final contact points), draw AX and BY perpendicular to AB, meeting TT at X and Y.

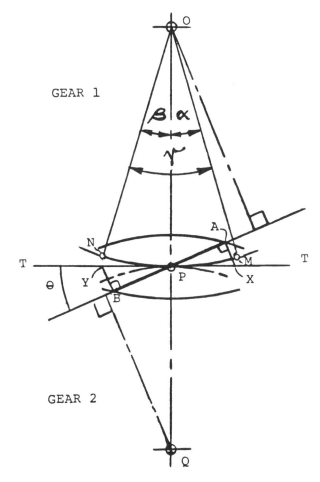

Figure 20.25   Arc of action.

3.   The required length of the rectified arc of action is obtained by
     scaling the line segment XY, where

$$PX = \text{length of arc of approach}$$

$$PY = \text{length of arc of recess}$$

This construction is based on the observation that in the right triangles
PAX and PAY,

$$\frac{PA}{PX} = \cos\theta = \frac{\text{path of approach}}{\text{arc of approach}}$$

and

$$\frac{PB}{PY} = \cos\theta = \frac{\text{path of recess}}{\text{arc of recess}}$$

Thus,

$$PX + PY = \text{arc of approach} + \text{arc of recess}$$

or

$$XY = \text{arc of action}$$

To obtain the true arcs of action, approach, and recess, as well as their
respective angles, lay off distances PM and PN on the pitch circle equal
to PX and PY found above. The points M and N, therefore, will define
the extremities of the arc of action. That is,

$$\text{Arc } MN = \text{arc of action}$$

where

$$\text{Arc } PM = \text{arc of approach}$$

and

$$\text{Arc } PN = \text{arc of recess}$$

Consequently, the angles of action are given by

$\angle MON$ = angle of action

$\angle MOP$ = angle of approach

$\angle NOP$ = angle of recess

Angles of Approach and Recess

Referring to Figure 20.24, angles of approach and recess for gear 1 may be determined more precisely from the following relationships:

Angle of approach (α_1), in radians

$$\alpha_1 = \frac{PA}{R_{B1}}$$

$$= \frac{\sqrt{(R_{P1} + A_1)^2 - R_{P1}^2 \cos^2 \theta} - R_{P1} \sin \theta}{R_{B1}}$$

$$= \frac{\sqrt{(R_{P1} + A_1)^2 - R_{B1}^2} - R_{B1} \tan \theta}{R_{B1}}$$

Angle of recess (β_1), in radians

$$\beta_1 = \frac{PB}{R_{B1}}$$

$$= \frac{\sqrt{(R_{P2} + A_2)^2 - R_{P2}^2 \cos^2 \theta} - R_{P2} \sin \theta}{R_{B1}}$$

$$= \frac{\sqrt{(R_{P2} + A_2)^2 - R_{B2}^2} - R_{B2} \tan \theta}{R_{B1}}$$

Geometrical Relationship Based on Pressure Angle

The pressure angle θ (see Figure 20.26) is a key parameter used in the study of a gear tooth system. Its size establishes a number of geometrical relationships between mating gears, such as the following:

$$\frac{\text{Base circle radius}}{\text{Pitch circle radius}} = \frac{R_B}{R_P} = \cos\theta$$

$$\frac{\text{Base pitch}}{\text{Circular pitch}} = \frac{P_B}{P_C} = \cos\theta$$

$$\frac{\text{Path of approach}}{\text{Arc of approach}} = \cos\theta$$

$$\frac{\text{Path of recess}}{\text{Arc of recess}} = \cos\theta$$

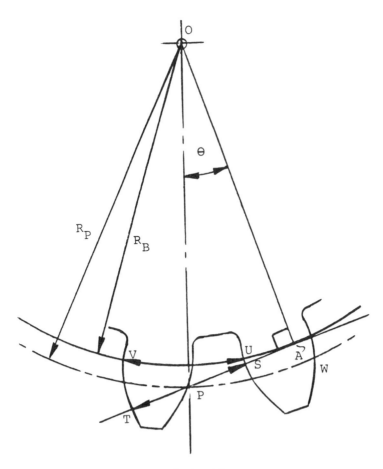

Figure 20.26 Geometric relationships.

$$\frac{\text{Path of contact}}{\text{Arc of contact}} = \cos\theta$$

Note that the denominators of these ratios are related to the pitch circle, whereas the numerators are base-circle-related.

Contact Ratio

The contact ratio may be defined as the average number of pairs of teeth in contact when two gears are in mesh. If smooth and continuous action between the gears is to be achieved, it is desirable to have more than one pair of teeth in engagement when the gears are meshing. The contact ratio may be computed as

$$\text{Contact ratio (CR)} = \frac{\text{angle of action}}{\text{pitch angle}}$$

where the angle of action, as previously seen, is the angle subtended by the arc of contact at the gear center, and the **pitch angle** is the angle subtended by the circular pitch at the gear center. That is,

$$CR = \frac{\text{path of contact}}{\text{base pitch}} = \frac{\text{path of contact}}{\text{normal pitch}}$$

where the **base pitch** (P_B) is the distance from a point on a tooth to the corresponding point on the next tooth, measured along the base circle. See Figure 20.26. The **normal pitch** (P_N) is the distance from the side of one tooth to the corresponding side of the next tooth, measured along the common normal. See Figure 20.26.

The above equation is based on the fact that the normal pitch is equal in length to the base circle, based on the construction of the involute.

Expressed algebraically, from Eqs 20.10 and 20.11,

$$CR = \frac{\sqrt{(R_{P1} + A_1)^2 - R_{P1}^2 \cos^2\theta} + \sqrt{(R_{P2} + A_2)^2 - R_{P2}^2 \cos^2\theta} - (R_{P1} + R_{P2})\sin\theta}{P_C \cos\theta}$$

$$= \frac{\sqrt{(R_{P1} + A_1)^2 - R_{B1}^2} + \sqrt{(R_{P2} + A_2)^2 - R_{B2}^2} - (R_{B1} + R_{B2})\tan\theta}{P_B}$$

If the contact ratio is 1, this means that one pair of teeth is in contact. If the contact ratio is 2, then two pairs of teeth are in contact. A contact ratio of 1.5, however, does not mean that 1.5 pairs of teeth are in contact. Rather, it means that there are alternately one pair and two pairs of teeth in contact and, on a time basis, the average number is 1.5.

Because there must always be at least one pair of teeth in contact for continuous gear action, the theoretical minimum contact ratio is 1. However, in practice, 1.4 has been recommended as a minimum. The larger the contact ratio, the smoother the action of the gears. This ratio can be increased, within limits, by

increasing the addendums of the gears
increasing the pitch diameters
decreasing the pressure angle

20.5 LAYOUT OF INVOLUTE GEARS

Rack and Pinion

Although gears can be obtained readymade from manufacturers, it is nevertheless instructive for the student of mechanisms to experience firsthand the construction of a simple gear system on paper. Suppose that it is required to lay out a standard rack and pinion gear assembly having the following specifications:

$$\text{Diametral pitch } (P_D) = 1$$

$$\text{Pressure angle } (\theta) = 20°$$

$$\text{Number of pinion teeth } (T) = 18$$

PROCEDURE (see Figure 20.27)

1. Draw the gear centerlines and pitch circles and locate the pitch point P, noting the following:
    a. Pitch circle radius is given by

$$R_P = \frac{T}{\pi P_D}$$

$$= \frac{18}{\pi(1)} = 9 \text{ in.}$$

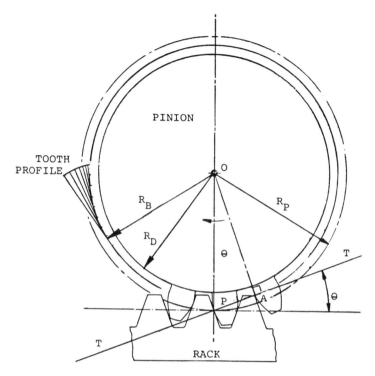

Figure 20.27   Rack and pinion layout.

    b.   The pitch circle for a rack is a straight line whose center is at infinity.

2.   Through P, draw the pressure line TT at the angle (θ) given. This line slopes outward from the center of the driver in the direction of rotation.

3.   Draw the base circle tangent to the pressure line. This can be done, first, by dropping a perpendicular from center 0 to meet TT at point A and then, with radius OA, describe the required base circle.

4.   Locate the addendum and dedendum circles using standard data for the required gear type. From Table 20.2, it is seen that for a 20°, full-depth tooth

$$\text{Addendum } (A) = \frac{1}{P_D} = 1 \text{ in.}$$

$$\text{Addendum radius } (R_A) = R_P + A$$

$$= 9 + 1 = 10 \text{ in.}$$

$$\text{Dedendum } (B) = \frac{1.157}{P_D} = 1.157 \text{ in.}$$

$$\begin{aligned} \text{Dedendum radius } (R_D) &= R_P - B \\ &= 9 - 1.157 = 7.84 \text{ in.} \end{aligned}$$

5. Construct an involute of the base circle (preferably on tracing paper for transfer to the layout) as follows:
   a. Using a small divider, divide a segment of the base circle into eight (or more) equal divisions.
   b. At each division, draw tangents to the base circle.
   c. Starting at the first division, mark off, along each tangent, distances equivalent to respective arc lengths along the base circle.
   d. Draw a smooth curve connecting the marked points on the tangents to obtain the required involute.
6. Using the involute curve developed in Step 5, construct the first tooth at the initial contact point A to standard dimensions (see Table 20.2). In this case,

$$\text{Circular pitch } P_C = \frac{\pi}{P_D}$$

$$= \frac{3.141}{1} = 3.141 \text{ in.}$$

$$\text{Tooth width} = \frac{P_C}{2} = 1.57 \text{ in.}$$

Tooth flanks are drawn radially from the base circle to the addendum circle, whereas tooth fillet and clearances are drawn to standard dimensions (see Table 20.2).

7. Construct the rack teeth, positioning the first tooth at the initial contact point A. Note that basic tooth dimensions, such as tooth width, addendum and dedendum, clearance and fillet, are the same as for the involute gear. Also, the contacting surface is drawn perpendicular to the pressure line, whereas the opposite surface is similarly sloped in the opposite direction to make the drive reversible.
8. Lay out the remaining two gear teeth in a similar manner to the above.

Gear and Pinion

To further illustrate the layout procedure, Figure 20.28 shows the layout of a typical gear and pinion assembly. See Appendix A.2 for design data calculation program for this type of assembly.

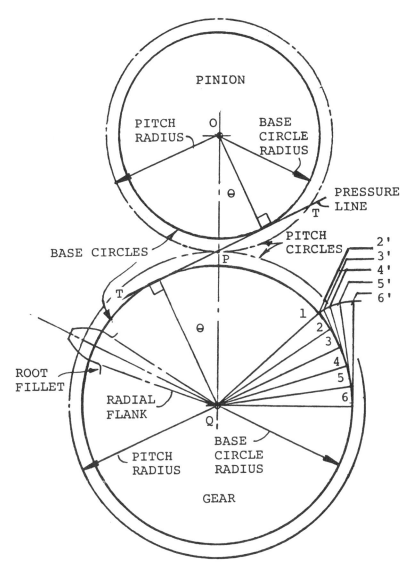

Figure 20.28   External gear and pinion layout.

20.6 INTERFERENCE

For uninterrupted gearing action, the meshing teeth must be clear of
each other as they approach and leave their designated contact positions. In the case of gears with involute tooth profiles, contact with any
portion of the tooth surface within the base circle, such as the flank,
fillet, or dedendum surface, is considered to be interference and should
be avoided. Interference then may be defined as the undesirable contact
between nonconjugate tooth surfaces of mating gears. Interference can
occur for various reasons. Fortunately, however, it can be easily de-
tected and avoided at the design stage with the aid of simple graphical
constructions. If we refer to Figure 20.29, the length CD represents the

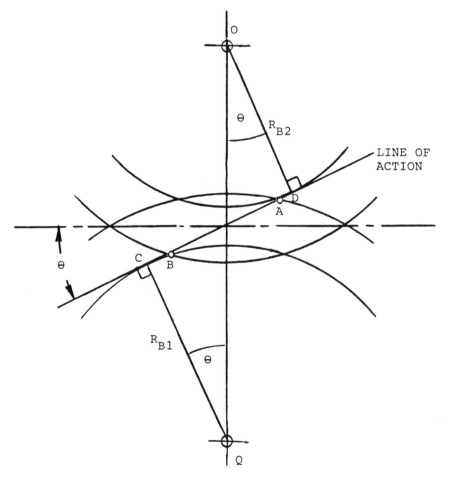

Figure 20.29   Interference conditions.

maximum possible length of contact the two mating gear teeth can have while maintaining involute contact (i.e., without interference). However, actual contact length is limited by points A and B due to the addendum circles of both gears. Therefore, to avoid interference, both points A and B must fall within the tangent points C and D (also known as interference points). In other words, a tangency point must not fall between the limits of the path of contact.

As an example of interference, when a gear having a small number of teeth meshes with another gear having a large number of teeth or a rack as depicted in Figure 20.30, this will result in interference. In this example, it can be seen that initial contact between the gears is at point D, which is outside the tangency point A, or the upper limit of the path of contact. Hence, the shaded portions of the rack teeth above the interference line are outside the range of the involute curve (or inside the base circle) of the pinion teeth.

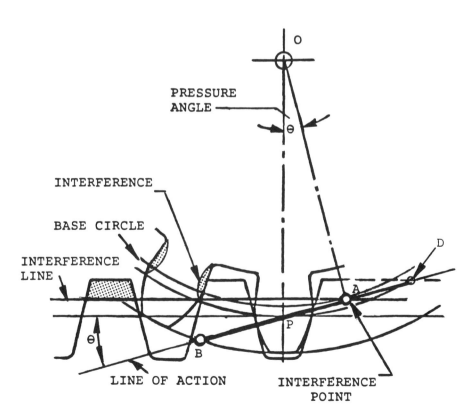

Figure 20.30   Rack and pinion interference.

   To remedy this situation, there are two options available, as follows:

1. Remove the metal in the flanks of the obstructing pinion teeth. This procedure is known as **undercutting**, which is usually not very desirable as it tends to weaken the teeth. Also, by undercutting, the contact ratio is reduced, resulting in noisier and rougher gear action.
2. Trim off the interfering (shaded) portions of the rack teeth, that is, shorten the addendum. This approach is generally preferred to removal of metal in the flanks of the teeth.

To avoid the problem of interference, the following corrective actions can be taken depending on the problem constraints.

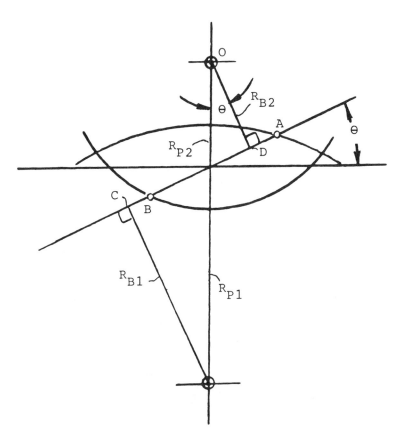

Figure 20.31   Effect of diametrical pitch on interference.

Effect of Diametral Pitch on Interference

In Figure 20.31, the pitch radii and pressure angle are assumed to be fixed. Since the addendum radius of gear 1 is much larger than that of gear 2, the addendum circle cuts the line of action at point B, outside CD. To avoid interference, the addendum radius must be reduced, which means that the diametral pitch must be increased. By increasing the diametral pitch, the number of teeth on each gear is increased, whereas the pitch radii are held constant.

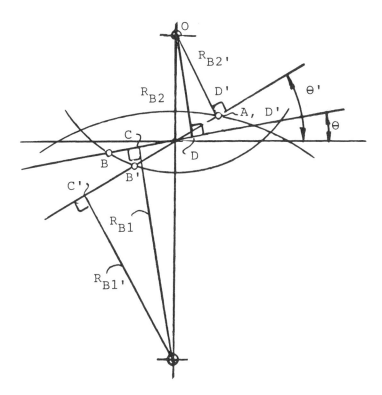

Figure 20.32   Effects of pressure angle on interference.

Effect of Pressure Angle on Interference

In Figure 20.32, pitch radii and diametral pitch are assumed to be fixed. With a pressure angle at 0, points A and B fall outside of tangency points C and C, indicating interference. By decreasing the base radius, the pressure angle is effectively increased from θ to θ' and the tangency points C and D are relocated to C' and D' outside the path of contact

AB, which eliminates interference. Thus, if interference is found to take place in a drive composed of standard gears of, say, 14 1/2° presure angle, this problem might be solved if similar gears of, say, 20° pressure angle were substituted.

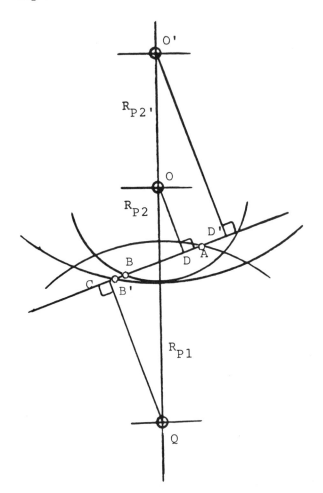

Figure 20.33   Effect of speed ratio on interference.

Effect of Speed Ratio on Interference

In Figure 20.33, the pressure angle and diametral pitch are assumed to be fixed. As in the case of the rack and pinion example discussed above, since gear 1 is much larger than gear 2, the tangency point C falls within the path of radius of gear 2 and is increased such that both C and D fall outside the interval AB. By increasing the pitch radius of the smaller gear, holding the diametral pitch constant, the center distance between the base circles must be increased.

Algebraic Approach

The criterion for avoiding interference may also be expressed algebraically, as follows:

For a gear and pinion:

$$\sqrt{R_{A2}^2 - R_{B2}^2} < (R_{B1} + R_{B2}) \tan \theta$$

For a rack and pinion:

$$A_R/\sin \theta < R_B \tan \theta$$

where

$A_R \equiv$ rack addendum
$R_{A2} \equiv R_{P2} + A_2$
$R_B \equiv$ pinion base radius

20.7 GEAR TOOTH PROPORTIONS

To obtain the benefits of interchangeability as well as economy of design and tooling, a limited number of standards for tooth proportions have been established by the American Gear Manufacturers Association (AGMA). Gears with any number of teeth, but having the same pitch and pressure angle, are interchangeable if their tooth profiles have been cut to the same standard tooth system.

The basic dimensions for certain standardized systems are given in Table 20.2. Note that all standardized systems employ the basic rack geometry to specify the tooth proportions and pressure angle (PA or θ). Also, note that the minimum number of teeth required on a pinion that will mesh with a rack in any standard involute system without interference or undercutting may be determined from the expression

$$T = 2 A \operatorname{cosec}^2 \theta$$

Table 20.2 Gear Tooth Propoportions

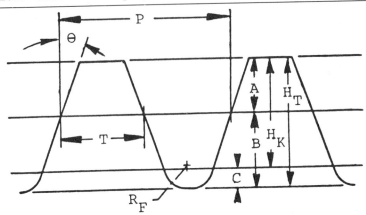

A = Addendum H_K = Working depth R_F = Fillet radius of basic rack
B = Dedendum H_T = Whole depth   T = Circular tooth thickness
C = Clearance   P = Circular pitch   θ = Pressure angle

Gear parameters	$14\frac{1}{2}°$ Full-depth involute	20° Full-depth involute	20° Stub involute
Pressure angle (θ)	$14\frac{1}{2}°$	20°	20°
Addendum (A)	$\dfrac{1.000}{P_D}$	$\dfrac{1.000}{P_D}$	$\dfrac{0.800}{P_D}$
Dedendum (B)	$\dfrac{1.157}{P_D}$	$\dfrac{1.157}{P_D}$	$\dfrac{1.000}{P_D}$
Working depth (H_K)	$\dfrac{2.000}{P_D}$	$\dfrac{2.000}{P_D}$	$\dfrac{1.600}{P_D}$
Whole depth (H_T)	$\dfrac{2.157}{P_D}$	$\dfrac{2.157}{P_D}$	$\dfrac{1.800}{P_D}$
Clearance (C)	$\dfrac{0.157}{P_D}$	$\dfrac{0.157}{P_D}$	$\dfrac{0.200}{P_D}$
Tooth width (T)	$\dfrac{1.5708}{P_D}$	$\dfrac{1.5708}{P_D}$	$\dfrac{1.5708}{P_D}$
Fillet radius (R_F)	$\dfrac{0.209}{P_D}$	$\dfrac{0.235}{P_D}$	$\dfrac{0.200}{P_D}$

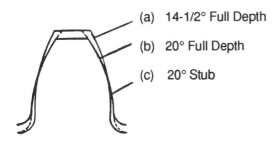

(a) 14-1/2° Full Depth

(b) 20° Full Depth

(c) 20° Stub

Figure 20.34 Comparisons of tooth profiles: (a) 14½° full-depth, (b) 20° full-depth, (c) 20° stub.

where

A = addendum (for 1 diametral pitch)

θ = pressure angle

If there is no interference between a pinion and rack, there would be no interference between the pinion and a gear the same size as or larger than the pinion.

The 14 1/2° full-depth involute system consists of teeth having involute profiles, except for portions that fall inside the base circle, and a pressure angle of 14 1/2°. This system is usually used for gears with a large number of teeth, 40 or more. The smallest pinion that will mesh with a rack without interference has a minimum of 32 teeth.

The **20° full-depth involute system** was developed so that gears having fewer than 32 teeth could be used without interference. The larger the pressure angle, the smaller the base circle. Thus, the 20° PA gear has a base circle farther away from the roots of the teeth than that of the 14 1/2° PA gear. In addition, the 20° PA tooth is broader at the base and, hence, stronger in bending. A disadvantage of this gear type is that the higher pressure angle gives a smaller contact ratio on account of the smaller base circle. Minimum tooth number for this gear type is 18.

The 20° stub involute system, in effect, is a modification of the 20° full-depth involute system, where the addendum has been shortened. As a result, this gear system has the strongest teeth of the three systems listed. However, it has the lowest contact ratio, since the length of contact between the mating gears is the shortest. The stub involute system was developed mainly for heavy-duty industrial applications. It is also used extensively in automotive transmissions. Minimum tooth number for this gear type is 14.

Figure 20.34 shows a comparison of the profiles for the three systems discussed above.

21

Gear Train Fundamentals

21.1 TYPES OF GEAR TRAINS

A *gear train* is a system consisting of two or more meshing gears, designed to provide a desired angular speed ratio between the shaft of the input gear and that of the output gear. There are four basic types of gear trains: simple, compound, reverted, and planetary (or epicyclic).

A *simple gear train* is one in which each shaft carries only one gear, as shown in Figure 21.1. In this type of train, the centerlines of all shafts or axles are fixed relative to the housing or frame, and all gears have the same diametral pitch. Simple gear trains are used in applications where the speed ratio is comparatively low.

A *compound gear train* is one in which one or more shafts carry more than one gear, as shown in Figure 21.2. Like the simple gear train, the shafts or axles are also fixed relative to a frame. However, the diametral pitch may or may not vary from stage to stage. Compound gear trains are used in applications where large speed ratios are desired.

A *reverted gear train* is a compound gear train in which both input and output gears rotate about coaxial shafts, as shown in Figure 21.3. One of the most important advantages of this type of train is its compactness. Some common applications of the reverted gear train include automatic transmissions, industrial speed reducers, and clocks.

A *planetary (or epicyclic) gear train* is one in which one or more gears rotate about a moving axis as shown in Figure 21.4. Usually, this

438

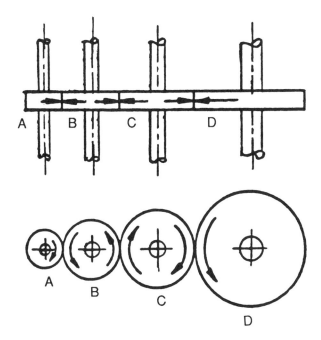

Figure 21.1   Simple gear train.

type of gear train consists of four elements: a central gear, the **sun gear**; a meshing pinion, the **planet gear**; a connecting arm, the **planet arm**; and an internal gear, the **ring gear**, inside of which the planet gear rotates. Three distinguishing characteristics of the planetary gear train are the following:

1. The input and output shafts are coaxial.
2. Either the input or output shaft turns with the planet arm.
3. Either the driver or follower must be stationary. Because of these characteristics, several combinations of input–output motions are possible. For example, we can consider:
   a. The planet arm is fixed, whereas the ring is free to rotate, in turn driving the planet and sun gears. (This is the case where the system behaves like a simple gear train.)
   b. The sun gear is fixed, whereas both the planet arm and planet gear are rotating.
   c. The ring gear is fixed, whereas the sun gear rotates, in turn driving the planet gear and planet arm.
   d. All gears rotate. In this case, the ring gear is externally driven.

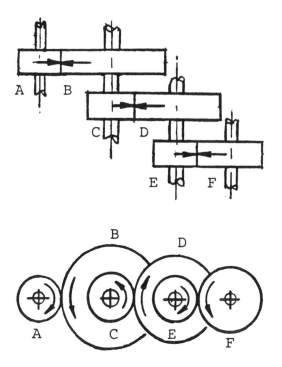

Figure 21.2   Compound gear train.

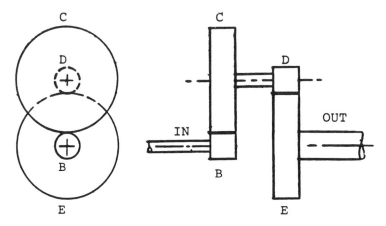

Figure 21.3   Reverted gear train.

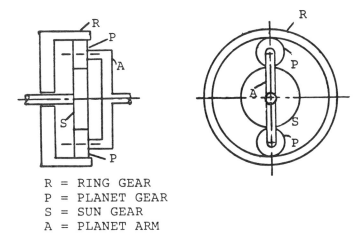

R = RING GEAR
P = PLANET GEAR
S = SUN GEAR
A = PLANET ARM

Figure 21.4   Planetary (Epicyclic) gear train.

The main advantage of the planetry gear train is its capability of providing large speed reductions with a small number of gears of moderate
size. Within recent years, the increasing use of planetary gears has been
seen in high-power transmissions systems, notably in turbo-generation
plants.

Practical Design Considerations

When designing a gear train, there are certain practical considerations
that the designer must take into account. They are as follows:

1. *Number and size of gears.* Although it is possible to select any
   number of gears to achieve a specified speed ratio, it is generally
   not feasible to obtain a large speed ratio in a single step, since not
   only would the gear be quite large and expensive, but space requirements would also be excessive. Therefore, a compromise must
   be made between economy and space requirements to arrive at an
   optimum system. Generally speaking, a greater number of smaller
   gears are usually preferred to a few larger gears.
2. *Tooth numbers.* Because tooth numbers must be integers and speed
   ratios commensurable numbers, sometimes it is impossible to obtain the precise ratio required for a gear train. In such cases, the
   design then involves investigation of possible tooth ratios that will
   give the best approximation.

3. **_Repetition of tooth contact._** In situations where cyclic loading is involved, it is particularly desirable that the same pair of teeth not be engaged at the peak loads. If this were to occur, tooth wear would be uneven and excessive. For example, if a gear with 21 teeth were to engage another gear with 42 teeth, the same pair would come into contact once every two revolutions of the first gear. However, if the 21-tooth gear were to engage a 43-tooth gear, a given pair of teeth would come into contact once in 903 (or 21 × 43) revolutions. Consequently, tooth wear will be more evenly distributed. It should be noted that the use of an odd number of teeth, in one of the gears, does change the speed ratio slightly. However, for many applications this is not important.

4. **_Tooth strength._** The required strength of a gear tooth is determined by the load it carries. When a large amount of horsepower is transmitted at a low speed, a large value of torque will exist and, as a result, a high load will be exerted on the gear tooth. Consequently, gears at the low-speed end of a gear train should use larger teeth than those at the high-speed end. Thus, the amount of load determines the maximum diametral pitch that may be used and, in turn, the tooth numbers.

21.2 SIMPLE GEAR TRAIN

Speed Ratio

Consider the simple gear train in Figure 21.1 consisting of gears A, B, C and D, in which gear A is the input or drive gear. Let it be required to find the speed ratio for the gear train. To begin the analysis, we recall from the law of gearing that if two gears 1 and 2 are in mesh, they behave like two rolling cylinders, and their speed ratio is inversely proportional to their tooth numbers. That is,

$$\frac{\omega_2}{\omega_1} = \frac{T_1}{T_2} \tag{21.1a}$$

Applying this principle to successive mating gears of the present gear train, we can write

For pair A and B:

$$\frac{\omega_B}{\omega_A} = \frac{T_A}{T_B}$$

For pair B and C:

$$\frac{\omega_C}{\omega_B} = \frac{T_B}{T_C}$$

For pair C and D:

$$\frac{\omega_D}{\omega_C} = \frac{T_C}{T_D}$$

Putting together these relationships, we obtain the overall speed ratio

$$\text{SR} = \frac{\omega_B \times \omega_C \times \omega_D}{\omega_A \times \omega_B \times \omega_C} = \frac{T_A \times T_B \times T_C}{T_B \times T_C \times T_D} \qquad (21.1b)$$

$$= \frac{\omega_D}{\omega_A} = \frac{T_A}{T_D} \qquad (21.1c)$$

Note that the intermediate gears B and C do not influence the overall ratio. For this reason, these gears are called **idlers**. Idlers are used when a large distance must be spanned between the input and output gears, and gears with large diameters are not permitted. They are also used to change the direction of rotation of the output gears.

Note also that the direction of rotation of the follower with respect to the driver is very important. This direction is easily determined by drawing arrows to indicate the rotation of each gear in the layout. When the driver and follower rotate in opposite directions, the speed ratio is considered negative and given a minus sign. When they rotate in the same direction, the speed ratio is considered positive and given a plus sign. A good practice is to place the appropriate sign before all tooth ratios to avoid errors. Thus,

$$\text{Speed ratio} = \frac{\text{output speed}}{\text{input speed}} = \frac{\text{tooth number of output}}{\text{tooth number of input}} \qquad (21.2)$$

EXAMPLE 21.1

Figure 21.5 shows a simple gear train in which the arm A is fixed. Determine the angular speed of the output gear R if the input gear rotates at 1800 rpm.

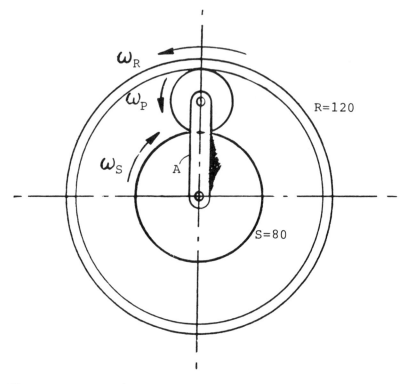

Figure 21.5   Example 21.1.

SOLUTION

From Eq. (21.2), the speed ratio of the simple gear train is

$$\frac{\omega_R}{\omega_S} = \frac{-T_S}{T_R} \tag{21.3a}$$

Substituting for ω_S, T_R, and T_S in the above equation, we obtain

$$\frac{\omega_R}{1800} = \frac{-80}{120}$$

$$\omega_R = -1200 \text{ rpm}$$

The negative sign indicates that R rotates opposite to S.

EXAMPLE 21.2

Consider the gear train shown in Figure 21.6. Arm A is fixed while gear E turns at 1 rpm. Determine the angular rotations of gears D, C, and B.

SOLUTION

1. The angular rotation of D is given by

$$\frac{\omega_D}{\omega_E} = \frac{T_E}{T_D} \qquad (21.3b)$$

$$\omega_D = \frac{1(140)}{(20)}$$

$$= 7 \text{ rpm}$$

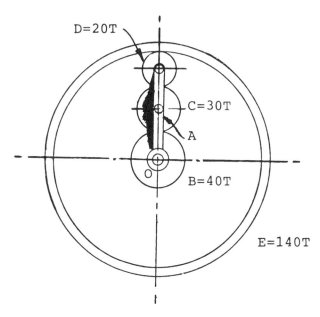

Figure 21.6   Example 21.2.

2. The angular rotation of C is given by

$$\frac{\omega_C}{\omega_E} = \frac{(+T_E)(-T_D)}{(T_D)(T_C)} \qquad (21.3c)$$

$$\omega_C = \frac{1(140)(-20)}{(20)(30)}$$

$$= \frac{-14}{3} \text{ rpm}$$

3. The angular rotation of B is given by

$$\frac{\omega_B}{\omega_E} = \frac{(+T_E)(-T_D)(-T_C)}{(T_D)(T_C)(T_B)} \qquad (21.3d)$$

$$\omega_B = \frac{1(+140)(-20)(-30)}{(20)(30)(40)}$$

$$= \frac{7}{2} \text{ rpm}$$

21.3 COMPOUND GEAR TRAIN

Speed Ratio

Consider the compound gear train shown in Figure 21.2, where the input gear is A and the output gear F. Let it be required to find the overall speed ratio for the gear train. Like the approach used in simple gear train analysis, we begin by writing the speed ratio successive mating pairs as follows:

For pair A nd B:

$$\frac{\omega_B}{\omega_A} = \frac{-T_A}{T_B}$$

For pair C and D:

$$\frac{\omega_D}{\omega_C} = \frac{-T_C}{T_D}$$

For pair E and F:

$$\frac{\omega_F}{\omega_E} = \frac{-T_E}{T_F}$$

Since

$$\omega_C = \omega_B \quad \text{(common shaft)}$$

and

$$\omega_E = \omega_D \quad \text{(common shaft)}$$

we can rewrite the above equations as follows:

$$\frac{\omega_B}{\omega_A} = \frac{-T_A}{T_B}$$

$$\frac{\omega_D}{\omega_B} = \frac{-T_C}{T_D}$$

$$\frac{\omega_F}{\omega_D} = \frac{-T_E}{T_F}$$

Putting together these relationships, we obtain

$$\frac{\omega_B}{\omega_A} \times \frac{\omega_D}{\omega_B} \times \frac{\omega_F}{\omega_D} = \frac{(-T_A)}{(T_B)} \frac{(-T_C)}{(T_D)} \frac{(-T_E)}{(T_F)} \qquad (21.4a)$$

or the overall speed ratio

$$\frac{\omega_F}{\omega_A} = \frac{T_A \times T_C \times T_E}{T_B \times T_D \times T_F} \qquad (21.4b)$$

Speed Ratio Diagram

As a quick check of Eq. (21.4b), it may be useful to note that the tooth ratios can be represented diagrammatically in the following manner:

$$\frac{\omega_F}{\omega_A} = \begin{array}{l} T_A \\ \quad | \\ T_B{-}T_C \\ \qquad | \\ \qquad T_D{-}T_E \\ \qquad\quad | \\ \qquad\quad T_F \end{array}$$

$$(21.4c)$$

The vertical lines in Eq. 21.4c imply that direct drive exists between the gears indicated, whereas the horizontal lines imply that the indicated gears are keyed to the same shaft. Thus,

$$\text{Speed ratio} = \pm \frac{\text{product of tooth numbers of driver gears}}{\text{product of tooth numbers of driven gears}}$$

(21.5)

Where the + sign indicates that the input and output gears rotate in the same direction, whereas the − sign indicates that they rotate in the opposite direction.

Note that this relationship is valid for all compound gear trains in which the gears have the same diametral pitch. For compound trains having gears with different diametral pitch, the tooth numbers must be expressed in terms of P_D and D_P. Recalling that $T = P_D \times D_P$, we can then rewrite Eq. (21.4b) as follows:

$$\frac{\omega_F}{\omega_A} = \frac{(P_D \times D_P)_A (P_D \times D_P)_C (P_D \times D_P)_E}{(P_D \times D_P)_B (P_D \times D_P)_D (P_D \times D_P)_F}$$

(21.6)

EXAMPLE 21.3

Figure 21.7 shows a diagram of a gear train consisting of various gear types: spur, bevel, internal, and helical. The tooth numbers for all gears are shown. Compute the output speed of gear H if the input gear A rotates at 1200 rpm.

SOLUTION

Using the diagrammatic approach we can express the speed ratio as

$$\frac{\omega_H}{\omega_A} = \begin{array}{c} T_A \\ | \\ T_B - T_C \\ | \\ T_D - T_E \\ | \\ T_F - T_G \\ | \\ T_H \end{array}$$

(21.7a)

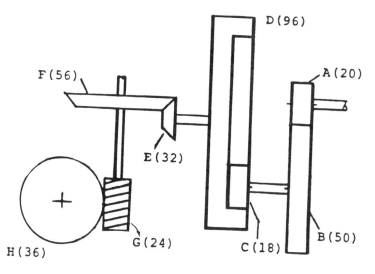

Figure 21.7   Example 21.3.

$$\frac{\omega_H}{\omega_A} = \left(\frac{T_A}{T_B}\right)\left(\frac{T_C}{T_D}\right)\left(\frac{T_E}{T_F}\right)\left(\frac{T_G}{T_H}\right)$$

$$\omega_H = 1200 \left(\frac{20}{50}\right)\left(\frac{18}{96}\right)\left(\frac{32}{56}\right)\left(\frac{24}{36}\right)$$

$$= 34.28 \text{ rpm}$$

(21.7b)

Note: Since the output gear H occurs at right angles to the input gear A, the sign convention is meaningless.

Determining the Tooth Numbers

In designing a compound gear train, the problem usually encountered is one in which the speed ratio is known and the tooth numbers are to be determined. One of the simplest methods for finding tooth numbers is a trial-and-error method that is the best explained by the following example.

EXAMPLE 21.4

Determine a set of teeth for a compound gear train that will produce an overall speed reduction ratio of 15 to 1 in two stages. No gear is to have more than 100 teeth or less than 18.

SOLUTION

To start the analysis, let us assume the two stages to be defined by gear pairs B and C, and D and E. Also assume the reduction is accomplished in two equal steps, namely,

$$\frac{T_C}{T_B} \times \frac{T_D}{T_E} = \frac{1}{15} \tag{21.8a}$$

Then the tooth ratio for each step will be

$$\frac{1}{\sqrt{15}} \left(\text{or } \frac{1}{3.88} \right)$$

so that

$$\frac{T_C}{T_B} \times \frac{T_D}{T_E} = \frac{1}{15} = \frac{1}{3.88} \times \frac{1}{3.88} \tag{21.8b}$$

To determine the actual tooth numbers, we need to convert the decimal ratios to a reasonably simple fraction consisting of integers only. Since 3.88 is very nearly 3.875 or 3⅞, we try 31/8. Using this value, we can alter the speed ratio equation as follows:

$$\frac{1}{15} = \frac{1}{(31/8)} \times \frac{(31/8)}{15}$$

$$= \left(\frac{8}{31} \right) \left(\frac{31}{120} \right) \tag{21.8c}$$

The tooth numbers represented in the above ratios will provide the required speed reduction. However, they are not acceptable since the tooth number 120 is larger than the permissible maximum and the tooth number 8 is smaller than the permissible minimum. Therefore, we should try a larger reduction for the first stage, say, 4 to 1. This gives

$$\frac{1}{15} = \frac{1}{4} \times \frac{4}{15} \tag{21.8d}$$

Now, it is evident that if we multiply the first ratio in the above equation by 18/18 and likewise the second ratio by 5/5, we will satisfy both minimum and maximum tooth number requirements. Thus,

$$\frac{1}{15} = \left(\frac{1}{4}\right)\frac{18}{18} \times \left(\frac{4}{15}\right)\frac{5}{5} = \frac{18}{72} \times \frac{20}{75} \tag{21.8e}$$

which makes

$$T_B = 18$$

$$T_C = 72$$

$$T_D = 20$$

$$T_E = 75$$

21.4 REVERTED GEAR TRAIN

Tooth Number Relationships

Figure 21.3 shows a schematic of a reverted gear train consisting of gears B (the input gear), C and D (the intermediate gears), and E (the output gear). As previosly defined, for this type of train, both the input and output shafts have the same axis. Therefore, the centerline distance C must be the same for both pairs of gears. That is,

$$C = \frac{D_B}{2} + \frac{D_C}{2} = \frac{D_D}{2} + \frac{D_E}{2} \tag{21.9a}$$

$$2C = \left(\frac{T}{P_D}\right)_B + \left(\frac{T}{P_D}\right)_C = \left(\frac{T}{P_D}\right)_D + \left(\frac{T}{P_D}\right)_E \tag{21.9b}$$

where D_B, D_C, D_D, and D_E are the pitch diameters and $(P_D)_B$, $(P_D)_C$, $(P_D)_D$, and $(P_D)_E$ are the diametral pitches for the respective gears.

Note that although the diametral pitches for gears B and C or D and E must be the same if these pairs are to mesh, these pitches need not be the same for both pairs. For example, if the diametral pitch for pair B and C differs from that for pair D and E, Eq. (21.9b) becomes

$$(P_D)_{D \text{ or } E}(T_B + T_C) = (P_D)_{B \text{ or } C}(T_D + T_E) \tag{21.10}$$

If the diametral pitch is the same in both pairs, which implies that all gears have the same pitch, then Eq. (21.9b) becomes

$$T_B + T_C = T_D + T_E \qquad\qquad (21.11)$$

Equations (21.10) and (21.11), which give the tooth number relationships, are sometimes referred to as the "tooth equations."

Speed Ratio

The speed ratio for the reverted gear train is determined in the same manner as that for the compound gear train. Thus,

$$\frac{\omega_E}{\omega_B} = \left(\frac{-T_B}{T_C}\right)\left(\frac{-T_D}{T_E}\right) \qquad\qquad (21.12a)$$

$$= \frac{T_B}{T_C} \times \frac{T_D}{T_E} \qquad\qquad (21.12b)$$

Determining the Tooth Numbers

EXAMPLE 21.5

In the reverted gear train in Figure 21.3, determine a set of tooth numbers that will produce an overall speed ratio of 1/48 to 1. All gears are assumed to have the same diametral pitch. No gear is to have less than 15 teeth.

SOLUTION

We begin by stating the two conditions that must be satisfied.

a. Since the diametral pitch for all gears is the same, the tooth relationship is

$$T_B + T_C = T_D + T_E \qquad\qquad (21.11)$$

b. The overall speed ratio may be expressed as

$$\frac{T_B}{T_C} \times \frac{T_D}{T_E} = \frac{1}{48} \qquad\qquad (21.13)$$

The objective then is to determine the tooth numbers that will satisfy Eqs. (21.11) and (21.13).

Let us now consider Eq. (21.13). The ideal value for each of the two ratios is $(1/\sqrt{48})$ or $(1/6.92)$. However, since ratios of integers are easier to handle than ratios of decimals, a better choice of ratios is $4/1$ and $6/1$. Thus, Eq. (21.13) can be rewritten as

$$\frac{T_B}{T_C} \times \frac{T_D}{T_E} = \frac{1}{6} \times \frac{1}{8} \tag{21.3}$$

which implies that

$$\frac{T_B}{T_C} = \frac{1}{6} \quad \text{and} \quad \frac{T_D}{T_E} = \frac{1}{8} \tag{21.4}$$

Suppose the numbers 6, 8, and 1 in the above ratios represented actual tooth numbers. Then the tooth number relationship in Eq. (21.11) would not be satisfied, since $(1 + 6)$ is not equal to $(1 + 8)$. The tooth equation requires that the sum of the numerator and denominator of one ratio must equal the sum of the numerator and denominator of the other. To satisfy this condition, while still maintaining the required ratios, we multiply and divide the first ratio by the sum of the numerator and denominator of the second ratio to obtain

$$\frac{T_B}{T_C} = \frac{1}{6} \times \frac{9}{9} = \frac{9}{54} \tag{21.17}$$

which makes

$$T_B = 9 \quad \text{and} \quad T_C = 54$$

Likewise, we multiply and divide the second ratio by the sum of the numerator and denominator of the first ratio to obtain

$$\frac{T_D}{T_E} = \frac{1}{8} \times \frac{7}{7} = \frac{7}{56} \tag{21.18}$$

which makes

$$T_D = 7 \quad \text{and} \quad T_E = 56$$

Note that the tooth ratios remain unchanged since

$$\frac{9}{54} = \frac{1}{6} \quad \text{and} \quad \frac{7}{56} = \frac{1}{8}$$

And the basic tooth equation (Eq. 21.11) is satisfied since

$$9 + 54 = 7 + 56 \tag{21.19}$$

However, since the smallest gear found ($T_D = 7$) has fewer teeth than the acceptable minimum (15), it is necessary to multiply the tooth numbers of all the gears by some factor so that the minimum took requirement is satisfied. In this case, we must multiply the tooth numbers found by 3, which gives

$$T_B = 27$$

$$T_C = 162$$

$$T_D = 21$$

$$T_E = 168$$

EXAMPLE 21.6

Consider the same reverted train as in the previous example, except assume that the diametral pitch of gears B and C is 3 and that of D and E is 4.

SOLUTION

Since the diametral pitches for the two gear pairs are not the same, we use the tooth relationship.

$$(P_D)_{D \text{ or } E}(T_B + T_C) = (P_D)_{B \text{ or } C}(T_D + T_E) \tag{21.10}$$

where

$$(P_D)_{D \text{ or } E} = 4 \text{ diametral pitch of gear } D \text{ or } E \tag{21.20}$$

and

$$(P_D)_{B \text{ or } C} = 3 \text{ diametral pitch of gear } B \text{ or } C \tag{21.21}$$

or

$$4(T_B + T_C) = 3(T_D + T_E) \tag{21.22}$$

And, as before, the tooth ratios are

$$\frac{T_B}{T_C} = \frac{1}{6} \tag{21.15}$$

$$\frac{T_D}{T_E} = \frac{1}{8} \tag{21.16}$$

Now our objective is to find a set of tooth numbers that will satisfy Eq. (21.22), as well as 21.15 and 21.16.

Again, if the numbers 8, 6, and 1 in the tooth ratios above were actual tooth numbers, then the tooth equation, Eq. (21.22), would not be satisfied since

$$4(6 + 1) \text{ is not equal to } 3(8 + 1)$$

or

28 is not equal to 27.

To satisfy the tooth equation, we multiply the numerator and denominator of the first ratio by 3 times the sum of the numerator and denominator of the second ratio to obtain

$$\frac{T_B}{T_C} = \frac{1}{6} \times \frac{27}{27} = \frac{27}{162} \tag{21.23}$$

which makes $T_B = 27$ and $T_C = 162$.

Likewise, we multiply the numerator and denominator of the second ratio by 4 times the sum of the numerator and denominator of the first ratio to obtain

$$\frac{T_D}{T_E} = \frac{1}{8} \times \frac{28}{28} = \frac{28}{224} \tag{21.24}$$

Thus, $T_D = 28$ and $T_E = 224$. Note that the tooth ratios remain unchanged, since

$$\frac{27}{162} = \frac{1}{6} \quad \text{and} \quad \frac{28}{224} = \frac{1}{8}$$

And the tooth equation is satisfied, since

$$4(27 + 162) = 3(28 + 224) \tag{21.25}$$

Alternative Solution

In this example, the tooth number relationship is given as

$$(P_D)_{D \text{ or } E}(T_B + T_C) = (P_D)_{B \text{ or } C}(T_D + T_E) \tag{21.10}$$

or

$$4(T_B + T_C) = 3(T_D + T_E) \tag{21.22}$$

And the tooth ratios are assumed to be

$$\frac{T_B}{T_C} = \frac{1}{6} \quad \text{and} \quad \frac{T_D}{T_E} = \frac{1}{8}$$

PROCEDURE (Equations 21.15 and 21.16)

1. Find the LCM of the following quantities
   a. diametral pitch of gears B and C (3)
   b. diametral pitch of gears D and E (4)
   c. sum of the numerator and denominator of the T_B/T_C ratio (or $1 + 6 = 7$)
   d. sum of the numerator and denominator of the T_D/T_E ratio (or $1 + 8 = 9$) LCM = 252
2. Equate each side of Eq. (21.10) to the LCM found in Step 1 and solve for the possible tooth numbers as follows:
   a. $(P_D)_{D \text{ or } E}(T_B + T_C) = 252$

$$4 \left(\frac{T_C}{6} + T_C \right) = 252$$

$$\frac{7T_C}{6} = 63$$

Therefore,

$$T_C = 54 \quad \text{(as before)}$$

$$T_B = 9 \quad \text{(as before)}$$

b. $(P_D)_{B \text{ or } C}(T_D + T_E) = 252$

$$3\left(\frac{T_E}{8} + T_E\right) = 252$$

$$\frac{9}{8} T_E = 84$$

Therefore,

$$T_E = \frac{224}{3} \quad \text{(theoretical)}$$

and

$$T_D = \frac{28}{3} \quad \text{(theoretical)}$$

Since tooth numbers cannot be fractional, the above values for T_E and T_D are not acceptable. However, practical values (or whole numbers) for these tooth numbers can be found in balancing the tooth equation, which requires that the theoretical values of T_E and T_D be multiplied by 3. Thus,

$$4(9 + 54) = 3\left(\frac{224}{3} + \frac{28}{3}\right) \tag{21.26}$$

$$4(27 + 162) = 3(224 + 28) \tag{21.27}$$

Hence, $T_B = 27$, $T_C = 162$, $T_D = 224$, and $T_E = 28$.

As previously noted, if the diametral pitch were the same as in Example (21.15) for all the gears, Eq. (21.10) would reduce to

$$T_B + T_C = T_D + T_E \tag{21.11}$$

in which case the LCM would be a smaller number: 63, and possible tooth numbers would be

$$T_B = 9$$

$$T_C = 54$$

$$T_D = 7$$

$$T_E = 56$$

Again, to satisfy the minimum tooth number requirement 15, we multiply all tooth numbers by 3 to give

$$T_B = 27$$

$$T_C = 162$$

$$T_D = 21$$

$$T_E = 168$$

Automotive Transmission

The automotive transmission mentioned earlier is a typical application of the reverted gear train. Figure 21.8 shows a schematic of a conventional three-speed automotive transmission. In this arrangement, gear A is driven by the engine, gears B and C can slide axially on a splined shaft connected to the drive wheels, gears D, E, F, and G rotate as a unit, and gear H is an idler.

In operation, all gears are in motion, and the input speed is constant since A and D are always in mesh. However, the output speed to the drive wheels depends on the position of B or C with respect to one of the lower gears.

For **low speed** [see Figure 21.8(b)], gear C is shifted to the left to mesh with gear F. As a result, the speed ratio is given by

$$
\frac{\omega_D}{\omega_C} = \left(\frac{-T_A}{T_D}\right)\left(\frac{-T_F}{T_C}\right)
$$

$$
= \left(\frac{-16}{29}\right)\left(\frac{-18}{27}\right)
$$

$$
= 0.367 \tag{21.28}
$$

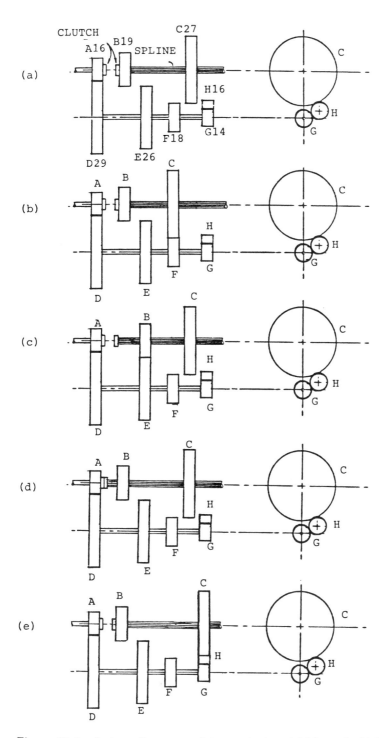

Figure 21.8 Automotive manual transmission: (a) Neutral, (b) low speed, (c) medium speed, (d) high speed, and (e) reverse.

For **medium speed** [see Figure 21.8(c)], gear B is shifted to the right to mesh with gear E. As a result, the angular speed ratio is given by

$$\frac{\omega_B}{\omega_A} = \left(\frac{-T_A}{T_D}\right)\left(\frac{-T_E}{T_B}\right)$$

$$= \left(\frac{-16}{29}\right)\left(\frac{-26}{19}\right)$$

$$= 0.754 \tag{21.29}$$

For **high speed** [see Figure 21.8(d)], gear B is shifted to the left to engage the clutch that directly couples the input and output shafts. Hence, the angular speed ratio between shafts is 1.

Finally, for **reverse speed** [see Figure 21.8(e)], C is shifted to the right to give a speed ratio of

$$\frac{\omega_C}{\omega_A} = \left(\frac{-T_A}{T_D}\right)\left(\frac{-T_G}{T_H}\right)\left(\frac{-T_H}{T_C}\right)$$

$$= \left(\frac{-16}{29}\right)\left(\frac{-14}{16}\right)\left(\frac{-16}{27}\right)$$

$$= 0.286 \quad \text{(reversed direction as expected)} \tag{21.30}$$

21.5 PLANETARY (OR EPICYCLIC) GEAR TRAIN

In studying planetary gearing problems, the usual objective is to determine the speed ratios between the various elements of the train, particularly the driver and follower. There are several methods by which the speed ratios may be determined. The first of the three methods selected for discussion here is based on the relative velocity method in which the key is to assume that one of the moving elements is held in a "fixed" position, while the motions of the other elements are viewed from that position.

The Formula Method

Consider the simple planetary gear train shown in Figure 21.9, where S, P, R, and A, respectively, designate the sun gear, planet gear, ring, and planet arm.

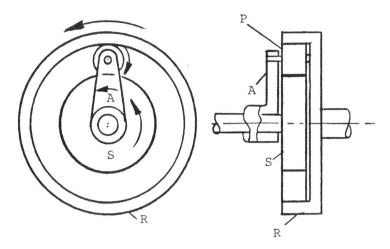

Figure 21.9   Planetary train indicating absolute motions.

Let us first assume that the planet arm is held fixed, while the sun, ring, and planet gears are in rotation. This then would reduce the system to that of a simple gear train consisting of three gears, one of which is internal. For this system, the speed ratio of ring to sun gear is

$$\frac{\omega_R}{\omega_S} = \left(\frac{-T_S}{T_P}\right) \times \left(\frac{+T_P}{T_R}\right) \quad \frac{\text{drivers}}{\text{followers}} \tag{21.31a}$$

or

$$\frac{\omega_R}{\omega_S} = \frac{-T_S}{T_R} \tag{21.31b}$$

Let us now suppose that the three components within the ring are rotating. Then to an observer rotating with the planet arm, the speeds of the other component would appear to be as follows:

Sun speed $\omega_S - \omega_A$

Ring speed $\omega_R - \omega_A$

Planet speed $\omega_P - \omega_A$

Arm speed $\omega_A - \omega_A = 0$

Since to the observer, the planet arm is fixed, the apparent speeds of the sun and ring gears must be related as in Eq. (21.31) above. That is,

$$\frac{\omega_R - \omega_A}{\omega_S - \omega_A} = \frac{-T_S}{T_R} \tag{21.32}$$

Likewise, to the observer (still on the fixed planet arm), the speeds of the planet and sun gears are related by

$$\frac{\omega_P - \omega_A}{\omega_S - \omega_A} = \frac{-T_S}{T_P} \tag{21.33}$$

Thus, the speeds of any two members F and D of a planetary train as viewed from the planet arm may be related by the general expression

$$\frac{\omega_F - \omega_A}{\omega_D - \omega_A} = (+/-) \frac{\text{product of driver teeth}}{\text{product of follower teeth}} \tag{21.34}$$

where

ω_F = angular velocity of the follower (F)

ω_D = angular velocity of the driver (D)

ω_A = angular velocity of the planet arm (A)

This general relationship may also be used to solve for the angular velocities or tooth numbers in planetary gear trains having one or two rotary inputs and a single rotary output. For a planetary train with one rotary input, one member must be fixed. The appropriate sign for the right-hand side of the equation depends on the signs of individual tooth ratios, determined by inspection.

Note that these signs do not necessarily indicate the absolute rotation of the member gears. Rather, they indicate the relative directions of rotation of the members as viewed from the "fixed" planet arm. This can be seen in Figure 21.10, where the absolute rotation of the ring gear is not the same as its rotation relative to the planet arm. To reduce the chance of errors, it is generally advisable to assume that the direction of the input velocity to the planetary is positive.

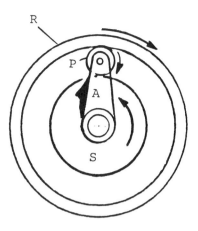

Figure 21.10   Planetary train indicating motion relative to the arm.

In a typical planetary gear train operation where any member could be held fixed, Eq. (21.34) can be conveniently used to define important speed relationships. For example, if the sun gear is held fixed, we have

$$\frac{\omega_R - \omega_A}{0 - \omega_A} = \frac{-T_S}{T_R} \tag{21.35a}$$

or

$$\frac{\omega_R}{\omega_A} = 1 + \frac{T_S}{T_R} \tag{21.35b}$$

This equation states that if the sun gear is fixed and the arm is given one revolution, then the ring gear will make one complete revolution plus an additional amount (due to the planet gear rotation) determined by the tooth ratio of the sun gear to the ring gear.

If the ring gear is held fixed, then

$$\frac{0 - \omega_A}{\omega_S - \omega_A} = \frac{-T_S}{T_R} \tag{21.36a}$$

or

$$\frac{\omega_S}{\omega_A} = 1 + \frac{T_R}{T_S} \tag{21.36b}$$

This equation states that if the ring gear is fixed and the arm is given one revolution, then the sun gear will make one complete revolution about its own axis plus an additional amount determined by the tooth ratio of the ring gear to the sun gear.

Note that the fact a gear is held fixed does not affect its function as a follower in the train. A follower gear may have any speed, including zero.

Mathematical Derivation

An alternate derivation of Eq. (21.32) employs a mathematical approach, as follows. Given the planetary gear train in Fig. 21.11, assume that ω_S and ω_A are known and both are clockwise. Then, applying the theory of relative motion, we obtain for point 1 on the planet gear,

$$V_1 = V_2 + V_{1/2} \tag{21.37}$$

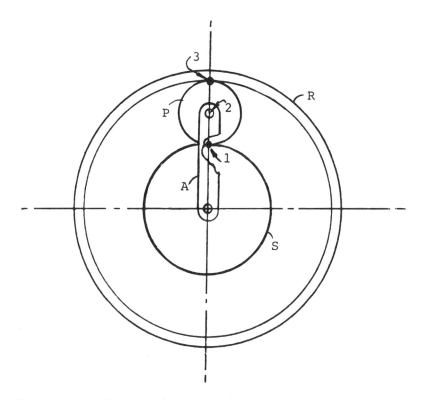

Figure 21.11 Planetary gear train.

and for point 3 on the planet gear,

$$V_3 = V_2 + V_{3/2} \tag{21.38}$$

Since points 1 and 3 rotate in opposite directions,

$$V_{3/2} = -V_{1/2} \tag{21.39}$$

or

$$V_3 - V_2 = -(V_1 - V_2) \tag{21.40}$$

Now, in terms of angular velocities, V_1, V_2, and V_3 are given by

$$V_1 = R_S\omega_S \tag{21.41}$$
$$V_2 = R_A\omega_A = (R_S + R_P)\omega_A \tag{21.42}$$
$$V_3 = R_R\omega_R \tag{21.43}$$

Substituting for V_1, V_2, and V_3 in Eq. (21.40), we get

$$R_R\omega_R - (R_S + R_P)\omega_A = -[R_S\omega_S - (R_S + R_P)\omega_A] \tag{21.44}$$

Expanding this equation, we have

$$R_R\omega_R - R_S\omega_A = -R_P\omega_A + R_S\omega_S + R_S\omega_A + R_P\omega_A \tag{21.45}$$

which reduces to

$$R_R\omega_R + R_S\omega_S = 2(R_S + R_P)\omega_A \tag{21.46}$$

This general equation describes the motion of the planetary gear train in terms of the angular velocities of the sun, planet, and ring gears as well as their respective radii. However, if we note that

$$R_R = R_S + 2R_P \tag{21.47}$$

or

$$T_R = T_S + 2T_P \tag{21.48}$$

Eq. (21.46) may be reduced to a more convenient form, in terms of the tooth numbers, as follows:

$$T_R\omega_R + T_S\omega_S = 2(T_S + T_P)\omega_A \tag{21.49}$$

$$= (2T_S + 2T_P)\omega_A$$

$$= (T_S + T_S + 2T_P)\omega_A$$

$$= (T_S + T_R)\omega_A$$

$$= T_S\omega_A + T_R\omega_A \tag{21.50}$$

$$T_R\omega_R - T_R\omega_A = -T_S\omega_S + T_S\omega_A \tag{21.51}$$

$$T_R(\omega_R - \omega_A) = T_S(\omega_S - \omega_A) \tag{21.52}$$

$$\frac{\omega_R - \omega_A}{\omega_S - \omega_A} = \frac{-T_S}{T_R} \quad \text{[same as Eq. (21.32)]}$$

Planetary Gear Train with One Input

EXAMPLE 21.7

Figure 21.12(a) shows a planetary gear train in which the ring gear R has 120 teeth and the planet gear 20 teeth. If the ring gear is fixed and the sun gear rotates counterclockwise at 1800 rpm, determine the angular speed of the planet arm A.

SOLUTION

First, find the sun tooth number. From the planetary geometry, T_S is given by

$$T_R = T_S + 2T_P \tag{21.48}$$

$$T_S = 120 - 2(20) = 80$$

Next, write the basic planetary relationship, based on the arm being considered fixed as in Figure 21.12(b)

$$\frac{\omega_R - \omega_A}{\omega_S - \omega_A} = \frac{-T_S}{T_R} \tag{21.32}$$

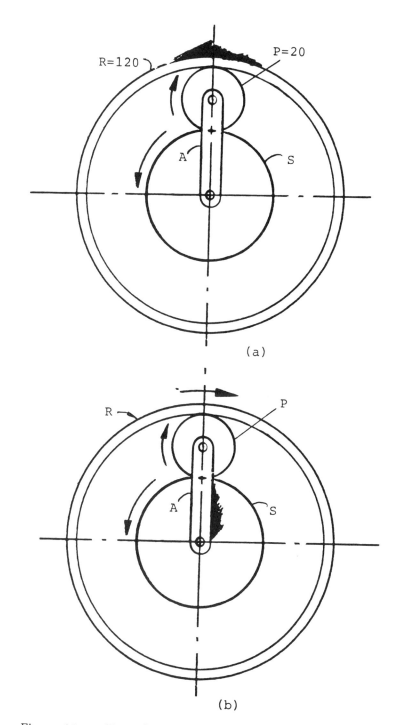

Figure 21.12 Example 21.7: (a) Ring given as fixed; (b) arm considered as fixed.

Substituting $\omega_R = 0$, we get

$$\frac{0 - \omega_A}{1800 - \omega_A} = \frac{-80}{120}$$

$$-120\omega_A = -80(1800 - \omega_A)$$

$$-200\omega_A = -144,000$$

$$\omega_A = 720 \text{ rpm}$$

The positive value of ω_A indicates that A rotates in the same direction as S.

Planetary Gear Train with More Than One Input

In some planetary gear train applications, none of the gears is fixed. In such a case, it might be necessary, for example, to find the angular velocity of the ring gear when the sun gear and planet arm are given independent input velocities, or to find the angular velocity of the planet arm when the sun and ring gears are given independent input velocities.

EXAMPLE 21.8

In the planetary gear train, shown in Figure 21.13, the sun gear and planet arm rotate at 750 to 200 rpm (CCW), respectively. Determine the angular speed and rotation of the ring gear.

SOLUTION

The problem may be solved using the general relationship in Eq. (21.34)

$$\frac{\omega_S - \omega_A}{\omega_R - \omega_A} = \frac{-T_R}{T_S} \tag{21.53}$$

where

$$\omega_S = 750 \text{ rpm}$$

$$\omega_A = 200 \text{ rpm}$$

$$T_S = 80$$

$$T_R = 120$$

Substituting these values in the above equation, we obtain

$$\frac{750 - 200}{\omega_R - 200} = \frac{-120}{80}$$

$$550 = \frac{-3}{2}(\omega_R - 200)$$

$$\frac{3}{2}\omega_R = 300 - 550$$

$$\omega_R = -166.7 \text{ rpm}$$

The negative sign indicates that the ring rotation is opposite that of the sun.

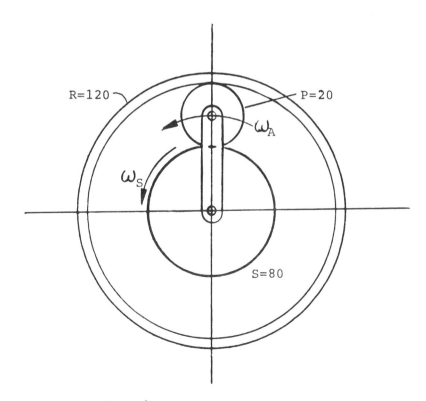

Figure 21.13   Example 21.8.

EXAMPLE 21.9

Assuming that the ring gear in Figure 21.14 has 150 teeth and is fixed, while the sun gear and planet arm rotate at 50 and 20 rpm CW, respectively, determine the tooth numbers for the sun and planet gears.

SOLUTION

By Eq. (21.34)

$$\frac{\omega_S - \omega_A}{\omega_R - \omega_A} = \frac{-T_R}{T_S}$$

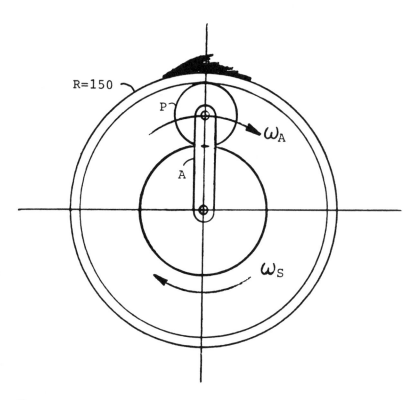

Figure 21.14   Example 21.9.

Given

$$\omega_A = 20 \text{ rpm}$$

$$\omega_S = 50 \text{ rpm}$$

$$T_R = 150$$

and substituting the values given for ω_A, ω_S and T_R in Eq. (21.53) above we get

$$\frac{50 - 20}{0 - 20} = \frac{-150}{T_S}$$

$$T_S = -150 \left(\frac{-20}{30} \right)$$

$$= 100$$

Now, for a planetary train, the teeth are related by

$$T_R = T_S + 2T_P \qquad\qquad (21.48)$$

or

$$150 = 100 + 2T_P$$

Therefore,

$$T_P = 25$$

EXAMPLE 21.10: Reverted Planetary Gear Train

In the reverted planetary gear train shown in Figure 21.15, the shaft M turns clockwise at 100 rpm while gear B is fixed. Determine the angular speed of shaft N.

SOLUTION

First, determine the number of teeth on gear E. From the train geometry,

$$D_D + D_E = D_B + D_C \qquad\qquad (21.54)$$

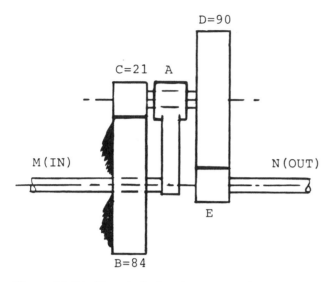

Figure 21.15 Reverted planetary gear train.

or

$$T_D + T_E = T_B + T_C \qquad (21.11)$$

Therefore,

$$T_E = 84 + 21 - 90$$

$$= 15$$

Next assume that the planet arm is "fixed." Then the velocity of E relative to the planet arm A can be written as

$$\frac{\omega_E - \omega_A}{\omega_B - \omega_A} = \frac{(-T_B)(-T_D)}{(T_C)(T_E)} \qquad (21.55)$$

$$= +\left(\frac{T_B}{T_C}\right)\left(\frac{T_D}{T_C}\right) \qquad (21.56)$$

or

$$\frac{\omega_E - 100}{0 - 100} = \frac{(84)(90)}{(21)(15)} = 24$$

from which

$$\omega_E - 100 = 24(-100)$$

$$\omega_E = -2300 \text{ rpm}$$

or

$$\omega_N = -2300 \text{ rpm}$$

The negative sign indicates the CCW rotation of E (opposite that of A).

Alternate Solution

The above example problem may also be solved in two steps as follows. First, consider the arm A with gears B and C as a subtrain, where C represents the planet gear orbiting about B, and find the rotational speed of C (ω_C). From the planetary gear equation (21.34)

$$\frac{\omega_C - \omega_A}{\omega_B - \omega_A} = \frac{-T_B}{T_C}$$

$$\frac{\omega_C - 100}{0 - 100} = \frac{-84}{21}$$

$$\omega_C = \frac{84}{21}(100) + 100$$

$$= 500 \text{ rpm} \tag{21.56}$$

Next, consider the arm A with gears D and E as a subtrain, where D represents the planet gear that drives E. Therefore, the planetary equation becomes

$$\frac{\omega_D - \omega_A}{\omega_E - \omega_A} = \frac{-T_E}{T_D} \tag{21.57}$$

Since gears C and D are keyed to the same shaft, their rotational speeds are the same. That is,

$$\omega_D = \omega_C = 500 \text{ rpm}$$

Thus, Eq. (21.57) becomes

$$\frac{500 - 100}{\omega_E - 100} = -\frac{15}{90}$$

$$\omega_E = -\frac{90}{15} (400) + 100$$

$$= -2300 \text{ rpm} \quad (\text{as before})$$

Planetary Gear Trains in Series

EXAMPLE 21.11

Let us now consider a planetary system as shown in Figure 21.16, where two trains similar to the one in Example 21.10 are joined in series. If shaft M turns at 2300 rpm, determine the output speed of shaft O.

SOLUTION

To solve this problem, the two trains are treated separately as two stages.

For stage 1:

$$\frac{\omega_E - \omega_A}{\omega_B - \omega_A} = \frac{(-T_B)(-T_D)}{(T_C)(T_E)} \tag{21.55}$$

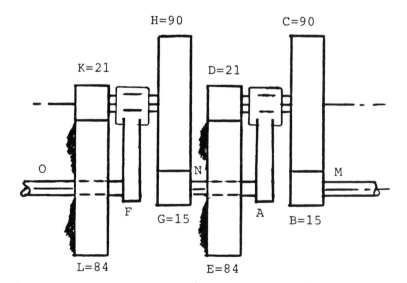

Figure 21.16   Two planetary trains in series.

Substituting the given data in the above equation gives

$$\frac{0 - \omega_A}{2300 - \omega_A} = \left(\frac{-15}{90}\right)\left(\frac{-21}{84}\right)$$

from which

$$-\omega_A = \frac{1}{24}(2300 - \omega_A)$$

$$-\frac{23}{24}\omega_A = \frac{2300}{24}$$

Therefore,

$$\omega_A = -100 \text{ rpm}$$

For stage 2:

$$\frac{\omega_L - \omega_F}{\omega_G - \omega_F} = \left(\frac{-T_G}{T_H}\right)\left(\frac{-T_K}{T_L}\right) \qquad (21.58)$$

where

$$\omega_G = \omega_A \quad \text{or input gear speed} \qquad (21.59a)$$

$$\omega_O = \omega_F \quad \text{or output gear speed} \qquad (21.59b)$$

Substituting the given values and the value found for ω_A in Eq. (21.58), we obtain

$$\frac{0 - \omega_F}{-100 - \omega_F} = \left(\frac{-15}{90}\right)\left(\frac{-21}{84}\right)$$

$$-\omega_F = \frac{1}{24}(-100 - \omega_F)$$

$$\frac{-23\omega_F}{24} = -\frac{100}{24}$$

Therefore,

$$\omega_F = 4.34 \text{ rpm}$$

Epicyclic Selective Speed Transmission

Consider the series gear train consisting of three epicyclic trains as shown in Figure 21.17. The driver gears S1, S2, and S3 are keyed to a common input shaft D, whereas the follower gears R1, R2, and R3, which turn freely on the same shaft, serve as the planet carriers or arm for planet gears P1, P2, and P3, respectively. The right end of shaft D is supported by bearings in shaft F, such that both shafts turn independently. Brake bands (not shown) are fitted to ring gears R1, R2, and R3, so that each ring gear can be held fixed as needed.

TOOTH NUMBERS

S1=S2=S3=36T
P1=P2=P3=18T
R1=R2=R3=72T

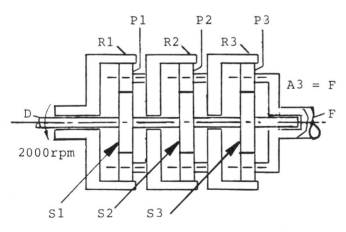

Figure 21.17   Epicyclic selective speed transmission.

Given the above mechanism, it will now be shown that by applying brakes selectively to the ring gear, three different output speeds will result.

Ring Gear R3 Held Fixed

Let R3 be held fixed. Then gear S3 becomes the driver, R3 the follower, and A3 (or F) the planet arm. Note that since trains 1 and 2 are upstream of the fixed ring gear R3, they do not affect the speed of F. Therefore, the speed of F is determined as follows:

$$\frac{\omega_{R3} - \omega_F}{\omega_{S3} - \omega_F} = \left(\frac{-T_{S3}}{T_{P3}}\right)\left(\frac{T_{P3}}{T_{R3}}\right)$$

$$\frac{0 - \omega_F}{2000 - \omega_F} = \left(\frac{-36}{18}\right)\left(\frac{18}{72}\right) = \frac{-1}{2}$$

$$-\omega_F = \frac{-1}{2}(2000 - \omega_F)$$

$$\omega_F = 666.6 \text{ rpm} \qquad (21.60)$$

Ring Gear R2 Held Fixed

Let R2 be held fixed. Then gear S2 becomes the driver, R2 the follower, and R3 the planet arm. To find the speed of F, we need to (1) determine the speed of arm R3 and (2) using this speed as the input to train 3, solve for the speed of F. Again, note that since train 1 is upstream of the fixed gear F2, it does not affect the speed of F.

Finding ω_{R3}

$$\frac{\omega_{R2} - \omega_{R3}}{\omega_{S2} - \omega_{R3}} = \left(\frac{-T_{S2}}{T_{P2}}\right)\left(\frac{T_{P2}}{T_{R2}}\right)$$

$$\frac{0 - \omega_{R3}}{2000 - \omega_{R3}} = \frac{-1}{2}$$

$$\omega_{R3} = 666.6 \text{ rpm}$$

Finding ω_F

$$\frac{\omega_{R3} - \omega_F}{\omega_{S3} - \omega_F} = \left(\frac{-T_{S3}}{T_{P3}}\right)\left(\frac{T_{P3}}{T_{R3}}\right)$$

$$\frac{2000/3 - \omega_F}{2000 - \omega_F} = \frac{-1}{2}$$

$$-\omega_F = \frac{-1}{2}(2000 - \omega_F) - \frac{2000}{3}$$

$$\omega_F = \frac{5000}{3}\left(\frac{2}{3}\right)$$

$$\omega_F = 1111.1 \text{ rpm} \tag{21.61}$$

Ring Gear $R1$ Held Fixed

Let $R1$ be held fixed. Then gear $S1$ becomes the driver, $R1$ the follower and $R2$ the planet arm. To find the speed of F, we need to (1) determine the speed of arm $R2$, (2) use the speed of arm $R2$ as the input to train 2 to determine the speed of arm $R3$, and (3) use the speed of arm $R3$ as the input to train 3 to determine the speed of F.

Finding ω_{R2}

$$\frac{\omega_{R1} - \omega_{R2}}{\omega_{S1} - \omega_{R2}} = \left(\frac{T_{S1}}{T_{P1}}\right)\left(\frac{-T_{P1}}{T_{R1}}\right)$$

$$\frac{0 - \omega_{R2}}{2000 - \omega_{R2}} = \frac{-1}{2}$$

$$-\omega_{R2} = \frac{-1}{2}(2000 - \omega_{R2})$$

$$\omega_{R2} = 666.6 \text{ rpm}$$

Finding ω_{R3}

$$\frac{\omega_{R2} - \omega_{R3}}{\omega_{S2} - \omega_{R3}} = \left(\frac{-T_{S2}}{T_{P2}}\right)\left(\frac{T_{P2}}{T_{R2}}\right)$$

$$\frac{2000/3 - \omega_{R3}}{2000 - \omega_{R3}} = \frac{-1}{2}$$

$$-\omega_{R3} = \frac{-1}{2}(2000 - \omega_{R3}) - \frac{2000}{3}$$

$$\omega_{R3} = 1111.1 \text{ rpm}$$

Finding ω_F

$$\frac{\omega_{R3} - \omega_F}{\omega_{S3} - \omega_F} = \left(\frac{-T_{S3}}{T_{P3}}\right)\left(\frac{T_{P3}}{T_{R3}}\right)$$

$$\frac{10,000/9 - \omega_F}{2000 - \omega_F} = \frac{-1}{2}$$

$$-\omega_F = \frac{-1}{2}(2000 - \omega_F) - \frac{10,000}{9}$$

$$\omega_F = \frac{19,000}{9}\left(\frac{2}{3}\right)$$

$$\omega_F = 1407.4 \text{ rpm} \tag{21.62}$$

Thus, by selectively applying brakes to the ring gears, three different output speeds can be obtained without stopping the motor or shifting any gears.

EXAMPLE 21.12

In the planetary gear train shown in Figure 21.18, the sun gear has 40 teeth, the planet gears 20 teeth, and the ring gear 80 teeth. If the ring gear is assumed to be fixed, while the sun gear rotates at 100 rpm, determine the angular velocity of the planet arm.

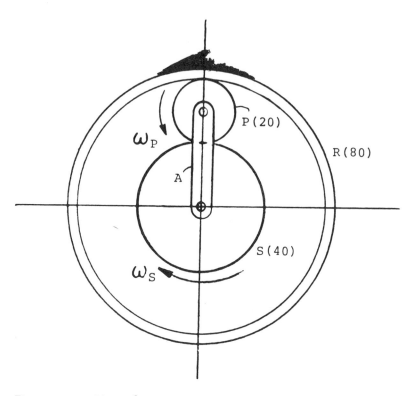

Figure 21.18   Example 21.12.

SOLUTION

By Eq. (21.34)

$$\frac{\omega_S - \omega_A}{\omega_R - \omega_A} = \frac{(T_R)(-T_P)}{(T_P)(T_S)} \tag{21.63}$$

Given

$$\omega_S = 100 \text{ rpm}$$

$$T_R = 80$$

$$T_S = 40$$

$$T_P = 20$$

and substituting the above data in Eq. (21.63), we get

$$\frac{100 - \omega_A}{0 - \omega_A} = \left(\frac{80}{20}\right)\left(\frac{-20}{40}\right)$$

$$-\frac{100}{\omega_A} + 1 = -2$$

$$\omega_A = 33.3 \text{ rpm}$$

The positive value of ω_A indicates that the direction of rotation of A is the same as that of the sun gear S. If the value were negative, this would indicate an opposite reotation to that of ω_S.

The Superposition Method

The second method, referred to as the **superposition method**, has also as its basis the relative motion concept, but in this case, the relative angular displacement is considered. In principle, the absolute displacement of the planet gear is equal to the sum of the absolute displacement of the planet arm and the displacement of the planet gear relative to the planet arm. That is,

$$\theta_A = \theta_B + \theta_{A/B} \tag{21.64}$$

EXAMPLE 21.13

To illustrate the superposition method, let us consider the reverted planetary gear system shown in Figure 21.15. Gear B has 84 teeth, C has 21, D has 90, and E has 15. Let it be required to find the speed of rotation of gear E if gear B is fixed and the planet arm A rotates CW at 100 rpm.

PROCEDURE

1. Assume that the entire gear train is locked and the assembly, with the planet arm, is rotated one turn in the direction specified. Then tabulate the displacement of all the gears and the planet arm as shown in Table 21.1.

Table 21.1   Angular Displacements For Example 21.13

	Gear B	Gear C	Gear D	Gear E	Arm
Gears locked	+1	+1	+1	+1	+1
Gears unlocked	−1	$\dfrac{+84}{21}$	$\dfrac{+84}{21}$	$\dfrac{-84}{21} \times \dfrac{90}{15}$	0
Arm fixed					
Total θ	0	5	5	−23	+1

2. Assume now that the planet arm is fixed, the gears are unlocked, and the gear B is rotated one turn in the opposite direction so that it is returned to its original position. Then record in the table the displacement of all the other gears relative to the fixed arm. These relative displacements are easily determined, if we consider that the system is now reduced to that of an ordinary reverted gear train. In this example, the relative displacements are

$$\frac{\theta_C}{\theta_B} = \frac{-T_B}{T_C} \qquad \text{same as } \frac{\omega_C}{\omega_B} \qquad (21.65)$$

$$\frac{\theta_D}{\theta_B} = \frac{-T_B}{T_D} \qquad \text{same as } \frac{\omega_D}{\omega_B} \qquad (21.66)$$

$$\frac{\theta_E}{\theta_B} = \left(\frac{T_B}{T_C}\right)\left(\frac{T_D}{T_E}\right) \qquad \text{same as } \frac{\omega_E}{\omega_B} \qquad (21.67)$$

3. Finally, sum the displacements recorded in the tabulation for each of the gears. This summation, which represents the absolute displacements of all the components in the train, can then be used to determine the required velocity ratios.

In this example, the velocity ratio between gears E and the planet arm A is computed as

$$\frac{\omega_E}{\omega_A} = \frac{\theta_E}{\theta_A} = -23 \qquad (21.68)$$

Since

$$\omega_A = 100 \text{ rpm}$$

$$\omega_E = 100(-23) = -2300 \text{ rpm}$$

the negative sign indicates that gear E rotates CCW (opposite to the arm).

EXAMPLE 21.14

In the planetary gear train shown in Figure 21.19, gear B turns at 5 rpm about fixed axis 0, while arm A turns independently about the same axis, carrying gears C and D. Gear D is meshed to a fixed gear E. Determine the angular velocity of arm A.

SOLUTION

1. Assume the entire gear train is locked and the planet arm is given one turn. Tabulate all displacements as $+1$ in Table 21.2.
2. Now hold the planet arm fixed and rotate gear E one turn in the opposite direction to return it to its original or fixed position. Re-

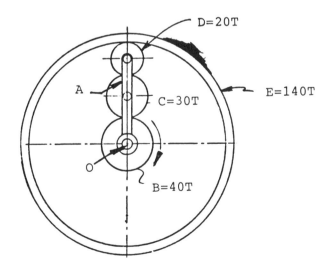

Figure 21.19   Example 21.13.

Table 21.2   Angular Displacements For Example 21.14

	Arm	Gear E	Gear D	Gear C	Gear B
Gear locked	+1	+1	+1	+1	+1
Gears unlocked	0	−1	−7	$+\dfrac{14}{3}$	$\dfrac{-7}{2}$
Arm fixed					
Total θ	+1	0	−6	$+\dfrac{17}{3}$	$\dfrac{-5}{2}$

cord this displacement as -1 in Table 21.2. Also record the relative displacements of gears D, C, and B as follows:

$$\frac{-\theta_D}{\theta_E} = \left(\frac{-140}{20}\right) = -7$$

$$\frac{-\theta_C}{\theta_E} = \left(\frac{-140}{20}\right)\left(\frac{-20}{30}\right) = \frac{14}{3}$$

$$\frac{-\theta_B}{\theta_E} = \left(\frac{-140}{20}\right)\left(\frac{-20}{30}\right)\left(\frac{-30}{40}\right) = -\frac{7}{2}$$

From Table 21.2, it is seen that

$$\frac{\omega_A}{\omega_B} = \frac{\theta_A}{\theta_B} = \frac{(+1)}{\left(\dfrac{-5}{2}\right)} = \frac{-2}{5}$$

Therefore,

$$\omega_A = -\omega_B\left(\frac{2}{5}\right)$$

$$= -5\left(\frac{2}{5}\right)$$

$$= -2 \text{ rpm}$$

The negative sign indicates that the angular velocity of arm A is opposite that of gear B.

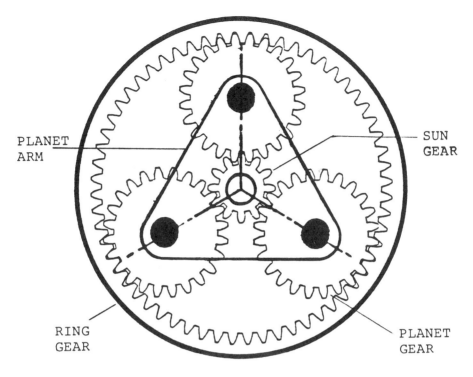

PLANET
ARM

SUN
GEAR

RING
GEAR

PLANET
GEAR

Figure 21.20   Planetary gear train.

Planetary Gear Train in Automotive Transmission

A planetary gear train can be used to step up or reduce the speed of an output shaft. It can also be used as a direct or reverse drive. The automatic transmission makes full use of these features of the planetary gear train.

The planetary gear train is compact. Its gears are in constant mesh. It is quiet and strong. Figure 21.20 illustrates a typical planetary gear, whereas Figure 21.21 illustrates its four basic modes of operation.

Low Speed [Figure 21(a)]

If the ring gear is held and power applied to the sun gear, the planet gears are forced to turn and walk around in the ring gear. This action causes the planet arm to revolve at a lower speed than the sun gear.

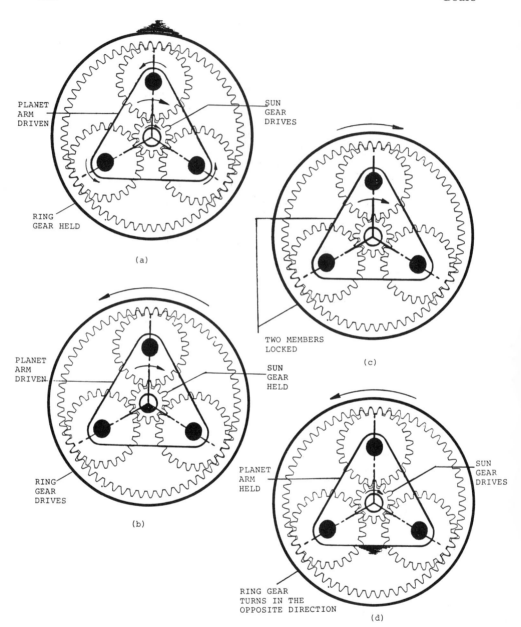

Figure 21.21   Planetary gear train operation: (a) low speed; (b) medium speed; (c) direct drive; (d) reverse.

Medium Speed [Figure 21(b)]

If the sun gear is held and the ring gear driven, the planet gears must walk around the sun gear. This causes the planet arm to move more slowly than the ring gear.

High Speed or Direct Drive [Figure 21(c)]

When any two members of the train are locked together, planetary action ceases. Under these conditions, the gear train will revolve as a solid unit with no increase or decrease in speed.

Reverse Drive [Figure 21(d)]

By holding the planet arm and driving the sun gear, the planet gears are forced to rotate about their axles. This causes them to drive the ring gear in a reverse direction at a reduced speed.

SUMMARY

1. When there is an input and output, and no member is held, the result is *neutral*.
2. When the planet arm is the output and one member is held, the result is a *speed reduction*. Conversely, when the planet arm is the input and one member is held, the result is a *speed increase*.
3. When the planet arm is held, the result is the *reverse*.
4. When any two members are locked together, the result is *direct drive* or *high speed*.

Relative sizes of gears determine the amount of reduction of the train. When the input member is smaller than the output member, there is gear speed reduction.

21.6 PLANETARY BEVEL GEAR TRAINS

A class of planetary gear trains that deserves special mention is the planetary bevel gear train that finds use in machine tools, automotive transmissions, and other reducing gears. Two common examples are the Humpage epicyclic gear and the bevel gear differential.

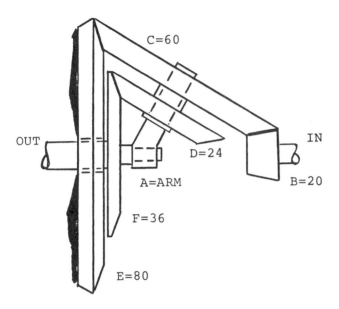

Figure 21.22   Humpage epicyclic gear.

The Humpage Epicyclic Gear

Figure 21.22 shows a Humpage epicyclic gear where the input is to
gear B and the output is from gear F. The gear E is fixed, while gears
C and D are one unit free to rotate on a planet arm A, which in turn
rotates freely on the output shaft.

   EXAMPLE 21.15

Given that the tooth numbers in Figure 21.22 are as shown and B turns
at 1200 rpm, find the angular speed of gear F.

   SOLUTION

This problem may be solved in two steps as follows:

1.   Consider the gears B, C, and E and solve for the arm speed ω_A

$$\frac{\omega_E - \omega_A}{\omega_B - \omega_A} = \left(\frac{-T_B}{T_C}\right)\left(\frac{+T_C}{T_E}\right) \qquad (21.69)$$

$$= \frac{-T_B}{T_E} \qquad (21.70)$$

$$\frac{0 - \omega_A}{1200 - \omega_A} = \frac{-20}{80} = \frac{-1}{4}$$

$$-\omega_A = \frac{-1}{4}(1200 - \omega_A)$$

$$\omega_A + \frac{\omega_A}{4} = 300$$

$$\omega_A = +300\left(\frac{4}{5}\right)$$

$$= +240 \text{ rpm}$$

2. Consider the gears B, C, D, and F and solve for the output speed ω_F using the arm speed ω_A found in Step 1.

$$\frac{\omega_F - \omega_A}{\omega_B - \omega_A} = \left(\frac{-T_B}{T_C}\right)\left(\frac{+T_D}{T_F}\right) \tag{21.71}$$

$$\frac{\omega_F - 240}{1200 - 240} = \frac{-20}{60} \times \frac{24}{36} = \frac{-2}{9}$$

$$\omega_F = \frac{-2}{9}(960) + 240$$

$$= +26.6 \text{ rpm} \tag{21.72}$$

The positive value indicates that the direction of rotation of F is the same as that of B.

Note that a system of determining the directions of motion of bevel gears using arrows is advisable. Generally, it is useful to assume that the motion is positive when the nearer side of the gear moves in a given direction, say, upward, in which case downward motion would be negative.

Alternative Solution

Alternatively, the above results can be obtained in one step, since it can be shown that by combining Eqs. (21.70) and (21.72), the overall speed ratio can be expressed as

$$\frac{\omega_F}{\omega_B} = \frac{T_B}{T_C} \left[\frac{T_C/T_E - T_D/T_F}{1 + T_B/T_E} \right] \tag{21.73}$$

Then, upon substitution,

$$\frac{\omega_F}{1200} = \frac{20}{60} \left[\frac{60/80 - 24/36}{1 + 20/80} \right]$$

$$= \frac{1}{3} \left[\frac{0.75 - 0.666}{1.25} \right]$$

$$\omega_F = 26.6 \text{ rpm}$$

21.7 BEVEL GEAR DIFFERENTIAL

The **differential gear**, first introduced by clock makers and now a standard feature on automobiles, solves the problem of cornering. In cornering, the outer wheels of a vehicle must travel farther than the inner wheels and therefore must turn faster.

Figure 21.23(a) shows a schematic of the bevel gear differential. Here, the input side gear B is taken as the sum gear, the spider gears D and E (mounted on the spider) as the planet gears, the spider A as the planet arm, and the output side gear C as the ring gear.

To understand the motion of the differential, it should be observed that the side gears B and C are capable of rotating only in one plane (the vertical plane) about the wheel axles, whereas the spider gears are capable of rotating about not only the wheel axles, but also their respective spindles.

To further illustrate the spider gear motion, not that each of the spindles extends through the spider gears a short distance. All power delivered to the wheel axles is received by these projections. For example, if one were to grasp the ends of these spindles and give the assembly a turn, both side gears will turn with the spider. That is, there will be no relative motion between the spider and side gears. However, if one of the side gears were to be held fixed while the spider was being turned, the spider gears will cause the other side gear to turn.

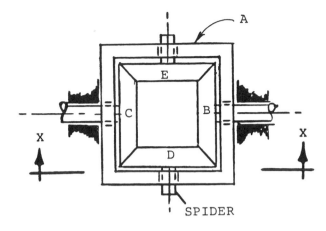

(a)

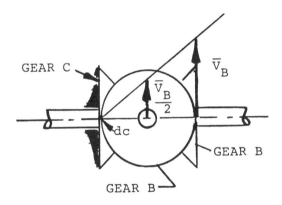

(b)

Figure 21.23   Velocity diagram of the differential: (a) Differential; (b) velocity diagram (viewed from X-X).

In summary, as long as the resistance to turning is equal for both side gears, they will turn together, and the spider gears will not rotate about their spindles. However, when the resistance is greater in one side gear than in the other, the gear with the greater resistance may turn more slowly or stand still, whereas the spider gears begin to turn on their spindles and drive the other side gear at a greater speed.

Velocity Analysis: Instant Center Method

Consider the output side gear C in Figure 21.23 fixed, while the input gear B rotates with an angular velocity ω_B, driving the spider gear D in turn. Let V_B be the instantaneous velocity of the contact point between gears B and D and dc designate the instant center for gears D and C. Then the velocity of the center (or spider arm) of gear D is $V_B/2$. See Figure 21.23(b). That is,

$$\omega_B = \frac{V_B}{R_B} \quad \text{and} \quad \omega_A = \frac{V_B}{2R_B} \tag{21.74}$$

Therefore,

$$\omega_B = 2\omega_A \tag{21.75}$$

Likewise, if gear B is fixed and gear C serves as the input to drive gear D, with V_C being the contact point velocity between gears D and C, the velocity of the center (or spider arm) of gear D is $V_C/2$.
    That is,

$$\omega_C = \frac{V_C}{R_C} \quad \text{and} \quad \omega_A = \frac{V_C}{2R_C} \quad (R_C = R_B = R) \tag{21.76}$$

Therefore,

$$\omega_C = 2\omega_A \tag{21.77}$$

Equations (21.75) and (21.77) indicate the arm rotation is twice that of the input gear when the output ear is fixed.
    Now, if both gears B and C are rotating, then the velocities $V_B/2$ and $V_C/2$ will add or subtract, depending on the directions of rotation of B and C. That is,

$$\omega_A = \frac{V_B/2 + V_C/2}{R} \tag{21.78}$$

$$= \frac{\omega_B + \omega_C}{2} \tag{21.79}$$

$$2\omega_A = \omega_B + \omega_C \tag{21.80}$$

This is the general equation of the bevel gear differential. Note that if ω_B and ω_C are equal in magnitude and sense, then

$$\omega_A = \omega_B = \omega_C \tag{21.81}$$

This means that the wheel axles and planet arm rotate as a unit at one speed.

Velocity Analysis: Relative Velocity Method

Consider again Figure 21.23 and, as before, let the output side gear C be fixed while B turns with an angular velocity ω_B. Using the planetary equation, we can write

$$\frac{\omega_C - \omega_A}{\omega_B - \omega_A} = \left(\frac{-T_B}{T_E}\right)\left(\frac{+T_D}{T_C}\right) \tag{21.82}$$

where

$$T_B = T_C$$
$$T_D = T_E$$

Then Eq. (21.82) becomes

$$\frac{\omega_C - \omega_A}{\omega_B - \omega_A} = -1 \tag{21.83}$$

Since C is fixed, $\omega_C = 0$ and Eq. (21.83) reduces to

$$\frac{0 - \omega_A}{\omega_B - \omega_A} = -1$$

$$\omega_A = \omega_B - \omega_A \tag{21.84}$$

$$\omega_B = 2\omega_A \quad \text{(as before)} \tag{21.75}$$

Similarly, when gear B is fixed, Eq. (21.83) reduces to

$$\omega_C = 2\omega_A \quad \text{(as before)} \tag{21.77}$$

Also, if both gears B anc C are rotating, Eq. (21.83) reduces to

$$\omega_C - \omega_A = -\omega_B + \omega_A$$

or

$$2\omega_A = \omega_B + \omega_C \quad \text{(as before)} \qquad (21.80)$$

Automobile Differential

As used in an automobile, the bevel gear differential is a mechanical arrangement that allows power from the engine to be transmitted to the wheel axles, while automatically allowing one rear wheel to slow down as the other speeds up.

Figure 21.24 shows a typical automobile differential, in which the ring gear A acts as the planet arm, gears D and E act as the planet gears, and gears B and C as the sun or ring gear, depending on their relative angular speeds.

When an automobile turns a corner, the outer rear wheel must rotate faster than the inner rear wheel in order that no slipping occurs. The use of the differential makes this action possible.

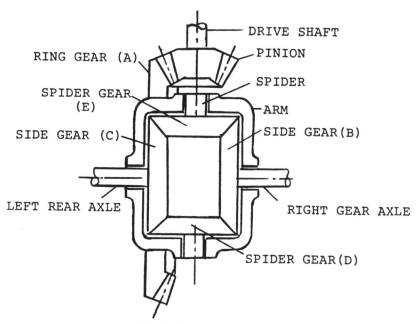

Figure 21.24   Automobile differential.

V

CAMS

Cams play an important part in most machines. As machine members, cams are used to convert rotary to oscillatory motions through direct contact with another member called the follower. As a cam rotates, the cam profile causes the follower to move along its axis or its pivot to bring a machine part into proper position at the proper time. This combination of cam and follower has made the cam probably the most versatile element of a machine since, by its use, complicated motions, which are impossible to achieve by any other means, can be produced. The cam is, therefore, extremely useful in performing a variety of machine functions. A familiar example of cam application is in the internal combustion engine where the cam is used for the opening and closing of the valves. Other examples can be found in bottling, canning and textile machines, printing presses, typewriters, sewing machines, machine tools, and household appliances.

22

Cam Fundamentals

22.1 CAM TERMINOLOGY

Before proceeding with a discussion of cams, one should have a clear understanding of the terminology used in cam analysis and design. Figure 22.1 illustrates some of the components and features defined in the list below.

Cam. A rotating or sliding machine member whose function is to impart a predetermined motion to another part, *the follower*, that rolls or slides along its surface.

Pitch curve. The path of the trace point.

Prime Circle. The smallest circle, centered on the cam axis, that touches the pitch curve.

Pitch point. The point on the pitch curve where the pressure angle is maximum.

Base circle. The smallest circle that can be drawn tangent to the cam profile.

Pressure angle or jamming angle. The angle between the normal to the pitch curve and the direction of the follower at the point of contact.

Guide. A bearing surface designed to maintain a constant axis of follower motion.

Spring. A means to maintain follower contact with the cam surface.

Cam inversion. When the roles of the cam and follower are reversed such that the follower drives the cam.

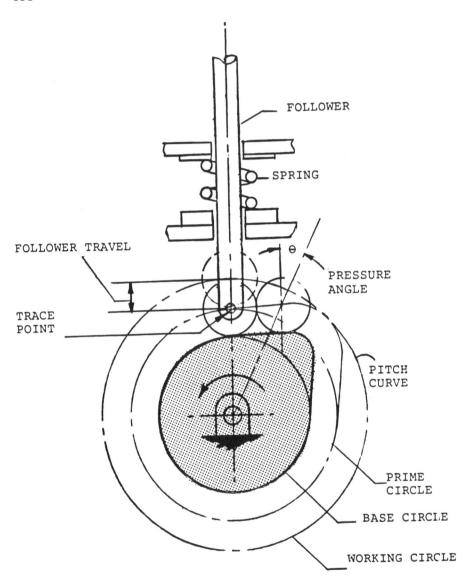

Figure 22.1   Cam terminology.

Jam. The tendency of two surfaces, ordinarily in sliding contact, to
    seize due to friction from side thrust.
Cam profile. The actual shape of the cam.
Trace point. The end point on a knife-edge follower.

MOTION TERMINOLOGY

Displacement. A measure of the net change in position of a point.
Velocity. The rate of change of position of a point with respect to time,
    or displacement per unit time.
Acceleration. The rate of change of velocity with respect to time, or
    the rate of speed-up.
Jerk. The rate of change in acceleration per unit time.
Cycle. One complete revolution of a cam.
Period. The time required for a motion to complete one cycle.
Lift. The maximum displacement of the follower.
Dwell. A period in the motion cycle during which the follower neither
    rises nor falls.
Return. That period of the motion cycle during which the follower falls
    or returns from its maximum displacement to its lowest (or home)
    position.
Transition point. Point of maximum velocity where the acceleration
    changes from positive to negative.
Stroke. The greatest distance through which the follower moves.

Cam-Follower Operation

The rotational cycle of a cam-follower mechanism is typified by the
rise, fall, and dwell motions of the follower. During the rise motion,
the follower is driven away from the cam axis by the increasing radius
of the cam profile at the point of contact. The follower acceleration
depends on the cam speed and instantaneous rate of change of the
profile radius with respect to the cam angle.

During the fall motion, the follower continues to maintain contact
with the cam profile whose radius is continually decreasing at the point
of contact. A closing force is essential to maintain the required contact
between the follower and cam. This force may be provided by gravity,
a spring, or some means of positive drive.

Occasionally, during the cam cycle, there may be periods where
the follower neither rises nor falls. Such periods are termed "dwells."

A dwell is obtained by making the appropriate section of the cam profile a circular arc concentric with the cam axis.

22.2 TYPES OF CAMS

Cams are made in various types. The most common types found in industry are the following:

1. *Translation or wedge cam*. The translation or wedge cam, shown in Figure 22.2(a), is the most rudimentary form of a cam. As simple as it appears, it serves to illustrate the basic principle of all cams. In simple form the wedge may be used to convert reciprocating motion in one plane into reciprocating or oscillatory motion in another plane.
2. *Disk or plate cam*. The disk or plate cam, shown in Figure 22.2(b), is one whose outline or periphery is shaped to produce the required motion of the follower as the cam rotates.
3. *Cylindrical cam*. The cylindrical or drum cam, shown in Figure 22.2(d), is another form of grooved cam designed for positive action. The surface of the cylinder is grooved such that as the cam rotates about its axis, the follower roller is guided within the groove. The cylindrical cam finds many applications, particularly in machine tools. One of the most common applications, however, is found level winding fishing reels.
4. *Positive-action cams*. A positive-action cam is one that employs a positive means to return the follower, without reliance on a spring or gravity. This type of cam is generally suitable for designs in which high speeds and heavy loads are to be considered.
5. *Grooved-plate cam*. The grooved-plate cam, shown in Figure 22.2(c), is shaped in the form of a disk and designed such that the follower is constrained to move within a groove cut in the face of the plate. This type of cam is very effective in producing positive action and is generally suitable for heavy loads and high speeds. However, it has the disadvantage that, in order for the roller to roll without slipping, some clearance has to be provided between the roller and groove. Consequently, considerable backlash can be developed, causing the follower motion to be inaccurate. This problem is generally solved using two rollers of slightly different diameters running on an eccentric groove.
6. *Constant-diameter cam*. The simplest form of a positive cam is the constant-diameter or *constant-breadth cam* illustrated in Fig-

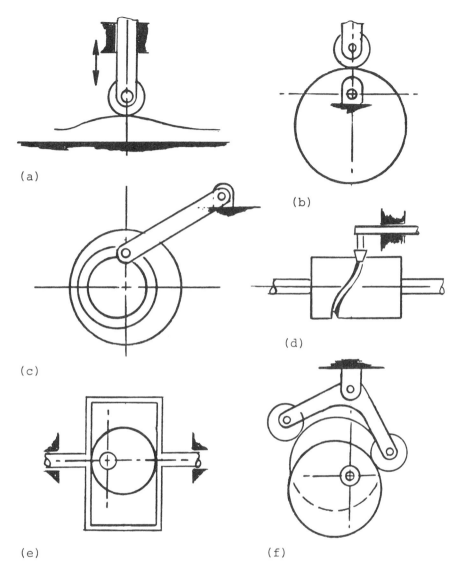

Figure 22.2   Principal types of cams: (a) Translation cam; (b) disk or plate cam; (c) grooved-plate cam; (d) cylindrical cam; (e) constant-diameter or constant breadth cam; (f) conjugate cam.

ure 22.2(e), where the follower surfaces are at a fixed distance apart and contact opposite sides of the cam. The fixed-distance requirement between the follower surfaces is based on the fact that, in order for the cam to drive back the follower, the motion during the first half of the cycle must be the same as that in the second half.

Or, the displacement must be the same in both directions. For this
to occur, there must be double contact between the follower and
cam throughout the cycle, which means that the width across the
cam, measured through the center, must be constant over 360° rotation. Hence, the term "constant diameter" or "constant breadth"
is frequently used to describe this type of cam. This cam has been
used successfully in movie cameras. Another form of this cam is
the **conjugate cam** with oscillating rollers, instead of two flat surfaces, as shown in Figure 22.2(f). Here the rollers are driven separately by two disks of identical profile, keyed to the same shaft.
Thus, the follower is always in contact with the cam surface. This
type of cam is generally used when the outward and return motions
of the follower are not the same.

22.3 CLASSIFICATION OF FOLLOWERS

Followers are classified according to their types, as well as their arrangements or paths with respect to the cam. There are four types of
followers, namely

1. **Roller follower** [see Figure 22.3(a)]. The use of this follower avoids
    the wear and stresses produced in the pointed follower and thus
    permits heavier loads and higher speeds. It, however, lacks the
    sensitivity and simplicity of the pointed follower.
2. **Pointed or knife-edge follower** [see Figure 22.3(b)]. This follower
    is best suited for intricate motions since the point is sensitive to
    abrupt changes in the cam profile. However, the point is subject to
    wear and produces large stresses. Therefore, the use of this follower
    is limited to very light loads and low speeds.
3. **Flat-face follower** [see Figure 22.3(c)]. This follower is suited only
    for cam profiles with very simple outlines, since compared to the
    pointed and roller followers, it is the least sensitive to abrupt
    changes in contour. However, it has the advantage that excessive
    steepness of the cam contour produces no adverse effects such as
    jamming that results from side thrust, as in the case of the pointed
    or roller follower.
4. **Round-edge or spherical follower** [see Figure 22.3(f)]. This fol-
    lower is simply a modified flat-faced follower. The round edge is
    generally employed to reduce the surface stresses and wear pro-
    duced in the flat-face follower.

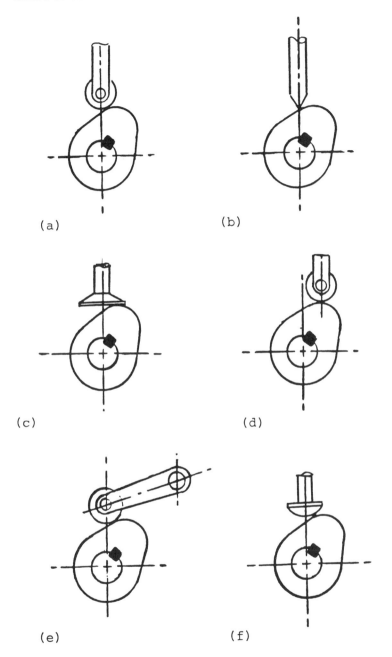

(a)

(b)

(c)

(d)

(e)

(f)

Figure 22.3   Types of followers: (a) Radial roller follower; (b) radial pointed follower; (c) radial flat-face follower; (d) offset roller follower; (e) pivoted roller follower; (f) radial spherical follower.

There are three basic arrangements as follows:

1. Radial
2. Offset
3. Pivoted or oscillating

When a follower is arranged such that its path is on a straight line passing through the axis of the cam, it is classified as a **radial follower**. Figure 22.3(a) illustrates a **radial roller follower**. When a follower is arranged such that its path does not pass through the cam axis, it is classified as an **offset follower**. Figure 22.3(d) illustrates an **offset roller follower**. When a follower is pinned such that its path is that of an arc, it is classified as a **pivoted or oscillating follower**. Figure 22.3(e) illustrates a **pivoted or oscillating roller follower**.

22.4 TYPES OF FOLLOWER MOTION

In designing a cam, the principal interest is the motion of the follower. The type of motion produced by a cam-follower mechanism is dependent on the cam profile. This profile is normally designed so that the follower displacement is as gradual as possible to prevent large acceleration or inertia forces in the mechanism. For low cam speeds, inertia forces in the follower system will cause little deflection of component parts. However, as speeds increase, these forces increase and are likely to create vibration. Vibration induces stresses that can result in failure of the system. Also, impact forces caused by separation between the cam and follower faces due to vibration or excessive clearances can cause structural damage.

Since inertia and impact forces are likely to develop when a cam operates at high speed, the acceleration characteristics of the follower motion are important. The most common types of follower motion employed in cam design are

1. Uniform velocity motion
2. Modified uniform motion
3. Uniformly accelerated (or parabolic) motion
4. Simple harmonic motion
5. Cycloidal motion

The characteristics of these motions will be discussed in the following sections.

Uniform Velocity Motion

Uniform velocity motion, depicted in Figure 22.4, is based on the principle that equal displacements are made in equal time periods. For this motion, displacement [see Figure 22.4(a)] is represented by the straight line AB having a constant slope, and hence the velocity [see Figure 22.4(b)] is constant at all times. Also, there is no acceleration, except at the beginning and end of the period where it is infinitely large in the positive and negative directions, respectively [see Figure 22.4(c)]. Because such acceleration characteristics tend to produce too abrupt

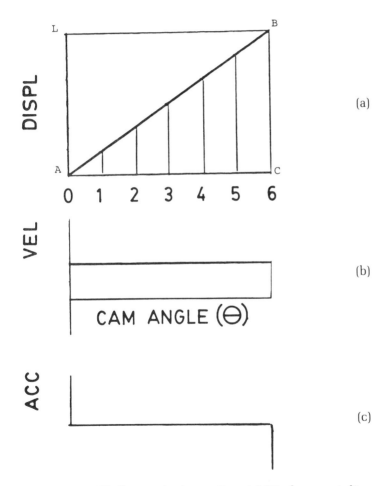

Figure 22.4   Uniform velocity motion: (a) Displacement diagram; (b) velocity diagram; (c) acceleration diagram.

changes in the follower motion, uniform velocity motion is seldom used, except for low-speed and nonprecision applications.

Displacement Curve Development

The graphical procedure for obtaining the displacement curve is as follows:

PROCEDURE [see Figure 22.4(a)]

1. Divide the displacement axis AL in equal parts, the same as the cam rotation axis AC.
2. Project points horizontally from the divisions on AL to intersect respective ordinates of the cam rotation axis AC.
3. Draw a straight line AB through the points of intersection found in Step 2 to obtain the required displacement curve AB.

Motion Equations

Let the curve shown in Figure 22.4(a) represent a typical uniform velocity motion where the maximum displacement L is attained over the period β. Since this curve is a straight line, the displacement, velocity, and acceleration can be expressed in terms of angle θ as follows:

For range $0 < \theta < \beta$

$$S = \left(\frac{L}{\beta}\right)\theta \tag{22.1}$$

$$V = \frac{dS}{d\theta} = \frac{L}{\beta} \tag{22.2}$$

$$A = \frac{dV}{d\theta} = 0 \tag{22.3}$$

Modified Uniform Velocity Motion

The uniform velocity curve is most effective when combined with other curves to improve the acceleration and deceleration conditions that occur at its beginning and end points. One such modification, seen in Figure 22.5, is the use of circular arcs to round off the ends of the curve. This results in a more gradual change in velocity or acceleration of the follower motion at the critical points and, hence, a smoother operation. [See Figures 22.5(b) and (c).]

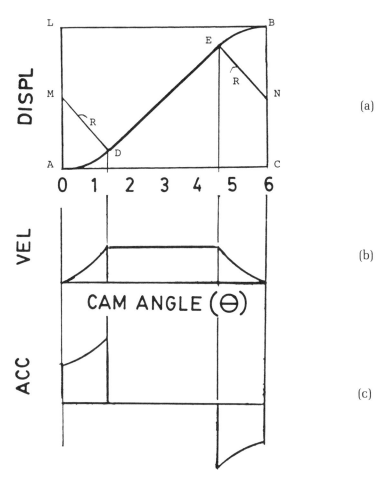

Figure 22.5   Modified uniform velocity motion: (a) Displacement diagram; (b) velocity diagram; (c) acceleration diagram.

Displacement Curve Development

The graphical procedure for obtaining the modified displacement curve is as follows:

PROCEDURE [see Figure 22.5(a)]

1. Using points M and N as the centers and the distance MD or NE as the radius, describe two arcs passing through A and B, the beginning and end points, respectively, of the rise cycle.
2. Draw a straight line tangential to both arcs to obtain the required curve.

It should be noted that, besides circular arcs, the use of other curves such as simple harmonic, parabolic, cycloidal, etc. for similar modifications is also possible.

Motion Equations

Figure 22.6 shows a typical modified motion curve where the entire period is divided into three regions, namely

$0 < \theta < \theta 1$ region 1 defined by arc at start of stroke

$\theta 1 < \theta < \theta 2$ region 2 defined by linear curve

$\theta 2 < \theta < \bar{\theta}$ region 3 defined by arc at end of stroke

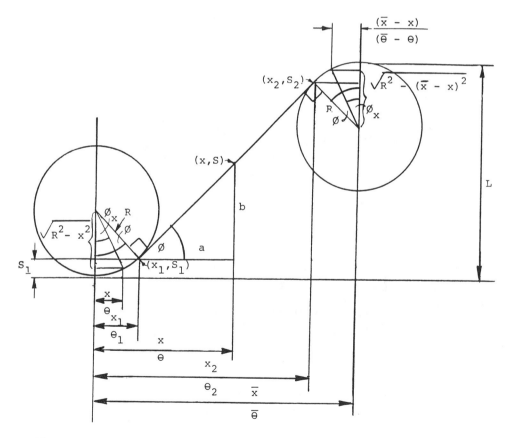

Figure 22.6   Geometric parameters.

In this figure, let

L = total lift

$\bar{\theta}$ = period (in degrees)

$\bar{x}$ = linear distance corresponding to period $\bar{\theta}$

R = radius of the circular arc at the beginning (region 1) and end (region 3) of the period

ϕ = angle subtended by the arc in regions 1 and 3

From this geometry, it can be shown that the angle θ and distance x are related by the proportionality

$$\frac{\theta}{\bar{\theta}} = \frac{x}{\bar{x}} \quad \text{in region 1} \tag{22.4}$$

and

$$\frac{(\bar{x} - x)}{\bar{x}} = \frac{(\bar{\theta} - \theta)}{\bar{\theta}} \quad \text{in region 3} \tag{22.5}$$

which means that $\theta 1$ and $\theta 2$, the cam angles where the points of tangency occur, and their respective distance $x1$ and $x2$ are given by

$$\frac{\theta 1}{\bar{\theta}} = \frac{x1}{\bar{x}} \quad \text{where } x1 = R \sin \phi \tag{22.6}$$

and

$$\frac{\theta 2}{\bar{\theta}} = \frac{x2}{\bar{x}} \quad \text{where } x2 = x - x1 \tag{22.7}$$

Also, the angle ϕ can be expressed as

$$\phi = \tan^{-1}\left[\frac{L - 2(R - R \cos \phi)}{x - 2R \sin \phi}\right] \tag{22.8}$$

$$= \tan^{-1}\left(\frac{-B \pm \sqrt{B^2 - 4AC}}{2A}\right) \tag{22.9a}$$

where A, B, and C are constants given by

$$A = \bar{x}^2 - 4R^2 \tag{22.9b}$$

$$B = -2\bar{x}(L - 2R) \tag{22.9c}$$

$$C = L^2 - 4RL \tag{22.9d}$$

With the geometric parameters of the curve completely defined, the resulting expressions for displacement, velocity, and acceleration for regions 1, 2, and 3 can be written as follows:

For region 1: $0 < \theta < \theta_1$,

Displacement:

$$S(\theta) = R - \sqrt{R^2 - \left(\frac{\theta\bar{x}}{\bar{\theta}}\right)^2} \tag{22.10}$$

Velocity:

$$V(\theta) = \frac{\theta \left(\frac{\bar{x}}{\bar{\theta}}\right)^2}{\sqrt{R^2 - \left(\frac{\theta\bar{x}}{\bar{\theta}}\right)^2}} \tag{22.11}$$

Acceleration:

$$A(\theta) = \frac{\left(\frac{\bar{x}}{\bar{\theta}}\right)^2 R^2}{\left[R^2 - \left(\frac{\theta\bar{x}}{\bar{\theta}}\right)^2\right]^{3/2}} \tag{22.12}$$

For region 2: $\theta_1 < \theta < \theta_2$,

Displacement:

$$S(\theta) = S(\theta_1) + \left(\frac{\theta\bar{x}}{\bar{\theta}} - R\sin\phi\right)\tan\phi \tag{22.13}$$

Velocity:

$$V(\theta) = \frac{\bar{x}}{\bar{\theta}}\tan\phi \quad \text{(constant)} \tag{22.14}$$

Acceleration

$$A(\theta) = 0 \tag{22.15}$$

For region 3: $\theta 2 < \theta < \bar{\theta}$,

Displacement:

$$S(\theta) = L - R + \sqrt{R^2 - \bar{x}^2 \left(1 - \frac{\theta}{\bar{\theta}}\right)^2} \qquad (22.16)$$

Velocity:

$$V(\theta) = \frac{\left(\frac{\bar{x}^2}{\bar{\theta}}\right)\left(1 - \frac{\theta}{\bar{\theta}}\right)}{\sqrt{R^2 - \bar{x}^2 \left(1 - \frac{\theta}{\bar{\theta}}\right)^2}} \qquad (22.17)$$

Acceleration:

$$A(\theta) = \frac{-\left(\frac{\bar{x}}{\bar{\theta}}\right)^2 R^2}{\left[R^2 - \bar{x}^2 \left(1 - \frac{\theta}{\bar{\theta}}\right)^2\right]^{3/2}} \qquad (22.18)$$

Uniformly Accelerated (or Parabolic) Motion

This motion depicted in Figure 22.7 is based on the principle that a body having uniform acceleration has a displacement $S = 1/2\ at^2$, as represented graphically by Figure 22.7(a). For this motion, the velocity increases linearly to a maximum at midperiod and then decreases linearly to zero at the end of the period. As a result, the acceleration [see Figure 22.7(c)] is positive and uniform over the first half of the period, up to the point of maximum velocity where it instantaneously reverses directions to become negative and uniform over the remaining half of the period. Compared to modified velocity motion, uniformly accelerated motion is smoother and hence more desirable. However, because of the sudden reversal of acceleration that occurs at maximum velocity, such motion is not suitable for high-speed applications, since in that type of situation the follower tends to leave the cam surface momentarily, only to return with a resulting shock to the mechanism with which it regains contact. The main advantage of this kind of motion is that a given displacement in a given length of time is accomplished with minimum acceleration.

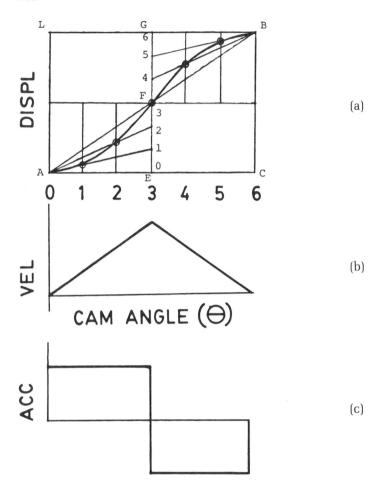

Figure 22.7   Uniform acceleration motion: (a) Displacement diagram; (b) velocity diagram; (c) acceleration diagram.

Displacement Curve Development

The graphical procedure for obtaining the displacement diagram is as follows:

PROCEDURE [see Figure 22.7(a)]

1. Divide the total cam lift and corresponding cam rotation angle into equal parts. Let EF and FG represent the first half and second half of the lift and AE the period corresponding to EF.

2. Erect ordinates from division points on *AE* to intersect their respective straight lines drawn from *A* to the division points on *EF*.
3. Draw a smooth curve through the points of intersection found in Step 2 to obtain the required curve for the first half of the period.
4. Repeat Steps 1, 2, and 3 for distance *EG* to obtain the remaining half of the required curve *AB*.

Motion Equations

Let the curve shown in Figure 22.7(a) represent a typical uniform acceleration (or parabolic) motion, where L is the total displacement over a period β. Given that this parabolic curve is inverted at $\beta/2$, the displacement, velocity, and acceleration can be expressed in terms of angle θ as follows:

For range $0 < \theta < \beta/2$

$$S = \left(\frac{2L}{\beta^2}\right) \theta^2 \tag{22.19}$$

$$V = \frac{dS}{d\theta} = \left(\frac{4L}{\beta^2}\right) \theta \tag{22.20}$$

$$A = \frac{dV}{d\theta} = \frac{4L}{\beta^2} \tag{22.21}$$

For range $\beta/2 < \theta < \beta$

$$S = L - \frac{2L}{\beta^2} (\beta - \theta)^2 \tag{22.22}$$

$$V = \frac{dS}{d\theta} = \frac{4L}{\beta} \left(1 - \frac{\theta}{\beta}\right) \tag{22.23}$$

$$A = \frac{dV}{d\theta} = -\frac{4L}{\beta^2} \tag{22.24}$$

Simple Harmonic Motion

This motion depicted in Figure 22.8 is the same motion produced by a Scotch yoke, where the displacement S [see Figure 22.8(a)] is the projection of a point on a circle to a reference line. For this motion, the maximum velocity [see Figure 22.8(b)] occurs in the middle of the period, whereas the acceleration [see Figure 22.8(c)] is greatest at both

ends of the period and least in the middle. Although these characteristics are more desirable than those of the uniformly accelerated motion,
they are generally unsuitable for high-speed operation unless the rise
and fall periods are both 180°, in which case the motion is smooth
throughout the cycle.

Displacement Curve Development

The graphical procedure for obtaining the displacement diagram is as
follows:

PROCEDURE [see Figure 22.8(a)]

1. Construct a semicircle on the distance AL as the diameter.
2. Divide the semicircle into the same number of equal parts as the
    cam rotation angle AC.
3. Draw horizontal lines from the division points on the semicircle
    to intersect respective ordinates of the cam rotation angle.
4. Draw a smooth curve through the points of intersection found in
    Step 3 to obtain the required curve AB.

Motion Equations

Let the curve in Figure 22.8(a) represent a typical simple harmonic
motion, where the total displacement L is attained over a period β. For
this motion, L is the diameter of the reference circle. Given this geometry, the displacement, velocity, and acceleration can be expressed
in terms of angle θ as follows:

For range $0 < \theta < \beta$

$$S = \frac{L}{2}\left(1 - \cos\frac{\pi\theta}{\beta}\right) \tag{22.25}$$

$$V = \frac{dS}{d\theta} = \frac{\pi L}{2\beta}\sin\frac{\pi\theta}{\beta} \tag{22.26}$$

$$A = \frac{dV}{d\theta} = \frac{\pi^2 L}{\beta^2}\cos\frac{\pi\theta}{\beta} \tag{22.27}$$

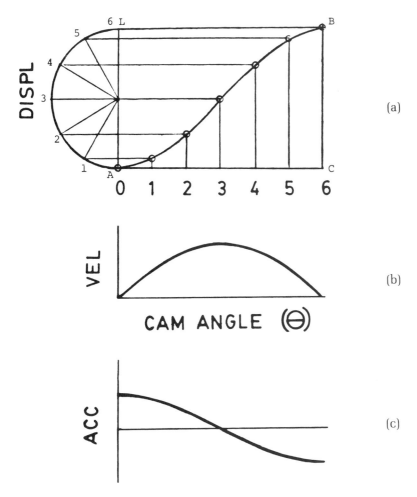

Figure 22.8   Simple harmonic motion: (a) Displacement diagram; (b) velocity diagram; (c) acceleration diagram.

Cycloidal Motion

This motion depicted in Figure 22.9, owes its name to the geometric curve [see Figure 22.9(a)] that is produced by a point on the circumference of a circle as it rolls without slipping on a straight line. For this motion, there is no abrupt change in the velocity profile and the acceleration is zero at the beginning and end of the period. See Figures 22.9(b) and (c). This makes cycloidal motion ideal for high-speed applications, since shock and vibration are eliminated. Hence, its dynamic performance is superior to that of simple harmonic and uniformly accelerated motions.

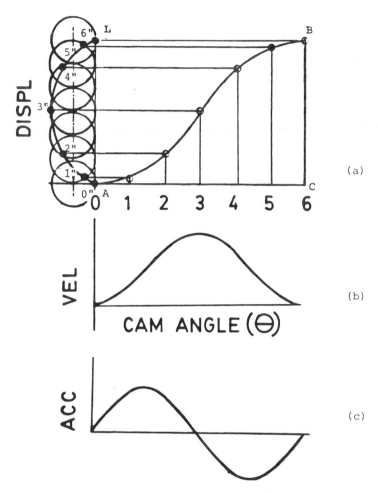

Figure 22.9   Cycloidal motion: (a) displacement diagram; (b) velocity diagram; (c) acceleration diagram.

Displacement Curve Development

The graphical procedure for obtaining the displacement diagram for this motion is as follows:

PROCEDURE [see Figures 22.9(a) and 22.10]

1. Using the cam angle axis as the centerline, construct a reference circle, whose circumference is equal to the total life AL, or whose radius is equal to $AL/2\pi$.
2. Divide the circumference of the circle into the same number of equal parts as the cam angle. Number these divisions $1'$, $2'$, $3'$, etc.

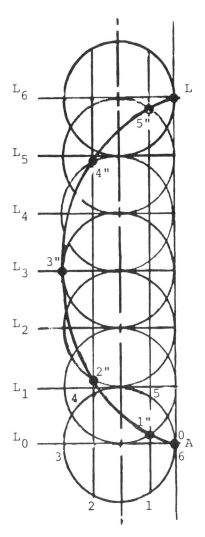

Figure 22.10   Cycloid construction.

3.   Divide the lift axis AL into the same number of equal parts as the circumference of the circle and through these division points construct lines L_1, L_2, L_3, etc., parallel to the cam angle axis AC.

4.   To locate the required points of the cycloid, (Fig. 22.10)

   a.   Project a line from point 1 on the reference circle, parallel to AL, to meet the circle centered on L_1 at point $1''$ that corresponds to point 1 on the reference circle.

   b.   Project a line from point 2 on the reference circle, parallel to AL, to meet the circle centered on L_2 at point $2''$ that corresponds

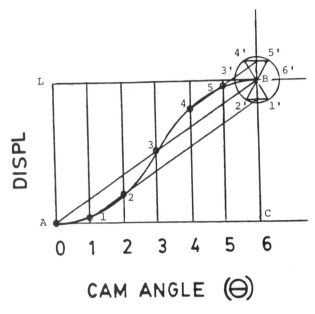

Figure 22.11   Alternate construction for cycloidal motion displacement.

       to point 2 on the reference circle.

   c. Continue the process above until all points are located.

5. Draw a smooth curve connecting points 1″, 2″, 3″, etc., to obtain the cycloid.

6. From points 1″, 2″, 3″, etc., project lines parallel to the cam angle axis to intersect with corresponding ordinates.

7. Draw a smooth curve through the intersections in Step 6 to obtain the required displacement curve AB.

Alternate Procedure

1. Draw two intersecting vertical and horizontal axes to represent cam lift and cam angle, respectively. (See Fig. 22.11) Define the origin as A.

2. Define lift AL on the vertical axis and cam angle AC on the abscissa.

3. At C erect an ordinate to meet a horizontal line drawn from L. Define this meeting point B, or end point of the rise cycle.

4. With the center at B and radius $L/2\pi$, draw a circle whose circumference is equal to lift AL.

5. Divide the circle and abscissa into an equal number of parts and number the divisions as shown.

6. Erect an ordinate at each division of the abscissa.
7. Join AB with a straight line.
8. Project each point of the circle horizontally until it intersects the ordinate CB (extended). Then from that point continue the project line parallel to AB until it intersects the ordinate projection line parallel to AB until it intersects the ordinate corresponding to the numbered division of the abscissa.
9. Draw a smooth curve through the intersection points in Step 8 to obtain the required curve.

Motion Equations

Let the curve in Figure 22.9 represent a typical cycloidal motion, where the total displacement L is attained over a period β. For this motion, L is also equal to the circumference of a circle (defined by $R = L/2\pi$) that rolls on the displacement axis. Given this geometry, the displacement, velocity, and acceleration can be expressed in terms of angle θ as follows:

For range $0 < \theta < \beta$

$$S = L \left(\frac{\theta}{\beta} - \frac{1}{2\pi} \sin \frac{2\pi\theta}{\beta} \right) \tag{22.28}$$

$$V = \frac{dS}{d\theta} = \frac{L}{\beta} \left(1 - \cos \frac{2\pi\theta}{\beta} \right) \tag{22.29}$$

$$A = \frac{dV}{d\theta} = \frac{2\pi L}{\beta^2} \sin \frac{2\pi\theta}{\beta} \tag{22.30}$$

22.5 LAYOUT OF CAM PROFILES

In order to lay out a cam profile, it is necessary to first define the following:

- Follower motion requirements, so that the displacement diagram can be developed
- Type and size of the follower
- Base circle radius on which the cam is to be built.

Once the above data are established, the construction of the profile can be accomplished in just a few basic steps. For the cam layouts given in this chapter, we will assume, for the purpose of illustration, that

a.   The follower motion, in each case, is as depicted in Figure 22.12. That is, starting from its home position at 0°, it rises by simple harmonic motion to attain a total lift L over a period of 180°. It then dwells for 60°, falls by simple harmonic motion for 90°, and dwells again for 30° before returning to its home position.

b.   The base circle radius in each case is the same.

To simplify the layout and reduce drafting time, we will use 30° increments to plot the displacements. If greater accuracy is desired, smaller increments must be used. Also, in plotting the displacement, we will apply the inversion principle where, instead of considering the follower frame as fixed while the cam is in rotation, we will consider the cam as fixed while the follower is in rotation. In this way, the relative motion between the cam and follower will remain unchanged.

Disk Cam with Radial Pointed Follower

PROCEDURE (See Figure 22.13)

1.   Draw the base circle.
2.   Draw the follower in contact with the base circle such that its centerline passes through the center of the base circle.
3.   Starting at the follower centerline and proceeding in the direction opposite to cam rotation, divide the base circle into angular segments corresponding to the displacement diagram. Each radial line will represent the angular position of the follower and its path relative to the cam. Number all radial lines corresponding to the division lines in the displacement diagram.
4.   Using the base circle as a reference, transfer to each follower position on the cam the displacement for the corresponding angle in the displacement diagram. This may be done by a projection of points from the diagram as shown, or by plotting values obtained from measurement. The points located in this manner will define the contact points of the follower on the cam surface.
5.   Draw a smooth curve connecting the projected points on the follower to obtain the required cam profile.

Disk Cam with Radial Flat-Faced Follower

PROCEDURE (See Figure 22.14)

1.   Draw the base circle.
2.   Draw the follower in contact with the base circle such that its centerline passes through the center of the base circle.

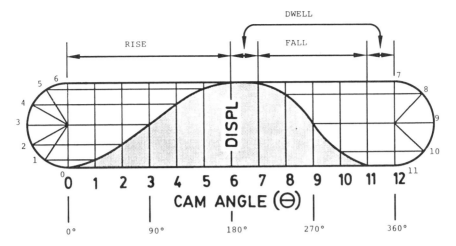

Figure 22.12   Displacement diagram.

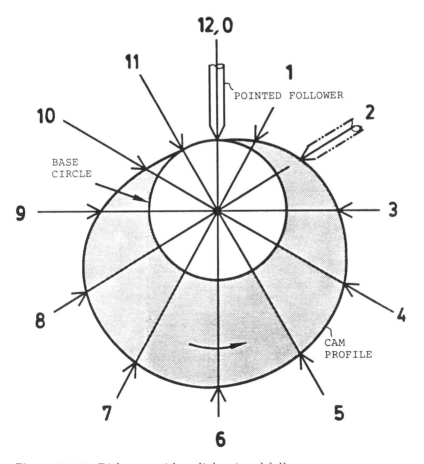

Figure 22.13   Disk cam with radial pointed follower.

521

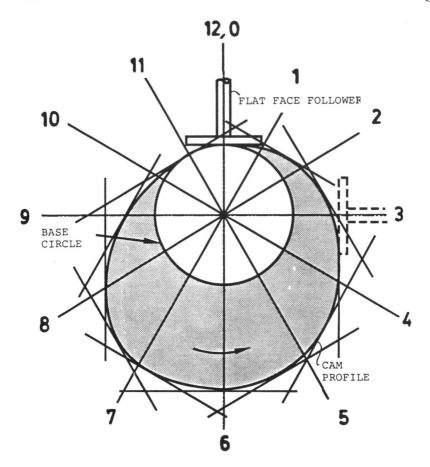

Figure 22.14   Disk cam with flat-faced follower.

3. Starting at the follower centerline and proceeding in the direction opposite to cam rotation, divide the base circle into angular segments corresponding to the displacement diagram. Each radial line will represent the angular position of the follower and its path relative to the cam. Number all radial lines corresponding to the division lines in the displacement diagram.

4. Using the base circle as a reference, transfer to each follower position on the cam the displacement for the corresponding angle in the displacement diagram. This may be done by a projection of points from the diagram as shown, or by plotting values obtained from measurement. The points located in this manner do not necessarily define the contact points of the follower on the cam surface. Rather, they merely indicate the displacements of the flat surface of the follower at various angular positions.

5. To locate the contact points, draw outlines of the follower (or straight lines to represent the flat faces) at each follower position around the cam. These lines will define the envelope for the desired cam profile.
6. Draw a smooth tangent to the lines on the faces of the envelope. This curve will define the contact points on the follower at the various angular positions, as well as the desired cam profile. It is important that the curve is drawn to touch all the surfaces of the envelope without undercutting.

Disk Cam with Radial Roller Follower

PROCEDURE (See Figure 22.15)

1. Draw the base circle.
2. Draw the follower such that its centerline passes through the axis of the base circle and the roller is tangent to the same circle. This point of tangency will define the "home" or 0° position of the follower.
3. Draw the prime circle. This circle is concentric with the base circle and passes through the axis of the roller.
4. Starting with the follower centerline and proceeding in the direction opposite to cam rotation, draw a radial-line rotation dividing the prime circle into angular segments corresponding to the divisions in the displacement diagram. The radial lines will represent the angular position of the follower and its path relative to the cam.
5. Using the prime circle as reference, transfer to each follower position on the cam the appropriate displacements from the displacement diagram. The displacements located in this manner will define the path of the roller axis about the cam.
6. Define the envelope for the desired cam profile by describing circular arcs to represent the follower outline at each angular position.
7. Draw a smooth curve tangent to the circular arcs of the envelope to obtain the desired profile.

Disk Cam with Offset Roller Follower

PROCEDURE (See Figure 22.16)

1. Draw the base circle.
2. Draw the offset circle concentric with the base circle and having a radius equal to that of the base circle less the required offset distance.

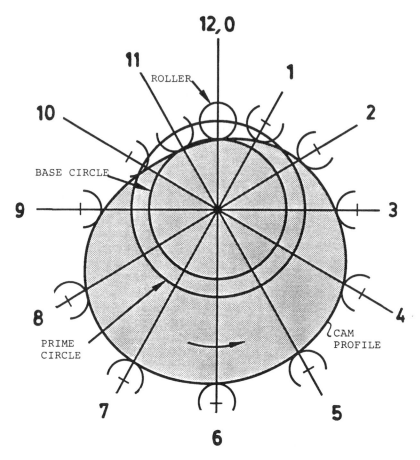

Figure 22.15   Disk cam with radial roller follower.

3. Draw the follower such that the roller surface just touches the base
    circle, while the centerline is tangent to the offset circle. This point
    of tangency corresponds to the "home" or 0° position of the fol-
    lower.
4. Starting with the point of tangency in Step 3, draw radial lines
    dividing the offset circle into angular segments corresponding
    to the divisions in the displacement diagram. Number the divi-
    sion points to correspond with the divisions in the displacement
    diagram.
5. At each division point in Step 4, draw tangents to represent the
    follower paths relative to the cam.

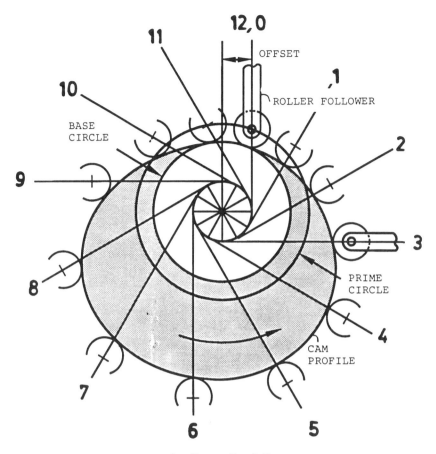

Figure 22.16 Disk cam with offset roller follower.

6. Through the center of the roller, draw the prime circle concentric with the base circle.
7. Using the prime circle as a reference, locate on the cam the appropriate displacements from the displacement diagram. The displacements located in this manner will define the path of the roller axis about the cam.
8. Define the envelope for the desired cam profile by describing circular arcs to represent the follower outline at each angular position.
9. Draw a smooth curve tangent to the circular arcs of the envelope to obtain the desired profile.

Disk Cam with Pivoted Roller Follower

PROCEDURE (See Figure 22.17)

1. Draw the base circle
2. Draw the pivoted follower in its "home" or 0° position, where the roller is tangent to the base circle.
3. Through the axis of the roller, draw the prime circle concentric with the base circle.
4. Starting with the point of tangency in Step 2 and proceeding in the

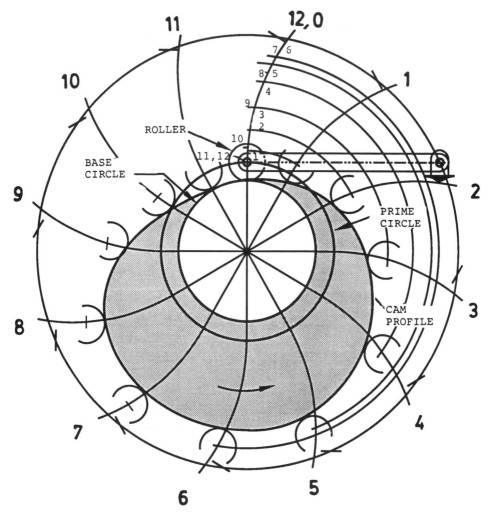

Figure 22.17   Disk cam with pivoted roller follower.

direction opposite to cam rotation, draw radial lines dividing the prime circle into angular segments corresponding to the divisions in the displacement diagram.

5. Through the follower pivot point, draw the pivot circle and locate the pivot points corresponding to the division points in Step 4.

6. Using each pivot point as the center, describe an arc directed outward from the corresponding division point on the prime circle.

7. Using the prime circle as a reference, transfer to each follower position (on the arcs) the appropriate displacements from the displacement diagram.

8. Define the envelope for the desired cam profile by describing circular arcs to represent the follower outline at each angular position.

9. Draw a smooth curve tangent to the circular arcs of the envelope to obtain the desired profile.

Other Cams with Roller Followers

Grooved-Plate Cam with Roller Follower

The layout of this cam profile follows the same procedure as that for disk cams, except that the cam surfaces are defined by two smooth curves drawn tangent to the roller circles on both sides of the pitch curve.

Cylindrical Cam with Roller Follower

The layout of this cam profile is more readily accomplished than that of the radial cam, since the displacement diagram is a true development of the cam surface. However, like the grooved-plate cam, the groove is defined by two smooth curves drawn tangent to the circular or roller outlines on both sides of the pitch curve.

EXAMPLE 22.1

a. Develop displacement, velocity, and acceleration vs cam angle curves for the motion cycle of a cam-follower having uniform motion modified by circular arcs as follows:

Rise:	*Fall*:
Lift (L) = 2 in.	Lift (L) = 2 in.
Period (θ) = 180°	Period (θ) = 90°
Distance ($\bar{x}$) = 4 in.	Distance ($\bar{x}$) = 2 in.
Arc radius (R) = 1 in.	Arc radius (R) = 0.5 in.

b. Using the above data and assuming intermediate dwells between rise and fall, design a 2.55 in.-diameter base circle cam for a roller follower whose roller is $\frac{1}{2}$ in. in diameter.

SOLUTION (FOR THE RISE)

First, we determine the angle ϕ, using Eq. (22.9a), where

$$A = 4^2 - 4(1)^2 = 12$$

$$B = -2(4)(2 - 2) = 0$$

$$C = 2^2 - 4(1)(2) = -4$$

$$\phi = \tan^{-1}\left[\frac{+0 \pm \sqrt{0 - 4(12)(-4)}}{2(12)}\right] = 30°$$

Since we know the angle ϕ, the distance's x1 and x2, as well as the corresponding angles θ1 and θ2, are found from Eqs. (22.6) and (22.7) as follows:

$$x1 = 1 \sin 30° = 0.5 \text{ in.}$$

$$x2 = 4 - 0.5 = 3.5 \text{ in.}$$

$$\theta1 = 180° \frac{0.5}{4} = 22.5°$$

$$\theta2 = 180° \frac{(3.5)}{4} = 157.5°$$

Substituting the above values into the appropriate displacement, velocity, and acceleration expressions for the required angles, one obtains the results in Table 22.1 and the plotted curves in Figure 22.18 (a–c).

SOLUTION (FOR THE FALL)

First, we find the angle ϕ, using Eq. (22.9a), where

$$A = 2^2 - 4(0.5)^2 = 3$$

$$B = -2(2)(2 - (2)0.5) = -4$$

$$C = 2^2 - 4(0.5)(2) = 0$$

$$\phi = \tan^{-1}\left[\frac{+ 4 \pm \sqrt{4^2 - 4(3)(0)}}{2(3)}\right] = 53.13°$$

Table 22.1   Results of Example 22.1

Region	Angle θ (deg.)	Displacement S (in.)	Velocity V (10^{-3} in./deg.)	Acceleration A (10^{-3} in./deg.^{2})
Rise Period				
$0° < \theta < 22.5°$	0	0	0	0.494
	10	0.025	0.507	0.533
	20	0.120	0.102	0.686
	22.5	0.133	1.282	0.761
$22.5° < \theta < 157.5°$				
	30	0.229	1.282	0
	60	0.612	1.282	0
	90	0.996	1.282	0
	120	1.379	1.282	0
	150	1.762	1.282	0
$157.5 < \theta < 180°$				
	157.5	1.858	1.282	-0.761
	160	1.875	1.107	-0.686
	170	1.975	0.506	-0.533
	180	2.0	0	-0.494
Fall Period				
$0° < \theta < 180°$	0	2.0	0	-0.988
	5	1.987	-0.507	-1.066
	10	1.94	-1.102	-1.376
	15	1.87	-1.986	-2.389
	18	1.8	-2.962	-4.573
$18° < \theta < 72°$				
	30	1.377	-2.962	0
	60	0.48	-2.962	0
$72° < \theta < 90°$				
	72	0.2	-2.962	4.573
	75	0.127	-1.986	2.389
	80	0.052	-1.102	1.376
	85	0.012	-0.507	1.066
	90	0	0	0.988

Since we know the angle ϕ, the distances x1 and x2, as well as the corresponding angles θ1 and θ2, are found from Eqs. (22.6) and (22.7) as follows:

$$x1 = 0.5 \sin 53.13° = 0.4 \text{ in.}$$

$$x2 = 2 - 0.4 = 1.6 \text{ in.}$$

$$\theta1 = 90 \frac{(0.4)}{2} = 18°$$

$$\theta2 = 90 \frac{(1.6)}{2} = 72°$$

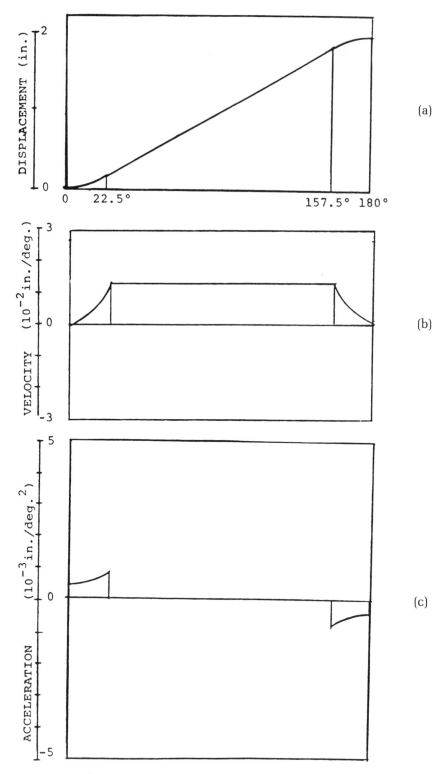

Figure 22.18   Rise: (a) Displacement; (b) velocity; (c) acceleration.

As in the rise case, substituting the above values into the appropriate motion equations for the required angles, one obtains the results in Table 22.1 and Figures 22.19(a–c). For the cam layout, see Figures 22.20(a) and (b).

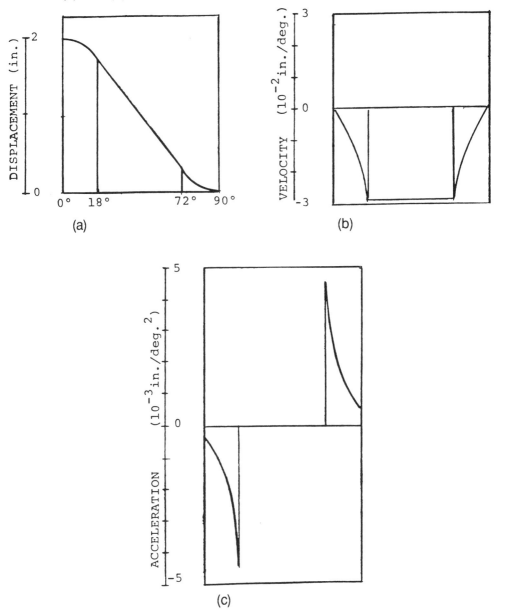

Figure 22.19   Fall: (a) Displacement; (b) velocity; (c) acceleration.

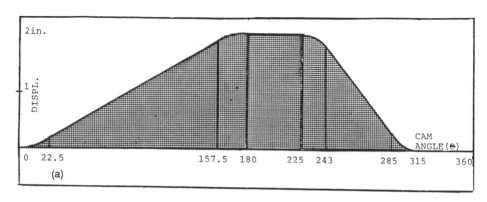

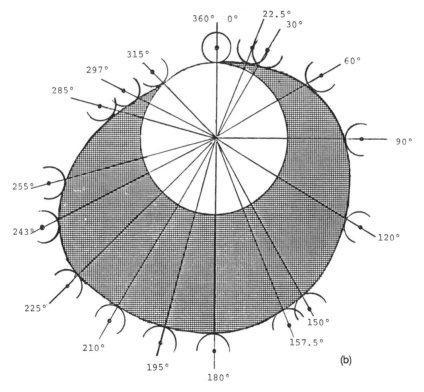

Figure 22.20 Cam layout: (a) Displacement vs cam angle; (b) cam profile.

Cam Displacement Tables

For accurate cam layouts, it is generally necessary to make use of standard cam displacement tables. Table 22.2, for example, provide nondimensional displacement values for uniform acceleration, simple harmonic and cycloidal curves. In each case, the table gives a listing of

Table 22.2   Cam Displacement for Uniform Acceleration, Simple Harmonic, and Cycloidal Motions

Cam displacement	Uniform acceleration	Simple harmonic	Cycloidal
Cam angle divisions (θ)	Follower displacement	Follower displacement	Follower displacement
0	0	0	0
0.05	0.005	0.006	0.0008
0.10	0.02	0.024	0.006
0.15	0.045	0.055	0.021
0.20	0.08	0.096	0.049
0.25	0.125	0.146	0.091
0.30	0.18	0.206	0.148
0.35	0.245	0.273	0.221
0.40	0.32	0.346	0.306
0.45	0.405	0.422	0.401
0.50	0.500	0.500	0.500
0.55	0.595	0.578	0.599
0.60	0.68	0.655	0.694
0.65	0.755	0.727	0.779
0.70	0.82	0.794	0.852
0.75	0.875	0.854	0.909
0.80	0.92	0.905	0.951
0.85	0.955	0.946	0.979
0.90	0.98	0.976	0.994
0.95	0.995	0.994	0.9992
1.00	1.000	1.000	1.000

fractional displacements corresponding to 120 equal divisions of a lift angle, where the total lift is considered to be one displacement unit. The following example will illustrate the use of these tables:

EXAMPLE 22.2

Suppose uniform acceleration motion has been selected and a total lift of 3 in. is required in 60°. Determine the displacements at each 15° interval.

Table 22.2 gives a lift of 1.00 in. in 20 equal angular divisions. This implies that for a 60° angular range, one division (0.05) in the

Table 22.3   Example 22.2

Deg.	Division	3(Y value)	Actual displacement
15°	5	3(0.125)	0.375 in.
30°	10	3(0.500)	1.500 in.
45°	15	3(0.875)	2.625 in.
60°	20	3(1.000)	3.000 in.

table represents 3° in actual cam rotation. Also, if we consider that the total lift in the table is 1 in., this then represents $\frac{1}{3}$ of the actual lift required.

Therefore, to find the actual displacements in the given period angles (60°), first divide this angle by 20 to obtain the cam angle per division in the table. Next, enter the table, obtain the Y value corresponding to the actual cam angle or division number, and multiply this value by 3 to obtain the actual lift. Thus, for angles 15, 30, 45, and 60°, the actual dispalcements are as shown in Table 22.3.

Pressure Angle

Every cam regardless of its shape tends to force the follower in the direction of the cam rotation, exerting a side thrust or pressure on the follower. The magnitude of this side thrust or pressure depends on the pressure angle on the angle at which the cam surface strikes the follower and on the magnitude of the load resistance. Pressure angle, therefore, may be defined as the angle between the normal to the pitch curve and the direction of the follower at the point of contact. See Figure 22.21(a). Alternatively, the pressure angle may be defined, in terms of the displacement diagram, as the slope of the curve at any given angle. As can be seen from the illustrations, the angle varies during the complete cycle of cam rotation, reaching both minimum and maximum values. If the angle is large, the follower will jam. This is because of friction resulting from excessive side thrust exerted by the cam on the follower. The side thrust, as shown in Figure 22.21(a), is the lateral component of the normal force at the point of contact. Thus, the side thrust increases with increasing pressure angle. It is difficult to set an absolute maximum value for the pressure angle of a cam, since this depends on many factors such as cam speed, friction, and type of follower. However, it is generally recommended that the limit on this angle not exceed 30°,

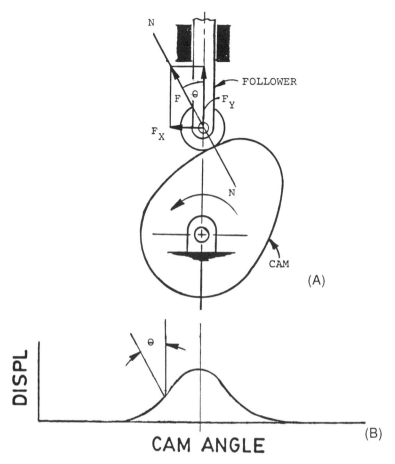

Figure 22.21   Pressure angle: (A) on cam follower; (B) on displacement diagram.

except for light loads where this limit may be extended to 47.5°. When jamming occurs, any of the following corrective actions may be taken:

1. Increase the base circle diameter. This, in effect, increases the cam size and makes the profile less steep.
2. Increase the cam rotation angle for a given follower displacement. This provides a lower maximum velocity as the slope of the displacement diagram (or cam profile) is reduced.
3. Decrease the total follower displacement.
4. Change the type of follower motion (i.e., uniform velocity, simple harmonic, cycloidal, etc.).

5. Change the amount of follower offset. Offsetting the follower in the proper direction effectively decreases the pressure angle during the rise. For example, if the rotation is clockwise, the offset should be to the left of the vertical cam centerline; if the rotation is counterclockwise, the offset should be to the right of the centerline.

Summary

The development of the cam profile can be summarized in five steps:

1. Draw the base and offset circles. For radial followers, these curves are one and the same, since the offset is zero.
2. Draw the follower in contact with the base circle.
   - For **radial followers**, the centerline passes through the cam axis.
   - For **offset followers**, the centerline is tangent to an offset circle.
3. Divide the base (or offset circle) into the same number of equal angles as the displacement diagram, and at each angle lay out the angular position of the follower and its path relative to the cam.
   - For the **radial follower**, the path is on the radial line of the cam.
   - For the **offset follower**, the path is on the tangent line to the offset circle.
   - For the **pivoted follower**, the path is an arc described by the pivot arm.
4. Proceeding in the direction opposite to cam rotation, transfer the displacements for individual angles in the displacement diagram to follower paths corresponding to those angles. Any of several methods may be used to transfer the displacements, provided that the follower is located in the proper position at the proper time when the cam is rotated through a specified angle.
5. Draw the outline of the follower surface in its proper position at each angle.
   - For the **roller follower**, these are the circular arcs of the roller.
   - For the **pointed follower**, these are the pointed tips.
   - For the **flat-face follower**, these are simply straight lines.

   These outlines will define the envelope for the required cam profile.
6. To obtain the required cam profile, draw a smooth curve as follows:
   - For the **roller follower**, the curve should be tangent to the circular arcs of the rollers that define the envelope.
   - For the **flat-face follower**, the curve should be tangent to the follower faces or straight lines that define the envelope.
   - For the **pointed follower**, the curve should simply connect the points of the envelope.

22.6   INDEXING MECHANISMS

Indexing mechanisms are those capable of converting a constant-speed rotary input motion to an intermittent rotary output motion. Packaging machines, press-feed tables, machine tools, switch gears, and feeding machines are but a few of the many machines found in industry in which indexing mechanisms are employed to produce the required intermittent motion. Two common indexing mechanism are:

1.   Ratchet and pawl mechanism
2.   Geneva mechanism

Ratchet and Pawl Mechanism

The **ratchet and pawl** is a special type of mechanism used extensively in certain classes of machinery where intermittent motion is desired. By means of this mechanism, reciprocating or oscillatory motion may be transformed into an intermittent form or rotary motion. Figure 22.22(a) shows a schematic of a simple ratchet and pawl assembly, where the toothed wheel W (or ratchet) is engaged by a pawl A (or follower) that is pinned to a lever arm L. When the lever arm is moved in the counterclockwise direction, the pawl is free to slide past the teeth of the ratchet. However, when the arm is moved in the clockwise direction, the pawl re-engages the teeth, causing the ratchet to turn in the direction of the arm. In this way, the alternating clockwise and counterclockwise motion of the arm produces an intermittent rotational motion of the ratchet. An additional pawl, B, is often provided in the mechanism to prevent reversal of direction of the ratchet. The number of indexing positions of the ratchet wheel depends on the number of teeth. That is, a large number of positions requires a large number of teeth. However, for any given size of ratchet, the larger the number of teeth, the smaller the size of each tooth, which means less strength.

　　　Two common applications of the ratchet and pawl mechanism are the ratchet wrench [similar to Figure 22.2(a)] and automotive jack [See Figure 22.22(b)].

Geneva Mechanism

The **Geneva mechanism** is a classical indexing mechanism used in automatic machines to give positive but intermittent motion to a drive

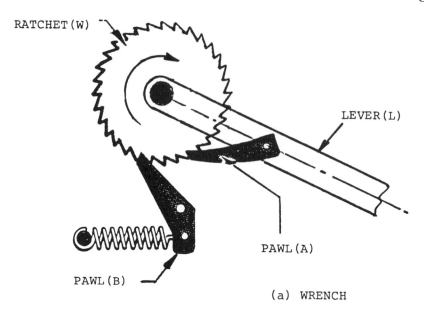

RATCHET(W)

LEVER(L)

PAWL(A)

PAWL(B)

(a) WRENCH

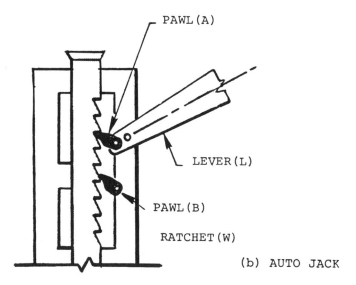

PAWL(A)

LEVER(L)

PAWL(B)

RATCHET(W)

(b) AUTO JACK

Figure 22.22   Ratchet and pawl mechanisms: (a) wrench; (b) auto jack.

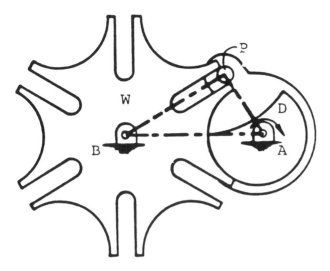

Figure 22.23   Geneva mechanism.

part. Figure 22.23 shows a 6-slot example of this mechanism, in which the partly circular disk D and arm AP are both keyed to an input shaft A that turns at a constant speed. The arm carries a pin P shown located in one of the slots of the wheel W that is keyed to an output shaft B.

When the drive is set in motion, the pin P enters one of the slots, and as the arm turns, the pin causes the wheel to turn as long a it remains engaged within the slot. To avoid jamming at the points of engagement and disengagement of the pin with the slot, the centerline of the arm AP must be perpendicular to that of the slot. At the instant the pin leaves the slot, the circular arc of disk D engages the matching surface of W between the lobes, thereby locking W in a stationary position until the pin P is rotated in position to enter the next slot. As the pin enters the slot, the circular surfaces of D and W are disengaged so as to again allow W to turn. The number of pins or the driver as well as the number of slots in the wheel may be varied, depending on the motion desired.

Problems

KINEMATIC TERMINOLOGY

1. Define (a) kinematics and (b) kinetics.
2. Distinguish between the terms "mechanism" and "machine."
3. How does a mechanism differ from a structure?
4. Name and give examples of the three types of plane motion.
5. Define (a) displacement, (b) velocity, and (c) acceleration.
6. Distinguish between the terms (a) "speed" and "velocity" and (b) "distance" and "displacement."
7. Explain mechanism inversion. How does it affect (a) relative motion and (b) absolute motion of the components?
8. How does rotation differ from curvilinear motion?
9. Distinguish between absolute motion and relative motion.
10. Define radian. Determine the number of radians in 30, 45, 150, and 330°.
11. What are the three basic modes of transmitting motion? Indicate, using sketches, one example in each case.
12. Distinguish between the terms "reciprocal motion" and "oscillatory motion."
13. Define (a) higher pair and (b) lower pair. Give an example of each, using sketches.
14. What is a kinematic chain? Describe the three types.

UNIFORMLY ACCELERATED MOTION

1. A train traveling at 50 mph speeds up to 70 mph in 1 min and 30 sec. Determine its acceleration and the distance traveled.

2. A flywheel turning at 200 rpm attains a speed of 300 rpm in 1 min with constant acceleration. Determine the acceleration and number of revolutions taken to attain the higher speed.

3. An engine crank pin has a linear velocity of 2400 ft/min while rotating at 150 rpm. What is the length of the crank?

4. Starting from rest, a body A held by a string, wrapped around a 12-in.-diameter pulley, falls 60 ft in 4 sec. Determine for the pulley the following:
   a. Number of revolutions
   b. Angular velocity after 4 sec
   c. Angular acceleration after 4 sec

5. An automobile accelerates from a speed of 20 mph to 55 mph in a distance of 300 ft. If the acceleration is constant, find the time taken.

6. The speed of an automobile is 55 mph. If the outside diameter of the tires is 27 in., determine the rpm of the wheels and angular speed in rad/sec.

7. A train traveling between two stations 5 miles apart takes 10 min. It uniformly accelerates to a maximum speed at 2 ft/sec^2 and uniformly decelerates at 6 ft/sec^2. What is the maximum speed of the train and the distance traveled at this speed? What are the distances covered during the first and last minutes of the train's motion?

8. Determine the minimum time for a car to travel between two stop signs, 1 mile apart, if its acceleration is 3 ft/sec^2, its deceleration 4 ft/sec^2, and its maximum speed 50 mph.

9. Develop a velocity-time curve to depict the motion of a body between two points, A and D, as folows:
   a. Passing point A, its velocity is 15 mph.
   b. During the next 30 sec, it accelerates uniformly to 50 mph to reach a point B.
   c. It then continues at 50 mph for 4 min to another point C.
   d. Finally, it comes to rest at D 6 min after passing point A.
   Determine also the total distance traveled between points A and D.

10. A rotating fan, 6 ft in diameter, comes to rest with uniform acceleration from a speed of 600 rpm. If it turns 15 revolutions while stopping, determine the time it takes to stop.

11. Two points, A and B, lie on a radial line of a rotating disk 2 in. apart. Determine the radius of rotation of each of these points if the velocity at A is 700 ft/min and at B 800 ft/min.

12. A 20-in.-diameter wheel turns at 200 rpm. Determine the:
    a. Angular velocity in rad/sec
    b. Linear velocity of a point on the rim
    c. Linear velocity of a point 12 in. from the center
    If the wheel speeds up to 300 rpm with uniform acceleration in 2 min, determine the:
    d. Angular acceleration
    e. Linear acceleration of a point on the rim

13. A car, initially traveling at 25 mph, accelerates at 5 miles/sec/sec for 5 sec. What is its speed at the end of the 5 sec of acceleration, and how far does it travel in this time?

14. An automobile accelerating from a standstill position reaches a speed of 50 miles in 10 sec. What is its average acceleration and the displacement in this time interval?

15. Referring to Figure P.1, how fast (i.e., angular velocity of gear G) must a cyclist pedal in order to travel at 15 mph? Determine the speed of the chain and linear velocity of the pedal (both with respect to the frame).

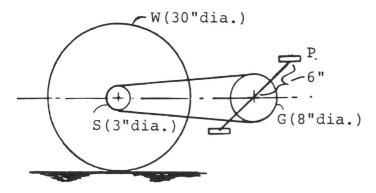

Figure P.1

VECTORS

1. Using data in Figure P.2, determine graphically the following:
   a. $\overline{A} + \overline{B}$
   b. $\overline{A} - \overline{B}$
   c. $\overline{A} + \overline{B} - \overline{C}$
   d. $\overline{A} + \overline{B} + \overline{C}$

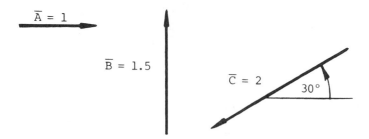

Figure P.2

2. Using the vector polygons given in Figure P.3, complete the following vector equations:
   a. $\overline{V} =$ $\overline{A} =$
   b. $\overline{B} =$ $\overline{D} =$
   c. $\overline{E} =$ $\overline{H} =$

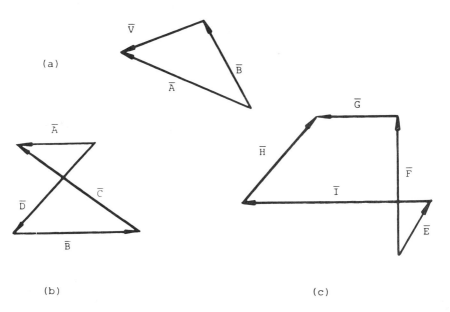

Figure P.3

3. A hiker desiring to go to a point northeast, because of various
   obstacles, goes $1\frac{1}{2}$ miles due east, then turns left 120°, and goes

straight to the point. How far was he originally from the point, and how far did he travel to arrive at the point?

4. Graphically determine the resultant of vectors $\overline{A} + \overline{B} + \overline{C}$ in Figure P.4. Then find the effective component of this vector along axes c-c, x-x, and y-y.

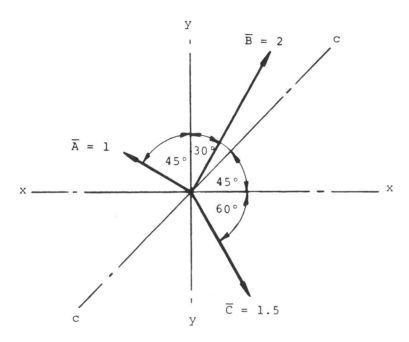

Figure P.4

5. Find the effective components of vector A in Figure P.5, along axes x-x, y-y, a-a, and b-b.

6. The effective component of a velocity vector $\overline{V}$ along the axis S-S (Figure P.6) is known. Locate this vector if it lies on the bisector of axes T-T and N-N. Also locate its effective components along axes T-T and R-R.

7. Determine:
    a. The sum of two vectors: one 10 units due north and the other 20 units northeast
    b. The resultant of 5 (at 90°) − 4 (at 180°)
    c. The value of a vector quantity when added to 100 units that due south gives 100 units northwest

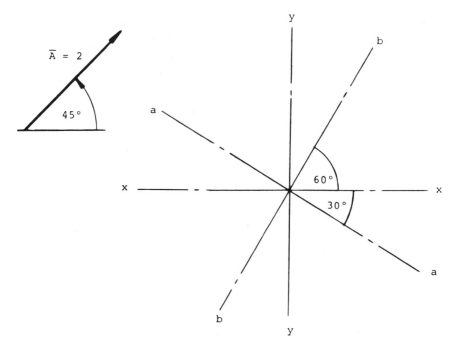

Figure P.5

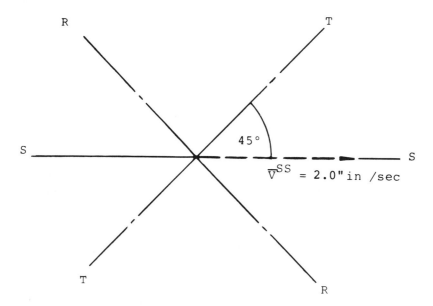

Figure P.6

8. In Figure P.7, $\overline{N}$ and $\overline{M}$ are effective components of a vector R, along axes n-n and m-m. Determine this vector.

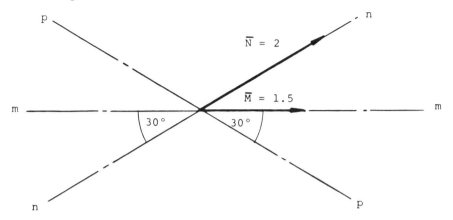

Figure P.7

9. Resolve the vector $\overline{A}$ in Figure P.8 into its components along axes b-b and c-c.

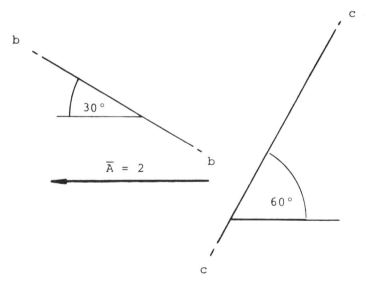

Figure P.8

10. Indicate the directions for all vectors shown in Figure P.9, based on the equations
   a. $\overline{R} = \overline{T} - \overline{S} + \overline{V} - \overline{U}$
   b. $\overline{A} = \overline{B} - \overline{C} + \overline{D} - \overline{E}$

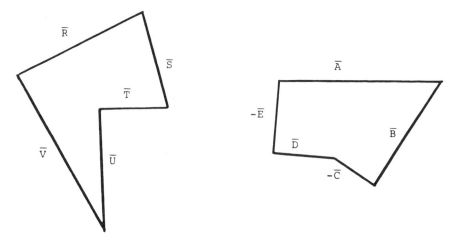

Figure P.9

11. The velocity of point B on the link BC in Figure P.10 is shown to act 60° with respect to the link centerline. Determine (a) the rotational and translational effects of this vector and (b) the effective component of the same vector along link BD oriented 15° with respect to link BC.

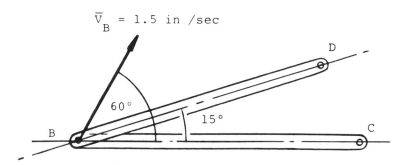

Figure P.10

12. From data given on vectors A, B, and C in Figure P.11, determine the magnitude and sense of all vectors when
     a. $\overline{C}$ is the resultant.
     b. $\overline{A}$ is the resultant.
     c. $\overline{B}$ is the resultant.

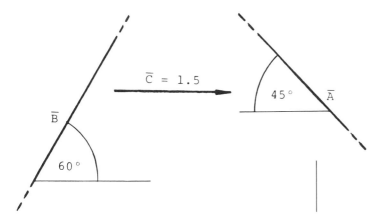

Figure P.11

13. An airplane is flying at an airspeed of 300 mph heading north at 45°. There is a tail wind from the west at 40 mph. What is the ground speed of the plane, and what is its actual flight direction?
14. In order to cross a stream flowing at 5 mph in a boat that travels at 12 mph, at what angle upstream should the boat be headed in order to reach a point directly opposite? What is the resultant speed of the boat?
15. A 20-ft ladder, supported at 45° by a horizontal floor and vertical wall, slides down the wall at 2 fps. What is the velocity of the bottom of the ladder and the relative velocity of one end to the other in magnitude and direction?
16. A boat with a speed of 12 mph in still water is to cross a stream 2 miles wide. The current of the stream is 5 mph. Determine the velocity of the boat and the time required to make the crossing.

GRAPHICAL TECHNIQUES: VELOCITY ANALYSIS

Effective Components

1. Using the effective component method, determine the linear and angular velocities of point C in Figure P.12 for any of the following conditions:
   a. $\theta = 30°$, $\omega = 1$ rad/sec (clockwise)
   b. $\theta = 30°$, $\omega = 1$ rad/sec (counterclockwise)
   c. $\theta = 60°$, $\omega = 1$ rad/sec (clockwise)
   d. $\theta = 60°$, $\omega = 1$ rad/sec (counterclockwise)

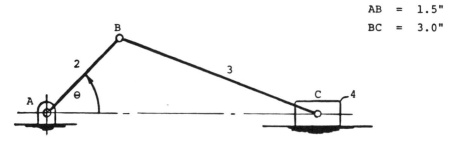

AB = 1.5"
BC = 3.0"

Figure P.12

2. Repeat Problem 1 but use Figure P.13.

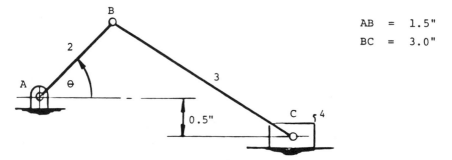

AB = 1.5"
BC = 3.0"

Figure P.13

3. Repeat Problem 1 but use Figure P.14.

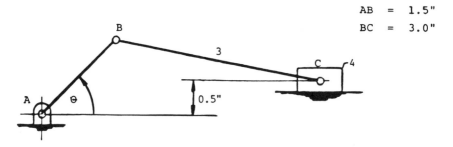

AB = 1.5"
BC = 3.0"

Figure P.14

4. Repeat Problem 1 but use Figure P.15.

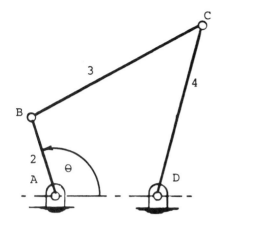

AB = 1.0"
BC = 2.5"
CD = 2.5"
AD = 1.5"

Figure P.15

5. Repeat Problem 1 but use Figure P.16.

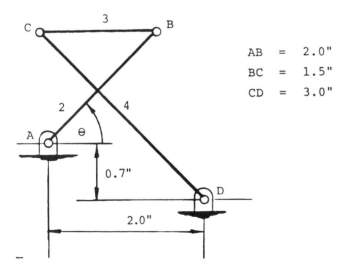

AB = 2.0"
BC = 1.5"
CD = 3.0"

Figure P.16

6. Repeat Problem 1 but use Figure P.17.

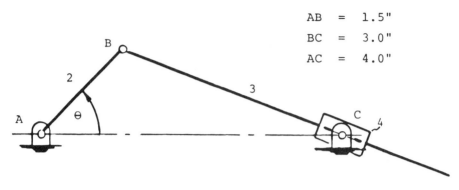

AB	=	1.5"
BC	=	3.0"
AC	=	4.0"

Figure P.17

7. Repeat Problem 1 but use Figure P.18.

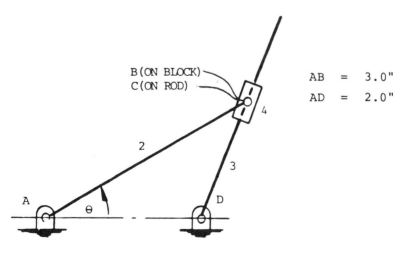

| AB | = | 3.0" |
| AD | = | 2.0" |

Figure P.18

8.   Repeat Problem 1 but use Figure P.19.

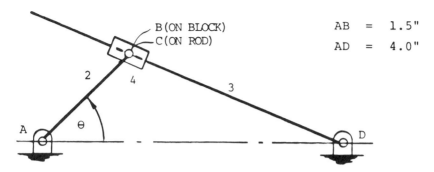

Figure P.19

9.   Repeat Problem 1 but use Figure P.20.

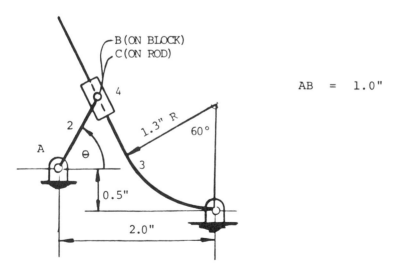

Figure P.20

10. Repeat Problem 1 but use Figure P.21.

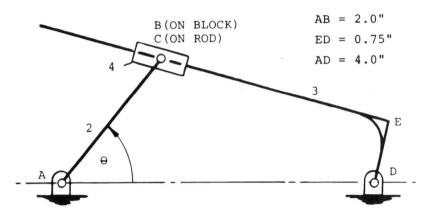

Figure P.21

11. Determine the linear velocities of points *C* and *E* in Figure P.22.

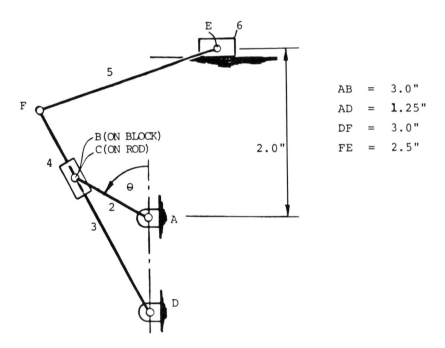

Figure P.22

12.   Repeat Problem 11 but use Figure P.23.

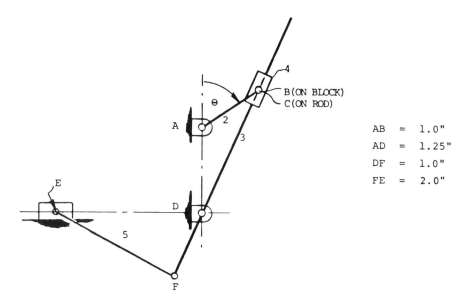

AB = 1.0"
AD = 1.25"
DF = 1.0"
FE = 2.0"

Figure P.23

13.   Repeat Problem 11 but use Figure P.24.

AB = 1.5"
BC = 3.0"
CF = 1.5"
FE = 2.0"
DE = 1.5"

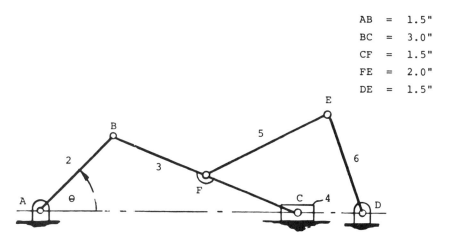

Figure P.24

14. Repeat Problem 11 but use Figure P.25.

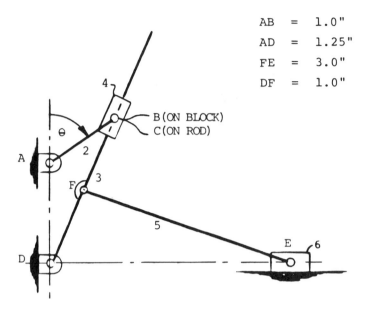

AB = 1.0"
AD = 1.25"
FE = 3.0"
DF = 1.0"

Figure P.25

15. Repeat Problem 11 but use Figure P.26.

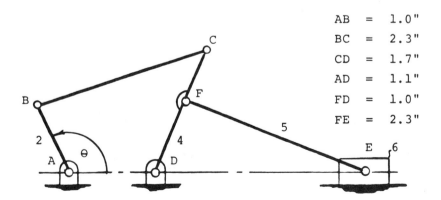

AB = 1.0"
BC = 2.3"
CD = 1.7"
AD = 1.1"
FD = 1.0"
FE = 2.3"

Figure P.26

16.   Repeat Problem 11 but use Figure P.27.

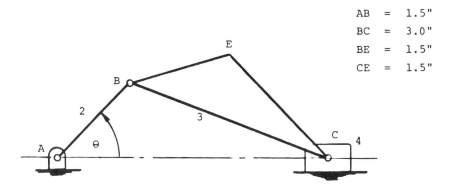

AB = 1.5"
BC = 3.0"
BE = 1.5"
CE = 1.5"

Figure P.27

17.   Repeat Problem 11 but use Figure P.28.

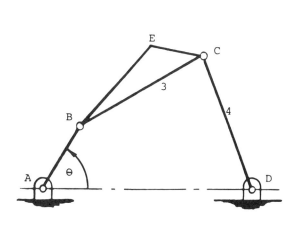

AB = 1.0"
BC = 2.0"
CD = 2.0"
AD = 3.0"
BE = 1.5"
CE = 0.75"

Figure P.28

18. Repeat Problem 11 but use Figure P.29.

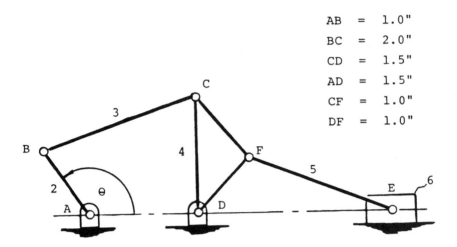

AB	=	1.0"
BC	=	2.0"
CD	=	1.5"
AD	=	1.5"
CF	=	1.0"
DF	=	1.0"

Figure P.29

19. Repeat Problem 11 but use Figure P.30.

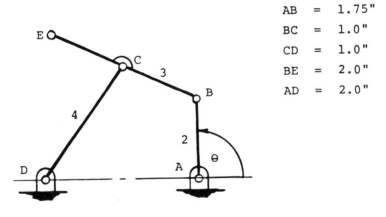

AB	=	1.75"
BC	=	1.0"
CD	=	1.0"
BE	=	2.0"
AD	=	2.0"

Figure P.30

20. Repeat Problem 11 but use Figure P.31.

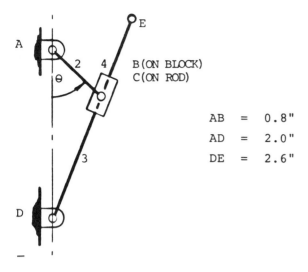

AB = 0.8"

AD = 2.0"

DE = 2.6"

Figure P.31

Instant Center

1. First, locate all the instant centers; then, using the instant center method, determine the linear and angular velocities of point C in Figure P.12 for any of the following conditions:
   a. $\theta = 30°$, $\omega = 1$ rad/sec (clockwise)
   b. $\theta = 30°$, $\omega = 1$ rad/sec (counterclockwise)
   c. $\theta = 60°$, $\omega = 1$ rad/sec (clockwise)
   d. $\theta = 60°$, $\omega = 1$ rad/sec (counterclockwise)
2. Repeat Problem 1 but use Figure P.13.
3. Repeat Problem 1 but use Figure P.14.
4. Repeat Problem 1 but use Figure P.15.
5. Repeat Problem 1 but use Figure P.16.
6. Repeat Problem 1 but use Figure P.17.
7. Repeat Problem 1 but use Figure P.18.
8. Repeat Problem 1 but use Figure P.19.
9. Repeat Problem 1 but use Figure P.20.
10. Repeat Problem 1 but use Figure P.21.
11. Locate all instant centers in Figure P.22.
12. Locate all instant centers in Figure P.23.

13. Locate all instant centers in Figure P.24.
14. Locate all instant centers in Figure P.25.
15. Locate all instant centers in Figure P.26.

Relative Velocity

1. Using the relative velocity method, determine the linear and angular velocities of point C in Figure P.12 for any of the following conditions:
    a. $\theta = 30°$, $\omega = 1$ rad/sec (clockwise)
    b. $\theta = 30°$, $\omega = 1$ rad/sec (counterclockwise)
    c. $\theta = 60°$, $\omega = 1$ rad/sec (clockwise)
    d. $\theta = 60°$, $\omega = 1$ rad/sec (counterclockwise)
2. Repeat Problem 1 but use Figure P.13.
3. Repeat Problem 1 but use Figure P.14.
4. Repeat Problem 1 but use Figure P.15.
5. Repeat Problem 1 but use Figure P.16.
6. Repeat Problem 1 but use Figure P.17.
7. Repeat Problem 1 but use Figure P.18.
8. Repeat Problem 1 but use Figure P.19.
9. Repeat Problem 1 but use Figure P.20.
10. Repeat Problem 1 but use Figure P.21.
11. Determine the linear velocities of points C and E in Figure P.22.
12. Repeat Problem 11 but use Figure P.23.
13. Repeat Problem 11 but use Figure P.24.
14. Repeat Problem 11 but use Figure P.25.
15. Repeat Problem 11 but use Figure P.26.
16. Repeat Problem 11 but use Figure P.27.
17. Repeat Problem 11 but use Figure P.28.
18. Repeat Problem 11 but use Figure P.29.
19. Repeat Problem 11 but use Figure P.30.
20. Repeat Problem 11 but use Figure P.31.

GRAPHICAL TECHNIQUES:
ACCELERATION ANALYSIS

Effective Components

1. Determine the linear acceleration of point C in Figure P.12 for any of the following conditions:
    a. $\theta = 45°$ rad, $\omega = 0.5$ rad/sec (clockwise), $\alpha = 1$ rad/sec^2 (clockwise)

b. $\theta = 45°$ rad, $\omega = 1$ rad/sec (counterclockwise), $\alpha = 0.5$ rad/ sec² (clockwise)

c. $\theta = 135°$ rad, $\omega = 0.5$ rad/sec (clockwise), $\alpha = 1$ rad/sec² (counterclockwise)

d. $\theta = 135°$ rad, $\omega = 1$ rad/sec (counterclockwise), $\alpha = 0.5$ rad/sec² (counterclockwise)

2.   Repeat Problem 1 but use Figure P.13.
3.   Repeat Problem 1 but use Figure P.14.
4.   Repeat Problem 1 but use Figure P.15.
5.   Repeat Problem 1 but use Figure P.16.
6.   Repeat Problem 1 but use Figure P.17.
7.   Repeat Problem 1 but use Figure P.18.
8.   Repeat Problem 1 but use Figure P.19.
9.   Repeat Problem 1 but use Figure P.20.
10.   Repeat Problem 1 but use Figure P.21.
11.   Determine the linear acceleration of points C and E in Figure P.22.
12.   Repeat Problem 11 but use Figure P.23.
13.   Repeat Problem 11 but use Figure P.24.
14.   Repeat Problem 11 but use Figure P.25.
15.   Repeat Problem 11 but use Figure P.26.
16.   Repeat Problem 11 but use Figure P.27.
17.   Repeat Problem 11 but use Figure P.28.
18.   Repeat Problem 11 but use Figure P.29.
19.   Repeat Problem 11 but use Figure P.30.
20.   Repeat Problem 11 but use Figure P.31.

Relative Acceleration

1.   Determine the linear acceleration of point C in Figure P.12 for any of the followng conditions:

a. $\theta = 45°$ rad, $\omega = 0.5$ rad/sec (clockwise), $\alpha = 1$ rad/sec² (clockwise)

b. $\theta = 45°$ rad, $\omega = 1$ rad/sec (counterclockwise), $\alpha = 0.5$ rad/ sec² (clockwise)

c. $\theta = 135°$ rad, $\omega = 0.5$ rad/sec (clockwise), $\alpha = 1$ rad/sec² (counterclockwise)

d. $\theta = 135°$ rad, $\omega = 1$ rad/sec (counterclockwise), $\alpha = 0.5$ rad/sec² (counterclockwise)

2.   Repeat Problem 1 but use Figure P.13.
3.   Repeat Problem 1 but use Figure P.14.
4.   Repeat Problem 1 but use Figure P.15.

5. Repeat Problem 1 but use Figure P.16.
6. Repeat Problem 1 but use Figure P.17.
7. Repeat Problem 1 but use Figure P.18.
8. Repeat Problem 1 but use Figure P.19.
9. Repeat Problem 1 but use Figure P.20.
10. Repeat Problem 1 but use Figure P.21.
11. Determine the linear acceleration of points C and E in Figure P.22.
12. Repeat Problem 11 but use Figure P.23.
13. Repeat Problem 11 but use Figure P.24.
14. Repeat Problem 11 but use Figure P.25.
15. Repeat Problem 11 but use Figure P.26.
16. Repeat Problem 11 but use Figure P.27.
17. Repeat Problem 11 but use Figure P.28.
18. Repeat Problem 11 but use Figure P.29.
19. Repeat Problem 11 but use Figure P.30.
20. Repeat Problem 11 but use Figure P.31.

Velocity Difference

1. Determine the linear acceleration of point C in Figure P.12 for any of the followng conditions:
   a. $\theta = 45°$ rad, $\omega = 0.5$ rad/sec (clockwise), $\alpha = 0$ rad/sec (clockwise)
   b. $\theta = 45°$ rad, $\omega = 1$ rad/sec (counterclockwise), $\alpha = 0$ rad/sec (clockwise)
   c. $\theta = 135°$ rad, $\omega = 0.5$ rad/sec (clockwise), $\alpha = 1$ rad/sec (counterclockwise)
   d. $\theta = 135°$ rad, $\omega = 1$ rad/sec (counterclockwise), $\alpha = 0.5$ rad/sec (counterclockwise)
2. Repeat Problem 1 but use Figure P.13.
3. Repeat Problem 1 but use Figure P.14.
4. Repeat Problem 1 but use Figure P.15.
5. Repeat Problem 1 but use Figure P.16.
6. Repeat Problem 1 but use Figure P.17.
7. Repeat Problem 1 but use Figure P.18.
8. Repeat Problem 1 but use Figure P.19.
9. Repeat Problem 1 but use Figure P.20.
10. Repeat Problem 1 but use Figure P.21.

GRAPHICAL TECHNIQUES: MISCELLANEOUS

1. Determine the linear velocity of point C (Figure P.32).

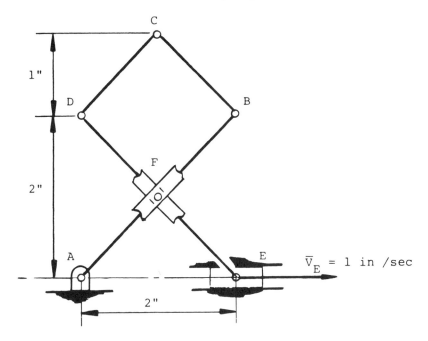

Figure P.32

2. Determine the linear velocity of point B (Figure P.33).

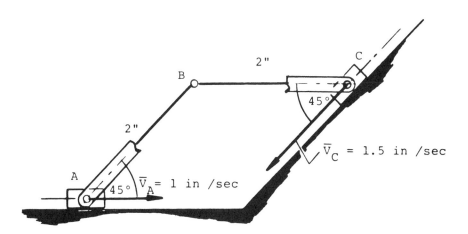

Figure P.33

3. Determine the linear velocities of points C and A (Figure P.34).

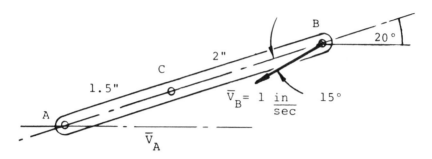

Figure P.34

4. Determine the linear velocities of points E and F (Figure P.35).

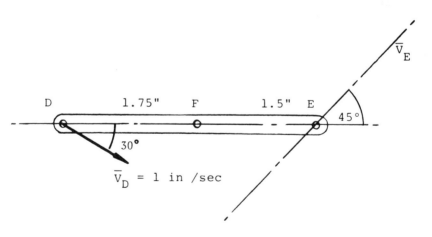

Figure P.35

5.   Determine the linear acceleration of point B (Figure P.36).

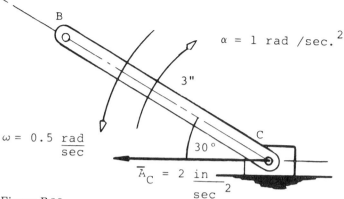

Figure P.36

6.   Determine the following (Figure P.37):
   a.   Linear acceleration of point B
   b.   Linear acceleration of point C
   c.   Linear acceleration of point B relative to point C

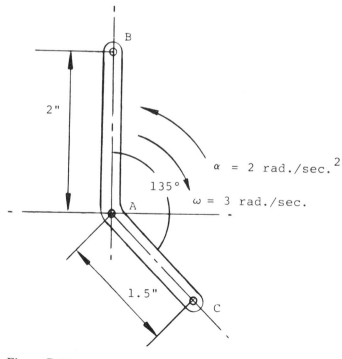

Figure P.37

7. Using instant center 24, determine the linear velocity of the pivot 34 (Figure P.38).

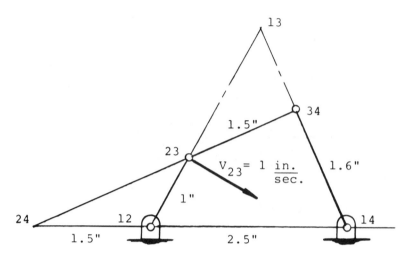

Figure P.38

8. Determine the linear velocities of points B and C and the linear velocity C relative to B (Figure P.39).

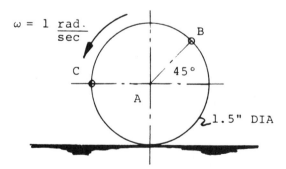

Figure P.39

9. Determine the linear acceleration of point C (Figure P.40) if the wheel rolls without slipping.

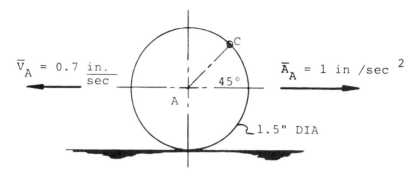

Figure P.40

10. Determine the linear acceleration of point P on the follower (Figure P.41), using the
    a. Relative acceleration method
    b. Equivalent linkage method

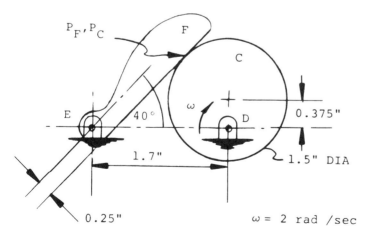

Figure P.41

11. Determine the following (Figure P.42):
    a. Linear acceleration of point B
    b. Angular acceleration of point C relative to point A
    c. Angular velocity of point C relative to point A

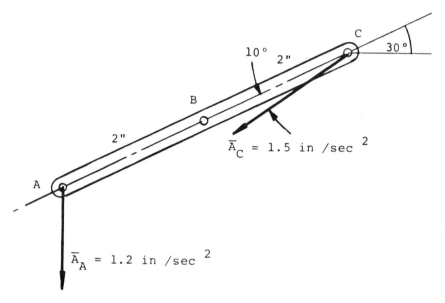

Figure P.42

12. Determine the following (Figure P.43):
    a. Linear acceleration of point B
    b. Angular acceleration of point A relative to point C
    c. Angular velocity of point A relative to point C

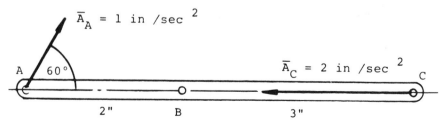

Figure P.43

13. Determine the following (Figure P.44):
    a. Linear acceleration of point C
    b. Angular acceleration of point A relative to point C
    c. Angular velocity of point A relative to point C

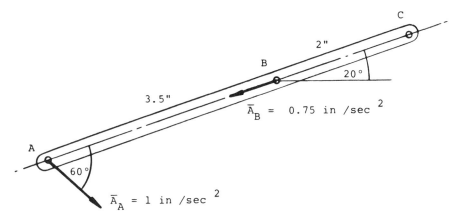

Figure P.44

14. Determine the linear acceleration of the cam follower, link 4 of Figure P.45, for the following positions:
    a. $\theta = 45°$
    b. $\theta = 75°$

15. Determine the linear acceleration of the cam follower, link 4 of Figure P.46, using the:
    a. Relative acceleration method
    b. Equivalent linkage method

16. a. Develop the linear velocity vs angular displacement curve for the complete cycle of the Scotch yoke mechanism (Figure P.47), based on a crank angular velocity of 2 rad/sec and crank displacement of 30°.
    b. From linear velocity vs angular displacement curve in part (a), develop the linear acceleration-displacement curve, using graphical differentiation.

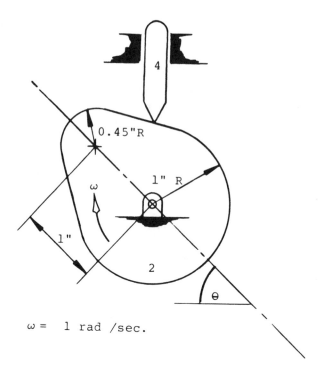

$\omega = 1$ rad /sec.

Figure P.45

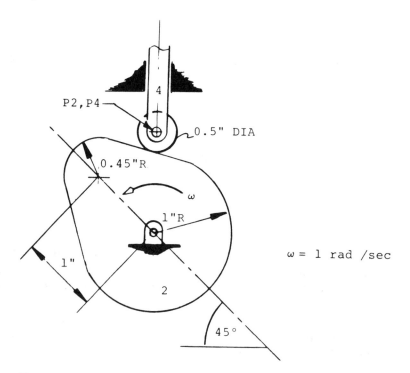

Figure P.46

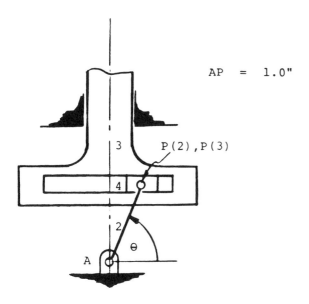

Figure P.47

17. a. Develop the complete linear velocity-time curve for point E
         on the slider-crank mechanism (Figure P.48), using convenient
         time intervals, based on a crank angular velocity of 6.28 rad/
         sec.
     b. From the linear velocity-time curve in part (a), develop the
         linear displacement-time curve, using graphical integration.

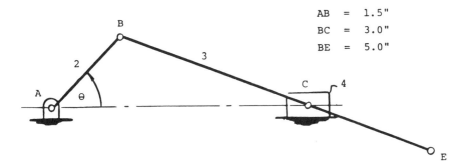

Figure P.48

18. Develop the linear acceleration vs angular displacement curve for
     a complete cycle of the drag-link mechanism (Figure P.49), based

on a crank angular velocity of 1 rad/sec and crank angular acceleration of 0.5 rad/sec.

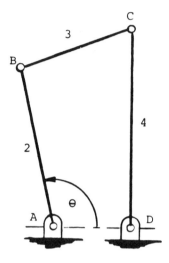

AB	=	2.0"
BC	=	1.5"
CD	=	2.5"
AD	=	1.0"

Figure P.49

19.   Develop the complete linear acceleration-time curve for the sliding
coupler mechanism (Figure P.50), based on a crank angular velocity of 30 rad/sec.

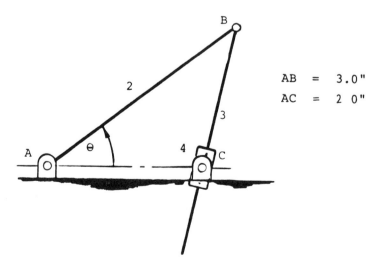

AB	=	3.0"
AC	=	2 0"

Figure P.50

20. A vehicle starting from rest is observed to have the following speeds at the times given:

Time (sec)	1	2	4	6	8	10	12
Velocity (mph)	1.5	3.0	8.5	16	21.5	25	26.5

Draw the velocity diagram and obtain from it the acceleration and displacement curves.

21. In Figure P.51, wheel W rolls without slipping. Determine the velocities of point D and D relative to C.

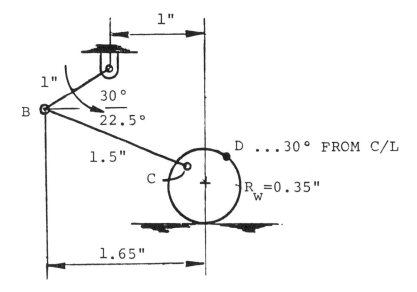

Figure P.51

22. In Figure P.52, wheels A and B roll without slipping as A moves to the right at 1 in./sec. Determine the:
    a. Angular velocity of B.
    b. Velocity of O relative to P

23. In Figure P.53, the slotted bar rotates at 0.5 rad/sec. $P2$ and $P3$ are coincident points on the rotating bar and the horizontal bar 3, respectively. Determine the linear acceleration of point P on link 3.

WHEEL RADIUS... 1.25"
R_S = R_R = 0.5"
S... 45° FROM WHEEL C/L
R... 75° FROM WHEEL C/L (HORZ.)

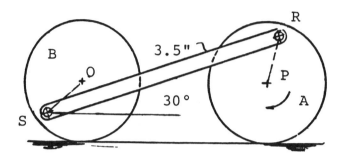

Figure P.52

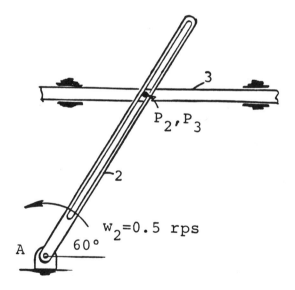

Figure P.53

24. In the belt and pulley system shown in Figure P.54, pulley 2 drives
    pulley 4 with an angular velocity of 0.25 rad/sec and angular
    acceleration of 1 rad/sec/sec. Determine the linear accelerations
    of points B, C, and D. Show all vectors.

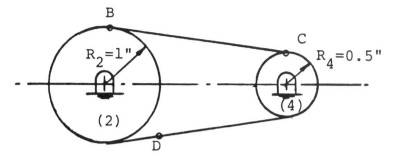

Figure P.54

25. In the planetary gear train shown in Figure P.55, the arm AB rotates at a constant angular speed of 1 rad/sec. Assuming there is no slipping, determine the velocity and acceleration of point P on the planet gear.

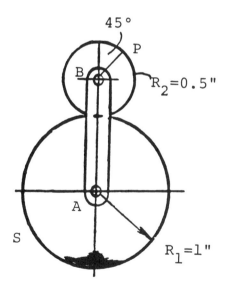

Figure P.55

26. In Figure P.56, wheels 1 and 2 roll without slipping. Using the effective component of the velocity vector for point B shown, determine the velocity of Point C. P and Q are fixed axes.

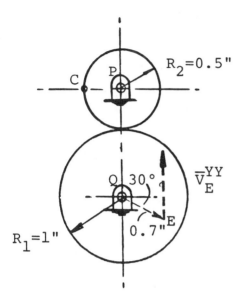

Figure P.56

27. In Figure P.57, wheel 2 rolls without slipping on wheel 1, which is fixed, and has an angular velocity of 2 rad/sec and angular acceleration of 1 rad/sec. Find the acceleration of point E.

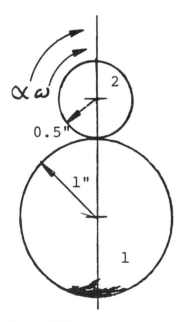

Figure P.57

28. The same as Problem 12, except wheel 2 rolls on the inside of wheel 1. (See Figure P.58.)

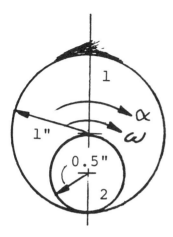

Figure P.58

29. The sames as Problem 12, except wheel 2 rolls on a straight track. (See Figure P.59.)

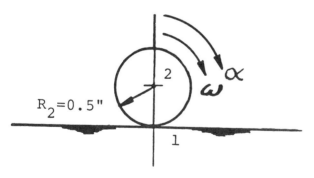

Figure P.59

30. Figure P.60 shows a ball bearing where the outer race 1 is stationary and the inner race 2 rotates wtih the shaft at 1600 rpm. Assuming pure rolling of the balls 3, determine the absolute velocity of the balls and angular velocity of the spacer ring.

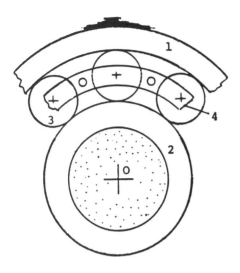

Figure P.60

ANALYTICAL TECHNIQUES: VELOCITY AND ACCELERATION

1. Using the simplified vector method equations, calculate the acceleration of point C and the acceleration of point C relative to B in Figure P.12 for any of the following conditions:
   a. $\theta = 45°$ rad, $\omega = 1$ rad/sec (clockwise), $\alpha = 0$ rad/sec² (clockwise)
   b. $\theta = 135°$ rad, $\omega = 1$ rad/sec (counterclockwise), $\alpha = 0.5$ rad/sec² (clockwise)
   c. $\theta = 225°$ rad, $\omega = 1$ rad/sec (clockwise), $\alpha = 0$ rad/sec² (counterclockwise)
   d. $\theta = 315°$ rad, $\omega = 1$ rad/sec (counterclockwise), $\alpha = 0.5$ rad/sec² (counterclockwise)

   Check the results using alternative equations, when possible, or a graphical method.
2. Repeat Problem 1 but use Figure P.13.
3. Repeat Problem 1 but use Figure P.14.
4. Repeat Problem 1 but use Figure P.15.
5. Repeat Problem 1 but use Figure P.16.
6. Repeat Problem 1 but use Figure P.17.
7. Repeat Problem 1 but use Figure P.18.
8. Repeat Problem 1 but use Figure P.19.

GEARS

1. Identify the following using letters or numbers given in Figure P.61:
   a. Pitch radius of driver D
   b. Pressure line
   c. Pressure angle
   d. Circular pitch
   e. Path of contact
   f. Pitch point
   g. Addendum distance of D
   h. Dedendum radius of D
   i. Clearance
   j. Tooth width of follower F

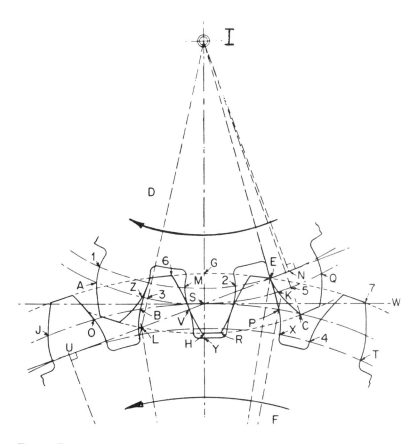

Figure P.61

2. The same as Problem 13, except identify the following:
    a. Arc of approach
    b. Tooth space on F
    c. Path of recess
    d. Arc of approach on F
    e. Arc of contact
    f. Pitch circle
    g. Base circle on D
    h. Base circle on F
    i. Addendum circle on D
    j. Addendum circle on F
3. Two standard 20° full-depth gears D and F are in mesh. Gear D turns clockwise about center O, driving gear F. The diametral pitch is 1.5. Gear F has 28 teeth. Compute the following values:
    a. Number of teeth on D
    b. Center distance
    c. Speed ratio
    d. Circular pitch
    e. Addendum
    f. Base diameter
    g. Normal pitch
    h. Arc of contact
    i. Path of contact
    j. Contact ratio
4. State the law of gearing.
5. Distinguish between the terms "circular pitch" and "diametral pitch."
6. Define "contact ratio."
7. Distinguish between the terms "backlash" and "clearance." Explain their importance in gear design.
8. Name the two basic gear tooth profiles. How are they formed? Using a base circle of 2.5-in. diameter, construct one of these profiles. Show the first five points of the curve.
9. Determine the pitch radii of a pair of mating gears for an angular velocity ratio of 3/4 and a center distance of 7 in.
10. A drive consists of two gears A and B. Gear A has 100 teeth and gear B 60 teeth. If either gear can rotate at 1000 rpm, determine the:
    a. Fastest output speed attainable
    b. Slowest output speed attainable
11. Gear A turns at 300 rpm, driving gear B at 750 rpm. The center

distance between the shafts is 12 in. and the diametral pitch 5. Determine the number of teeth on each gear

a. If they turn in opposite directions.
b. If they turn in the same direction.

12. A spur gear with 32 teeth and diametral pitch of 4 rotates at 1800 rpm, driving a gear at 750 rpm. Determine the:
   a. Number of teeth in the gear
   b. Center distance

13. A pair of standard of 20° full-depth involute spur gears have 18 and 20 teeth and a diametral pitch of 2. Compute the following:
   a. Circular pitch
   b. Base pitch
   c. Length of the path of contact (graphically)
   d. Contact ratio

14. A pair of standard 20° involute spur gears have 15 and 22 teeth and a diametral pitch of 2. The pinion rotates in a clockwise direction. Make a full-scale layout of two teeth in mesh and determine:
   a. Angle of approach for the pinion
   b. Angle of approach for the gear
   c. Angel of recess for the pinion
   d. Angle of recess for the gear

15. A 32-tooth spur gear has a pressure angle of 20°, diametral pitch of 4, and addendum of 0.2 in. Determine the number of teeth on the smallest pinion that will mesh the gear without interference.

16. A pair of meshing spur gears have 16 and 20 teeth, a diametral pitch of 4, and a 20° pressure angle.
   a. Determine the center distance.
   b. If the center distance is decreased 0.5 in., find the resultant pressure angle.

17. A pair of spur gears have 16 and 18 teeth, a diametral pitch of 2, and a pressure angle of $14\frac{1}{2}°$.
   a. Show that the gears have interference.
   b. Determine graphically the amount by which the addendum must be reduced to eliminate interference.
   c. Measure the length of the new path of contact.
   d. Compute the contact ratio.

18. A 50-tooth helical gear has a normal circular pitch of 0.75 and helix angle of 15°. Determine the:
   a. Normal diametral pitch
   b. Pitch diameter
   c. Diametral pitch in the traverse plane

19. A pair of helical gears have a center distance of 6 in. and helix angle of 15°. The smaller of the two gears has 20 teeth and the speed ratio is 261. Find the:
   a. Normal diametral pitch
   b. Transverse diametral pitch
   c. Pitch diameter of the two gears

20. Two helical gears on parallel shafts have a normal circular pitch of 0.5 in., pitch line velocity of 40 in./sec, and speed ratio of 1/8. If the pinion gear has 20 teeth and its pitch line speed is 60 rpm, find the:
   a. Center distance
   b. Helix angle
   c. Diametral pitch

21. A helical gear of normal diametral pitch has 40 teeth and a helix angle of 20°. Determine the:
   a. Pitch diameter
   b. Transverse diametral pitch
   c. Circular pitch in the normal plane

GEAR TRAINS

1. Determine the tooth numbers for a reverted gear train that will provide a speed reduction as near as possible to 24 to 1, given that the gears are to have a diametral pitch of 6 and the center distance between shafts is to be exactly 6 in.

2. In a simple planetary gear train, the sun gear has 90 teeth and the ring gear 150 teeth. The planet arm is rotated clockwise at 1000 rpm. Determine the speed and direction of rotation
   a. Of the ring gear if the sun gear is held fixed.
   b. Of the sun gear if the ring gear is held fixed.

3. Figure P.62 shows a schematic gear train in which gear A rotates at 1200 rpm and carrier C rotates in the same direction at 2000 rpm. All gears have the same diametral pitch. Determine the output speed of gear E. Use the formula method.

4. Repeat Problem 3, except use the superposition method.

5. Figure P.63 shows a schematic gear train in which gear A rotates at 1200 rpm and carrier C rotates in the same direction at 2000 rpm. All gears have the same diametral pitch. Determine the output speed of the planet carrier C. Use the formula method.

6. Repeat Problem 5, except use the superposition method.

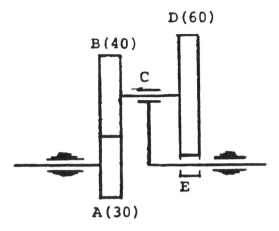

Figure P.62

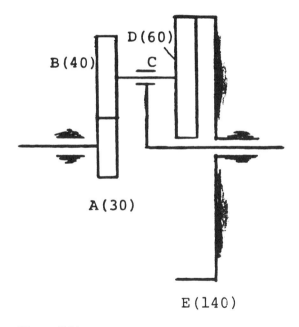

Figure P.63

7. Figure P.64 shows a schematic gear train in which gear A rotates at 1200 rpm and carrier C rotates in the same direction at 2000 rpm. All gears have the same diametral pitch. Determine the output speed of the planet carrier C. Use the formula method.

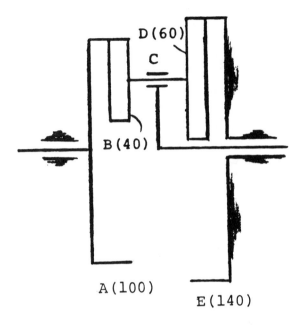

Figure P.64

8. Repeat Problem 7, except use the superposition method.
9. Figure P.65 shows a four-stage reverted gear train in which all gears turn with the shafts on which they are mounted and all shafts turn in fixed bearings. If the input speed of shaft M is 1800 rpm, determine the speed of the output shaft N.

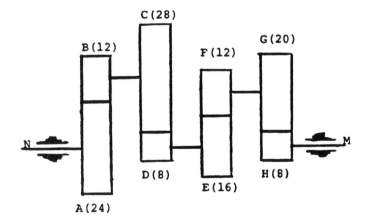

Figure P.65

10. In Figure P.66, shaft D turns at 3600 rpm (CW). Determine the speed of shaft F.

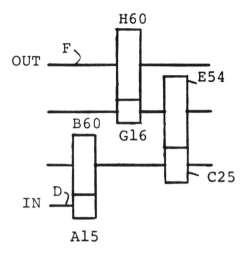

Figure P.66

11. In Figure P.67, gear D turns CW at 400 rpm, while planet arm A turns CCW at 200 rpm. Determine the absolute speed of gear F.

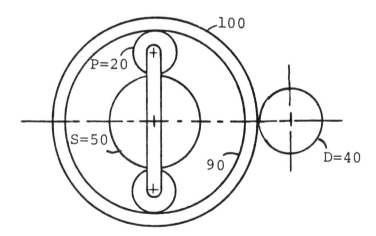

Figure P.67

12. In Figure P.68, gear D rotates at 200 rpm CCW and drives gear R, while the planet arm A rotates at 120 rpm CW. Determine the rotational speed and direction of gear S.

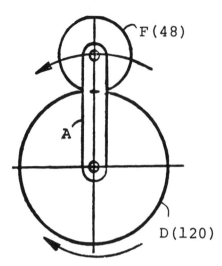

Figure P.68

13. The planetary gear train shown in Figure P.69 has a speed reduc-
     tion of 4 to 1 as the sun gear drives the planet arm. The sun gear
     rotates at 2000 rpm. The planet gear has 20 teeth and a diametral
     pitch of 10. Determine the tooth ratio between the ring and sun
     gears. What is the angular speed of the planet gear?

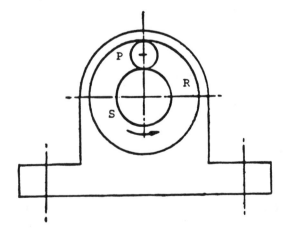

Figure P.69

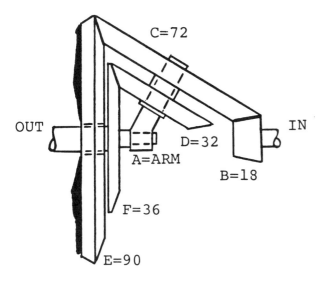

Figure P.70

14. For the Humpage epicyclic gear train shown in Figure P.70,
    a. Show that

$$\frac{\omega_F}{\omega_B} = \frac{T_B(T_C/T_E - T_D/T_F)}{T_C(1 + T_B/T_E)}$$

    b. Determine the angular velocity of the output shaft, if the input shaft rotates at 240 rpm.
15. For the planetary gear train shown in Figure P.71, what is the output–input speed ratio, using either shaft as the input.

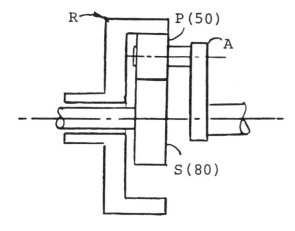

Figure P.71

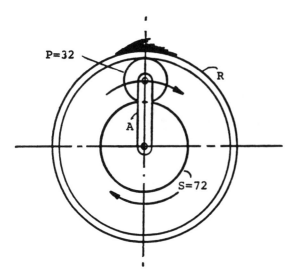

Figure P.72

16. In Figure P.72, gear S turns at 3600 rpm about fixed axis O, while planet arm A also turns about the same axis. Gear R is stationary. Determine the speed of arm A.

17. In Figure P.73, the planet arm turns at 200 rpm. Determine the speed of gear C.

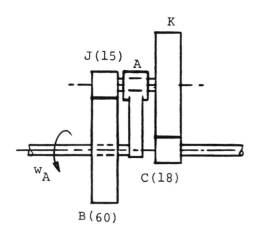

Figure P.73

18. In the series gear train shown in Figure P.74, determine the angular velocity of the planet arm A. The sun gear S and ring gear R turn at 400 and 600 rpm, respectively.

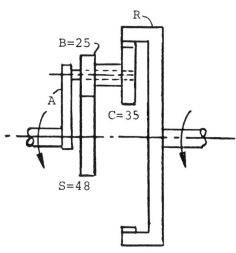

Figure P.74

19. Design a reverted gear train having a speed ratio of 1/15. The minimum number of teeth of any gear is 12. All gears have the same diametral pitch.
20. The same as Problem 19, except that the first pair has a diametral pitch of 5 and the second pair a diametral pitch of 6.
21. In Figure P.75, gear B is fixed, while arm A rotates at 60 rpm CW. Find the angular speed of gear E.

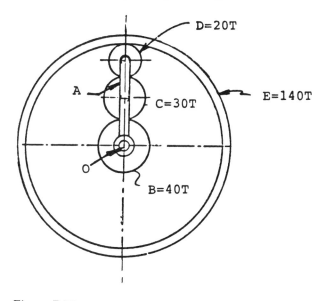

Figure P.75

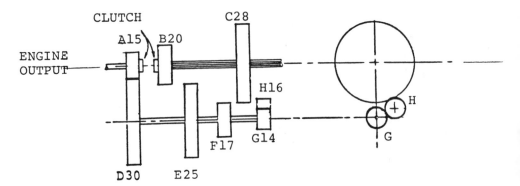

Figure P.76

22. Figure P.76 shows an automotive transmission schematic. A car
    equipped with this transmission has a differential ratio of 4/1 and
    tire outside diameter of 27 in. Determine the engine speed of the
    car as it travels in:
    a. Low gear at 25 mph
    b. High gear at 55 mph
    c. Reverse at 5 mph

23. In the series train shown in Figure P.77, the ring gears can be
    selectively held stationary as the input shaft D turns. If the input
    speed of D is 1800 rpm, determine the output speed of F,
    a. When R1 is held stationary.
    b. When R2 is held stationary.

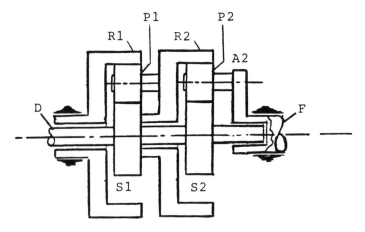

Figure P.77

Table P1 Cam Specifications for Problems 1–5

Cam angle	Rise/fall (in.)	Motion
0–18	1.5	Constant acceleration
180–210	Dwell	
210–330	1.5	Simple harmonic
330–360	Dwell	

CAMS

1. Make a layout of a 2-in. base diameter disk cam having a pointed follower to operate under the conditions given in Table P1 above.
2. Repeat Problem 1, except the follower is to be a radial roller follower, where the roller is 0.5 in. in diameter.
3. Repeat Problem 1, except the follower is to be an offset roller follower, where the offset is 0.5 in. and the roller 0.5 in. in diameter.
4. Repeat Problem 1, except the follower is to be a flat face follower 1 in. wide.
5. Repeat Problem 1, except the follower is to be an oscillating roller follower, where the oscillating arm is 2.5 in. long and horizontal when the roller is at the initial or home position of the cam cycle.

Table P2 Cam Specifications for Problems 6–10

Cam angle	Rise/fall (in.)	Motion
0–18	1.5	Constant acceleration
180–210	Dwell	
210–330	1.5	Simple harmonic
330–360	Dwell	

6. The same as Problem 1, except use Table P2.
7. The same as Problem 2, except use Table P2.
8. The same as Problem 3, except use Table P2.
9. The same as Problem 4, except use Table P2.
10. The same as Problem 5, except use Table P2.

11. Make a layout of a 2-in. base diameter roller follower disk cam for the following operating conditions:

Table P3   Cam Specifications for Problem 11

Cam angle	Lift (in.)	Motion
0–120	1.5	Modified constant velocity (arc R = 1 in.)
120–180	Dwell	
180–300	1.5	Constant acceleration
300–360	Dwell	

12. Make a layout of a 2-in. base diameter pointed follower disk cam for the following operating conditions:

Table P4   Cam Specifications for Problem 12

Cam angle	Lift (in.)	Motion
0–90	2.0	Constant velocity
90–180	Dwell	
180–270	1.0	Modified constant velocity (arc R = 0.5 in.)
270–360	1.0	Constant velocity

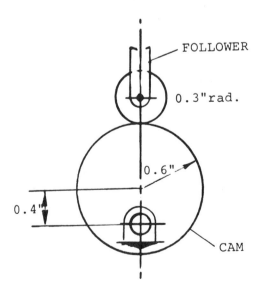

Figure P.78

13. Figure P.78 shows an eccentric cam with a radial roller follower. Draw the displacement vs cam angle curve and determine the maximum pressure angle.
14. Using the cam follower displacement Table 22.2, determine the follower displacement at 40° for a cycloidal cam, if the follower rises 1.5 in. during a cam rotation angle of 120°.
15. Repeat Problem 14, except consider a simple harmonic cam.
16. Repeat Problem 14, except consider a constant acceleration cam.

Appendix A

A.1 LINKAGES: FORTRAN PROGRAMS

The computer programs listed in this section were developed to analyze the linkages covered in Part III, based on the mathematical methods discussed in that section. The programs are written in Fortran language applicable to a Univac 1108 digital computer. To apply these programs, typically, the link lengths of the mechanism, crank angle, angular velocity, and angular acceleration must be known. In the slider-crank case, additional information on slider offset (or eccentricity) must also be known. From these data, the required velocities and accelerations can be computed for any specified angular position of the crank in its motion cycle.

   As illustrations of typical outputs obtained from these programs, printouts of example problem results are given following each program listing.

Four-Bar Linkage Analysis: Simplified Vector Method

```
C       AB     = CRANK
C       BC     = COUPLER
C       CD     = FOLLOWER
C       AD     = FRAME
C       C      = DEGREES/RADIAN
C       THETAA = THETA(A), POSITION ANGLE OF LINK AB
C       THETAB = THETA(B), POSITION ANGLE OF LINK BC
C       THETAC = THETA(C), POSITION ANGLE OF LINK CD
C       THETAX = THETA(A) + 90 DEGS.
C       THETAY = THETA(B) + 90 DEGS.
C       THETAZ = THETA(C) + 90 DEGS.
C       PHID   = PHI(D),ANGLE BETWEEN CB AND AD
C       GAMMAB = GAMMA(B)
C       GAMMAD = GAMMA(D)
C       OMEGAA = OMEGA(A),ANGULAR VELOCITY OF AB
C       ALPHAA = ALPHA(A),ANGULAR ACCELERATION OF AB
C       SMB    = THETA(B)-THETA(C)
C       SMC    = THETA(B)-THETA(A)
C       SMCB   = THETA(C)-THETA(A)
C       VB     = VELOCITY OF B
C       VC     = VELOCITY OF C
C       VCB    = VELOCITY OF C RELATIVE TO B
C       CVB    = VEL OF B (COMPLEX)
C       CVC    = VEL OF C (COMPLEX)
C       CVCB   = VEL OF C RELATIVE TO B (COMPLEX)
C       ABSVB  = ABSOLUTE VELOCITY OF B
C       ABSVC  = ABSOLUTE VELOCITY OF C
C       ABSVCB = ABSOLUTE VEL OF C RELATIVE TO B
C       ACCNB  = NORMAL ACC OF B (COMPLEX)
C       ACCTB  = TANGENTIAL ACC OF B (COMPLEX)
C       ACCNC  = NORMAL ACC OF C (COMPLEX)
C       ACCTC  = TANGENTIAL ACC OF C (COMPLEX)
C       ACCNCB = NORMAL ACC OF C REL. TO B(COMPLEX)
C       ACCTCB = TANG. ACC OF C REL.TO B(COMPLEX)
C       ATC    = TANGENTIAL ACC OF C
C       ATCB   = TANGENTIAL ACC OF C RELATIVE TO B
C       ACCB   = ACC OF B (COMPLEX)
C       ACCC   = ACC OF C (COMPLEX)
C       ACCCB  = ACC OF C RELATIVE TO B (COMPLEX)
C       ABSNB  = ABSOLUTE NORMAL ACC OF B
C       ABSTB  = ABSOLUTE TANGENTIAL ACC OF B
C       ABSNC  = ABSOLUTE NORMAL ACC OF C
C       ABSTB  = ABSOLUTE TANGENTIAL ACC OF C
C       ABSNCB = ABS. NORMAL ACC OF C RELATIVE TO B
C       ABSTCB = ABS. TANGENTIAL ACC OF C REL. TO B
C       ABSAB  = ABS. ACCELERATION OF B
C       ABSAC  = ABSOLUTE ACCELERATION OF C
C       ABSACB = ABSOLUTE ACC OF C RELATIVE TO B
C       PHZVB  = PHASE ANGLE OF VELOCITY OF B
```

```
C      PHZVC  = PHASE ANGLE OF VELOCITY OF C
C      PHZVCB = PHASE ANGLE OF VEL. OF C REL. TO B
C      PHZNB  = PHASE ANGLE OF NORMAL ACC OF B
C      PHZTB  = PHASE ANGLE OF TANGENTIAL ACC OF B
C      PHZNC  = PHASE ANGLE OF NORMAL ACC OF C
C      PHZTC  = PHASE ANGLE OF TANGENTIAL ACC OF C
C      PHZNCB = PHASE ANGLE OF NORMAL ACC OF C
C               RELATIVE TO B
C      PHZTCB = PHASE ANGLE OF TANGENTIAL ACC OF C
C               RELATIVE TO B
C      PHZAB  = PHASE ANGLE OF ABSOLUTE ACC OF B
C      PHZAC  = PHASE ANGLE OF ABSOLUTE ACC OF C
C      PHZACB = PHASE ANGLE OF ABSOLUTE ACC OF C
C               RELATIVE TO B
C
       COMPLEX ACCNB,ACCTB,ACCB,ACCNC,ACCTC,ACCC
       COMPLEX ACCNCB,ACCTCB,ACCCB
       COMPLEX CVB,CVC,CVCB
       C = 57.29578
C
       OPEN (UNIT=6, FILE='4BAR.LIS', STATUS='NEW')
       OPEN (UNIT=10, FILE='4BAR.DAT', STATUS='OLD')
C
    1  READ(10,*)AB,BC,CD,AD,THETAA,OMEGAA,ALPHAA
C 100  FORMAT(7F10.4)
       IF(AB.EQ.0.0) GO TO 999
       WRITE(6,104)
  104  FORMAT(1H1,9X,'FOUR BAR: SIMPLIFIED VECTOR M
      +ETHOD',//10X,'PROBLEM DATA')

       WRITE(6,105)
  105  FORMAT(/12X,'AB',10X,'BC',10X,'CD',10X,'AD')
       WRITE(6,106)AB,BC,CD,AD
       WRITE(6,1105)
 1105  FORMAT(/6X,'THETA(A)',4X,'OMEGA(A)',4X,'ALPH
      +A(A)')
       WRITE(6,106)THETAA,OMEGAA,ALPHAA
C
       THETAA= THETAA/C
       BD=SQRT(AB**2+AD**2-2.0*AB*AD*COS(THETAA))
       PHID= ASIN(AB/BD*SIN(THETAA))
       PHID =PHID*C
       S2=(BC+BD+CD)/2.0
C
       GAMMAB=ACOS((BC**2+BD**2-CD**2)/(2.0*BC*BD))
       GAMMAD=ACOS((BD**2+CD**2-BC**2)/(2.0*BD*CD))
C
       GAMMAB= GAMMAB*C
       GAMMAD= GAMMAD*C
C
```

```
      THETAA = THETAA*C
      THETAB = GAMMAB - PHID
      THETAC = 180. - PHID - GAMMAD
      WRITE(6,901)
  901 FORMAT(/12X,'BD',6X,'PHI(D)',4X,'GAMMA(B)'
     +,4X,'GAMMA(D)')

      WRITE(6,107)BD,PHID,GAMMAB,GAMMAD
      WRITE(6,1901)
 1901 FORMAT(/6X,'THETA(A)',4X,'THETA(B)',4X,
     +'THETA(C)')

      WRITE(6,107)THETAA,THETAB,THETAC
      SMB = THETAB - THETAC
      SMC = THETAB - THETAA
      SMCB = THETAC - THETAA
      SRB=SMB/C
      SRC=SMC/C
      SRCB=SMCB/C
      SNB = SIN(SRB)
      SNC = SIN(SRC)
      SNCB = SIN(SRCB)
      VB = OMEGAA * AB
      VC = VB*SNC/SNB
      VCB = VB*SNCB/SNB
      WRITE(6,903)
  903 FORMAT(/10X,'LINEAR VELOCITIES')
      WRITE(6,904)
  904 FORMAT(/10X,'V(B)',8X,'V(C)',7X,'V(CB)')
      WRITE(6,107)VB,VC,VCB
      THETAA = (THETAA/C)
      THETAB = (THETAB/C)
      THETAC = (THETAC/C)
C
      THETAY = (THETAB + 90.0/C)
      THETAX = (THETAA + 90.0/C)
      THETAZ = (THETAC + 90.0/C)
      X = COS(THETAX)
      Y = SIN(THETAX)
      CVB = VB * CMPLX(X,Y)
      X= COS(THETAZ)
      Y = SIN(THETAZ)
      CVC = VC* CMPLX(X,Y)
      X = COS(THETAY)
      Y = SIN(THETAY)
      CVCB = VCB * CMPLX(X,Y)
      ABSVB = CABS(CVB)
      PHZVB=C*ATAN2(AIMAG(CVB),REAL(CVB))
      ABSVC = CABS(CVC)
C     TEST ABSVC
```

```
      IF(ABSVC.GT. 0.0001)GO TO 80
      PHZVC=0.0
      GO TO 85
   80 PHZVC=C*ATAN2(AIMAG(CVC),REAL(CVC))
   85 CONTINUE
      ABSVCB = CABS(CVCB)
C     TEST ABSVCB
      IF(ABSVCB.GT. 0.0001)GO TO 90
      PHZVCB=0.0
      GO TO 95
   90 PHZVCB=C*ATAN2(AIMAG(CVCB),REAL(CVCB))
   95 CONTINUE
      WRITE(6,905)
  905 FORMAT(/10X,'REAL',8X,'IMAG',9X,'ABS',7X,
     +'PHASE')
      WRITE(6,301)CVB,ABSVB,PHZVB
      WRITE(6,302)CVC,ABSVC,PHZVC
      WRITE(6,303)CVCB,ABSVCB,PHZVCB
      D1 = VC**2*COS(THETAC)/CD
      D2 = -OMEGAA**2*AB*COS(THETAA)
      D3 = ALPHAA*AB*COS(THETAX)
      D4 =-VCB**2*COS(THETAB)/BC
      D5 = VC**2*SIN(THETAC)/CD
      D6 = -OMEGAA**2*AB*SIN(THETAA)
      D7 = ALPHAA*AB*SIN(THETAX)
      D8 =-VCB**2*SIN(THETAB)/BC
      C1 = D1+D2+D3+D4
      C2 = D5+D6+D7+D8
      A1 = COS(THETAZ)
      A2 = SIN(THETAZ)
      B1 =-COS(THETAY)
      B2 =-SIN(THETAY)
C
      ATC = (C1*B2 - C2*B1)/(A1*B2 - A2*B1)
      ATCB = (A1*C2 - A2*C1)/(A1*B2 - A2*B1)
C
      X = COS(THETAA)
      Y = SIN(THETAA)
      ACCNB = -OMEGAA**2*AB*CMPLX(X,Y)
      X=COS(THETAX)
      Y=SIN(THETAX)
      ACCTB = ALPHAA*AB*CMPLX(X,Y)
      ACCB = ACCNB + ACCTB
      X = COS(THETAC)
      Y = SIN(THETAC)
      ACCNC = -VC**2*CMPLX(X,Y)/CD
      X = COS(THETAZ)
      Y = SIN(THETAZ)
      ACCTC = ATC*CMPLX(X,Y)
      ACCC = ACCNC + ACCTC
```

```
      X = COS(THETAB)
      Y = SIN(THETAB)
      ACCNCB =-VCB**2*CMPLX(X,Y)/BC
      X = COS(THETAY)
      Y = SIN(THETAY)
      ACCTCB = ATCB * CMPLX(X,Y)
      ACCCB = ACCNCB + ACCTCB
C
      ABSNB=CABS(ACCNB)
      PHZNB = C*ATAN2(AIMAG(ACCNB),REAL(ACCNB))
      ABSTB=CABS(ACCTB)
C     TEST ABSTB
      IF(ABSTB.GT.0.0001)GO TO 10
      PHZTB = 0.0
      GO TO 15
   10 PHZTB = C*ATAN2(AIMAG(ACCTB),REAL(ACCTB))
   15 CONTINUE
      ABSNC=CABS(ACCNC)
C     TEST ABSNC
      IF(ABSNC.GT. 0.0001)GO TO 20
      PHZNC=0.0
      GO TO 25
   20 PHZNC = C*ATAN2(AIMAG(ACCNC),REAL(ACCNC))
   25 CONTINUE
      ABSTC=CABS(ACCTC)
C     TEST ABSTC
      IF(ABSTC.GT. 0.0001)GO TO 30
      PHZTC=0.0
      GO TO 35
   30 PHZTC = C*ATAN2(AIMAG(ACCTC),REAL(ACCTC))
   35 CONTINUE
      ABSNCB=CABS(ACCNCB)
C     TEST ABSNCB
      IF(ABSNCB. GT. 0.0001)GO TO 40
      PHZNCB=0.0
      GO TO 45
   40 PHZNCB = C*ATAN2(AIMAG(ACCNCB),REAL(ACCNCB))
   45 CONTINUE
      ABSTCB=CABS(ACCTCB)
C     TEST ABSTCB
      IF(ABSTCB.GT. 0.0001)GO TO 50
      PHZTCB=0.0
      GO TO 55
   50 PHZTCB = C*ATAN2(AIMAG(ACCTCB),REAL(ACCTCB))
   55 CONTINUE
      WRITE(6,906)
  906 FORMAT(/10X,'NORMAL ACCELERATIONS')
      WRITE(6,915)
  915 FORMAT(/10X,'REAL',8X,'IMAG',9X,'ABS',7X,
     +'PHASE')
```

```
      WRITE(6,301)ACCNB,ABSNB,PHZNB
      WRITE(6,302)ACCNC,ABSNC,PHZNC
      WRITE(6,303)ACCNCB,ABSNCB,PHZNCB
      WRITE(6,1906)
 1906 FORMAT(/10X,'TANGENTIAL ACCELERATIONS')
      WRITE(6,1916)
 1916 FORMAT(/10X,'REAL',8X,'IMAG',9X,'ABS',7X,
     +'PHASE')
      WRITE(6,301)ACCTB,ABSTB,PHZTB
      WRITE(6,302)ACCTC,ABSTC,PHZTC
      WRITE(6,303)ACCTCB,ABSTCB,PHZTCB
      WRITE(6,907)
  907 FORMAT(/10X,'ABSOLUTE ACCELERATIONS')
      WRITE(6,905)
      ABSAB = CABS(ACCB)
      PHZAB = C*ATAN2(AIMAG(ACCB),REAL(ACCB))
      ABSAC = CABS(ACCC)
C     TEST ABSAC
      IF(ABSAC.GT. 0.0001) GO TO 60
      PHZAC=0.0
      GO TO 65
   60 PHZAC = C*ATAN2(AIMAG(ACCC),REAL(ACCC))
   65 CONTINUE
      ABSACB = CABS(ACCCB)
C     TEST ABSACB
      IF(ABSACB.GT. 0.0001)GO TO 70
      PHZACB=0.0
      GO TO 75
   70 PHZACB = C*ATAN2(AIMAG(ACCCB),REAL(ACCCB))
   75 CONTINUE
      WRITE(6,301)ACCB,ABSAB,PHZAB
      WRITE(6,302)ACCC,ABSAC,PHZAC
      WRITE(6,303)ACCCB,ABSACB,PHZACB
      THETAX=THETAX*C
      THETAY=THETAY*C
      THETAZ=THETAZ*C
      WRITE(6,909)
  909 FORMAT(/12X,'A1',10X,'A2',10X,'B1',10X,'B2')
      WRITE(6,107)A1,A2,B1,B2
      WRITE(6,1909)
 1909 FORMAT(/12X,'C1',10X,'C2',9X,'ATC',8X,
     +'ATCB')
      WRITE(6,107)C1,C2,ATC,ATCB
  106 FORMAT(2X,8F12.4)
  107 FORMAT(2X,8F12.4)
  301 FORMAT(2X,4F12.4,8X,'B ')
  302 FORMAT(2X,4F12.4,8X,'C ')
  303 FORMAT(2X,4F12.4,8X,'CB')
      GO TO 1
  999 STOP
      END
```

Four-Bar: Simplified Vector Method

PROBLEM DATA

AB	BC	CD	AD
1.5000	3.0000	3.0000	4.0000

THETA(A)	OMEGA(A)	ALPHA(A)
30.0000	2.0000	1.0000

BD	PHI(D)	GAMMA(B)	GAMMA(D)
2.8032	15.5189	62.1478	62.1478

THETA(A)	THETA(B)	THETA(C)
30.0000	46.6289	102.3334

LINEAR VELOCITIES

V(B)	V(C)	V(CB)
3.0000	-1.0392	-3.4601

REAL	IMAG	ABS	PHASE	
-1.5000	2.5981	3.0000	120.0000	B
1.0152	0.2220	1.0392	12.3334	C
2.5152	-2.3761	3.4601	-43.3711	CB

NORMAL ACCELERATIONS

REAL	IMAG	ABS	PHASE	
-5.1962	-3.0000	6.0000	-150.0000	B
0.0769	-0.3517	0.3600	-77.6666	C
-2.7405	-2.9009	3.9907	-133.3711	CB

TANGENTIAL ACCELERATIONS

REAL	IMAG	ABS	PHASE	
-0.7500	1.2990	1.5000	120.0000	B
-10.7699	-2.3548	11.0244	-167.6666	C
-2.0064	1.8954	2.7601	136.6289	CB

ABSOLUTE ACCELERATIONS

REAL	IMAG	ABS	PHASE	
-5.9462	-1.7010	6.1847	-164.0363	B
-10.6930	-2.7065	11.0302	-165.7964	C
-4.7469	-1.0055	4.8522	-168.0402	CB

A1	A2	B1	B2
-0.9769	-0.2136	0.7269	-0.6867

C1	C2	ATC	ATCB
-8.7635	-4.2502	11.0244	2.7601

Four-Bar: Simplified Vector Method

```
PROBLEM DATA
```

AB	BC	CD	AD
2.0000	1.5000	2.5000	1.0000

THETA(A)	OMEGA(A)	ALPHA(A)
240.0000	1.0000	0.5000

BD	PHI(D)	GAMMA(B)	GAMMA(D)
2.6458	-40.8934	67.7923	33.7446

THETA(A)	THETA(B)	THETA(C)
240.0000	108.6857	187.1488

```
LINEAR VELOCITIES
```

V(B)	V(C)	V(CB)
2.0000	1.5332	1.6270

REAL	IMAG	ABS	PHASE	
1.7321	-1.0000	2.0000	-30.0000	B
0.1908	-1.5213	1.5332	-82.8512	C
-1.5413	-0.5213	1.6270	-161.3143	CB

```
NORMAL ACCELERATIONS
```

REAL	IMAG	ABS	PHASE	
1.0000	1.7321	2.0000	60.0000	B
0.9329	0.1170	0.9403	7.1488	C
0.5654	-1.6718	1.7648	-71.3143	CB

```
TANGENTIAL ACCELERATIONS
```

REAL	IMAG	ABS	PHASE	
0.8660	-0.5000	1.0000	-30.0000	B
0.1280	-1.0202	1.0282	-82.8512	C
-1.3705	-0.4635	1.4468	-161.3143	CB

```
ABSOLUTE ACCELERATIONS
```

REAL	IMAG	ABS	PHASE	
1.8660	1.2321	2.2361	33.4350	B
1.0609	-0.9032	1.3933	-40.4102	C
-0.8051	-2.1353	2.2820	-110.6594	CB

A1	A2	B1	B2
0.1244	-0.9922	0.9473	0.3204

C1	C2	ATC	ATCB
1.4985	-0.5567	1.0282	1.4468

Slider-Crank Mechanism Analysis: Simplified Vector Method

```
C       C      = DEGREES/RADIAN(CONSTANT)
C       AB     = CRANK
C       BC     = CONNECTING ROD
C       PHI    = PHI(3),ANGLE BETWEEN CONNECTING ROD
C                AND DEAD CENTER
C       THETA2 = POSITION ANGLE OF LINK AB
C       THETA3 = POSITION ANGLE OF LINK BC
C       THETAX = THETA(2)+90DEGS.
C       THETAY = THETA(3)+90DEGS.
C       OMEGA2 = ANGULAR VELOCITY OF LINK AB
C       ALPHA2 = ANGULAR ACC OF LINK AB
C       SNB    = 90 - THETA(DEGS)
C       SNC    = THETA(2)-THETA(3)
C       SNBC   = 90 - THETA(2)
C       VB     = VELOCITY OF B
C       VC     = VELOCITY OF C
C       VBC    = VELOCITY OF B RELATIVE TO C
C       CVB    = VEL OF B (COMPLEX)
C       CVC    = VEL OF C (COMPLEX)
C       CVBC   = VEL OF B RELATIVE TO C (COMPLEX)
C       ABSVB  = ABSOLUTE VELOCITY OF B
C       ABSVC  = ABSOLUTE VELOCITY OF C
C       ABSVBC = ABSOLUTE VEL. OF B RELATIVE TO C
C       ACCNB  = NORMAL ACC OF B (COMPLEX)
C       ACCTB  = TANGENTIAL ACC OF B (COMPLEX)
C       ACCNC  = NORMAL ACC OF C (COMPLEX)
C       ACCTC  = TANGENTIAL ACC OF C (COMPLEX)
C       ACCNBC = NORMAL ACC OF B REL TO C (COMPLEX)
C       ACCTBC = TANGENTIAL ACC OF B RELATIVE TO
C                C(COMPLEX)
C       ATC    = TANGENTIAL ACC OF C
C       ATBC   = TANGENTIAL ACC OF B RELATIVE TO C
C       ACCB   = ACC OF B (COMPLEX)
C       ACCC   = ACC OF C (COMPLEX)
C       ACCBC  = ACC OF B RELATIVE TO C (COMPLEX)
C       ABSNB  = ABSOLUTE NORMAL ACC OF B
C       ABSTB  = ABSOLUTE TANGENTIAL ACC OF B
C       ABSNC  = ABSOLUTE NORMAL ACC OF C
C       ABSTC  = ABSOLUTE TANGENTIAL ACC OF C
C       ABSNBC = ABSOLUTE NORMAL ACC OF B REL TO C
C       ABSTBC = ABSOLUTE TANGENTIAL ACC OF B
C                RELATIVE TO C
C       ABSAB  = ABSOLUTE ACC OF B
C       ABSAC  = ABSOLUTE ACC OF C
C       ABSABC = ABSOLUTE ACC OF B RELATIVE TO C
C       PHZVB  = PHASE ANGLE OF VELOCITY OF B
C       PHZVC  = PHASE ANGLE OF VELOCITY OF C
C       PHZVBC = PHASE ANGLE OF VEL. OF B REL. TO C
C       PHZNB  = PHASE ANGLE OF NORMAL ACC OF B
```

```
C       PHZTB  = PHASE ANGLE OF TANGENTIAL ACC OF B
C       PHZNC  = PHASE ANGLE OF NORMAL ACC OF C
C       PHZTC  = PHASE ANGLE OF TANGENTIAL ACC OF C
C       PHZNBC = PHASE ANGLE OF NORMAL ACC OF B
C                RELATIVE TO C
C       PHZTBC = PHASE ANGLE OF TANGENTIAL ACC OF B
C                RELATIVE TO C
C       PHZAB  = PHASE ANGLE OF ABSOLUTE ACC OF B
C       PHZAC  = PHASE ANGLE OF ABSOLUTE ACC OF C
C       PHZABC = PHASE ANGLE OF ABSOLUTE ACC OF B
C                RELATIVE TO C
        COMPLEX CVB,CVC,CVBC,ACCC,ACCNC,ACCTC
        COMPLEX ACCNB,ACCTB,ACCB,ACCNBC,ACCTBC,ACCBC
        C=57.29578
        OPEN (UNIT=6, FILE='SCRNK.LIS', STATUS='NEW')
        OPEN (UNIT=10, FILE='SCRNK.DAT', STATUS='OLD')
      1 READ(10,*)AB,BC,THETA2,OMEGA2,ALPHA2,E
     99 FORMAT(6F12.4)
        IF(AB.EQ.0.0) GO TO 999
        WRITE(6,899)
    899 FORMAT(1H1,9X,'SLIDER-CRANK: SIMPLIFIED VEC
       +TOR METHOD',
       +//10X,'PROBLEM DATA')
        WRITE(6,900)
    900 FORMAT(/12X,'AB',10X,'BC')
        WRITE(6,100)AB,BC
        WRITE(6,1900)
   1900 FORMAT(/6X,'THETA(A)',4X,'OMEGA(A)',4X,'ALPH
       +A(A)',6X,'ECC(E)')
        WRITE(6,100)THETA2,OMEGA2,ALPHA2,E
        THETA2=THETA2/C
        PHI=ASIN(AB/BC*SIN(THETA2)-E/BC)
        PHI=PHI*C
        THETA3=180.0-PHI
        WRITE(6,901)
    901 FORMAT(//8X,'PHI(3)',4X,'THETA(3)')
        WRITE(6,101)PHI,THETA3
        THETA2=THETA2*C
        SNB=90.0-THETA3
        SNC=THETA3-THETA2
        SNBC=90.0-THETA2
        SMB=SNB/C
        SMC=SNC/C
        SMBC=SNBC/C
        VB= AB*OMEGA2
        VC= VB*SIN(SMC)/SIN(SMB)
        VBC=VB*SIN(SMBC)/SIN(SMB)
        WRITE(6,903)
    903 FORMAT(/10X,'LINEAR VELOCITIES')
        WRITE(6,904)
    904 FORMAT(/10X,'V(B)',8X,'V(C)',7X,'V(BC)')
```

```
      WRITE (6,101) VB,VC,VBC
      THETAX= (THETA2+90.0) /C
      THETAY= (THETA3+90.0) /C
      THETA2=THETA2/C
      THETA3=THETA3/C
      X=COS (THETAX)
      Y=SIN (THETAX)
      CVB=VB*CMPLX (X,Y)
      ABSVB=CABS (CVB)
      PHZVB=C*ATAN2 (AIMAG (CVB) ,REAL (CVB) )
      WRITE (6,905)
  905 FORMAT (/10X,'REAL',8X,'IMAG',9X,'ABS',7X,
     +'PHASE')
      WRITE (6,301) CVB,ABSVB,PHZVB
      CVC=VC
      ABSVC=CABS (CVC)
C     TEST FOR ABS VEL OF C
      IF (ABSVC.GT.0.0001) GO TO 20
      PHZVC=0.0
      GO TO 25
   20 PHZVC=C*ATAN2 (AIMAG (CVC) ,REAL (CVC) )
   25 CONTINUE
      WRITE (6,302) CVC,ABSVC,PHZVC
      X=COS (THETAY)
      Y=SIN (THETAY)
      CVBC=VBC*CMPLX (X,Y)
      ABSVBC=CABS (CVBC)
C     TEST VEL OF B REL TO C
      IF (ABSVBC.GT.0.0001) GO TO 30
      PHZVBC=0.0
      GO TO 35
   30 PHZVBC=C*ATAN2 (AIMAG (CVBC) ,REAL (CVBC) )
   35 CONTINUE
      WRITE (6,303) CVBC,ABSVBC,PHZVBC
      A1=1
      A2=0
      B1=COS (THETAY)
      B2=SIN (THETAY)
      C1=- (OMEGA2**2) *AB*COS (THETA2) + (ALPHA2*AB)
     +*COS (THETAX) + (VBC**2/BC)
     +*COS (THETA3)
      C2=- (OMEGA2**2) *AB*SIN (THETA2) + (ALPHA2*AB)
     +*SIN (THETAX) + (VBC**2/BC)
     +*SIN (THETA3)
      ATC= (C1*B2-C2*B1) / (A1*B2-A2*B1)
      ATBC= (A1*C2-A2*C1) / (A1*B2-A2*B1)
      X=COS (THETA2)
      Y=SIN (THETA2)
      ACCNB=- (VB**2/AB) *CMPLX (X,Y)
      ABSNB=CABS (ACCNB)
```

```
      PHZNB=C*ATAN2(AIMAG(ACCNB),REAL(ACCNB))
      X=COS(THETAX)
      Y=SIN(THETAX)
      ACCTB=AB*ALPHA2*CMPLX(X,Y)
      ABSTB=CABS(ACCTB)
C     TEST ABS TAN ACC OF B
      IF(ABSTB.GT.0.0001) GO TO 40
      PHZTB=0.0
      GO TO 45
   40 PHZTB=C*ATAN2(AIMAG(ACCTB),REAL(ACCTB))
   45 CONTINUE
      ACCNC=0.0
      ABSNC=0.0
      PHZNC=0.0
      ACCTC=ATC
      ABSTC=CABS(ACCTC)
C     TEST FOR TAN ACC OF C
      IF(ABSTC.GT.0.0001)GO TO 90
      PHZTC= 0.0
      GO TO 95
   90 PHZTC=C*ATAN2(AIMAG(ACCTC),REAL(ACCTC))
   95 CONTINUE
      X=COS(THETA3)
      Y=SIN(THETA3)
      ACCNBC=-(VBC**2/BC)*CMPLX(X,Y)
      ABSNBC=CABS(ACCNBC)
C     TEST FOR ABS NOR ACC OF B REL TO C
      IF(ABSNBC.GT.0.0001) GO TO 50
      PHZNBC=0.0
      GO TO 55
   50 PHZNBC=C*ATAN2(AIMAG(ACCNBC),REAL(ACCNBC))
   55 CONTINUE
      X=COS(THETAY)
      Y=SIN(THETAY)
      ACCTBC=ATBC*CMPLX(X,Y)
      ABSTBC=CABS(ACCTBC)
C     TEST TAN ACC B REL TO C
      IF(ABSTBC.GT.0.0001) GO TO 60
      PHZTBC=0.0
      GO TO 65
   60 PHZTBC=C*ATAN2(AIMAG(ACCTBC),REAL(ACCTBC))
   65 CONTINUE
      WRITE(6,906)
  906 FORMAT(/10X,'NORMAL ACCELERATIONS')
      WRITE(6,915)
  915 FORMAT(/10X,'REAL',8X,'IMAG',9X,'ABS',7X,
     +'PHASE')
      WRITE(6,301)ACCNB,ABSNB,PHZNB
      WRITE(6,302)ACCNC,ABSNC,PHZNC
      WRITE(6,303)ACCNBC,ABSNBC,PHZNBC
```

```
      WRITE(6,1906)
 1906 FORMAT(/10X,'TANGENTIAL ACCELERATIONS')
      WRITE(6,916)
  916 FORMAT(/10X,'REAL',8X,'IMAG',9X,'ABS',7X,
     +'PHASE')
      WRITE(6,301)ACCTB,ABSTB,PHZTB
      WRITE(6,302)ACCTC,ABSTC,PHZTC
      WRITE(6,303)ACCTBC,ABSTBC,PHZTBC
      ACCB=ACCNB+ACCTB
      ABSAB=CABS(ACCB)
      PHZAB=C*ATAN2(AIMAG(ACCB),REAL(ACCB))
      WRITE(6,907)
  907 FORMAT(/10X,'ABSOLUTE ACCELERATIONS')
      WRITE(6,905)
      WRITE(6,301)ACCB,ABSAB,PHZAB
      ACCC=ACCNC+ACCTC
      ABSAC=CABS(ACCC)
C     TEST FOR ABS ACC OF C
      IF(ABSAC.GT.0.0001)GO TO 10
      PHZAC=0.0
      GO TO 15
   10 PHZAC=C*ATAN2(AIMAG(ACCC),REAL(ACCC))
   15 CONTINUE
      WRITE(6,302)ACCC,ABSAC,PHZAC
      ACCBC=ACCNBC+ACCTBC
      ABSABC=CABS(ACCBC)
C     TEST ABS ACC OF B REL TO C
      IF(ABSABC.GT.0.0001) GO TO 70
      PHZABC=0.0
      GO TO 75
   70 PHZABC=C*ATAN2(AIMAG(ACCBC),REAL(ACCBC))
   75 CONTINUE
      WRITE(6,303)ACCBC,ABSABC,PHZABC
      THETA2=THETA2*C
      THETA3=THETA3*C
      THETAX=THETAX*C
      THETAY=THETAY*C
      WRITE(6,909)
  909 FORMAT(/12X,'A1',10X,'A2',10X,'B1',10X,'B2')
      WRITE(6,101) A1,A2,B1,B2
      WRITE(6,1909)
 1909 FORMAT(/12X,'C1',10X,'C2',9X,'ATC',8X,'ATBC')
      WRITE(6,101)C1,C2,ATC,ATBC
  100 FORMAT(2X,6F12.4)
  101 FORMAT(2X,8F12.4)
  201 FORMAT(2X,4F12.4,12X,4F12.4)
  301 FORMAT(2X,4F12.4,8X,'(B) ')
  302 FORMAT(2X,4F12.4,8X,'(C) ')
  303 FORMAT(2X,4F12.4,8X,'(BC)')
      GO TO 1
  999 STOP
      END
```

Slider-Crank: Simplified Vector Method

PROBLEM DATA

AB	BC
1.5000	3.0000

THETA(A)	OMEGA(A)	ALPHA(A)	ECC(E)
150.0000	1.0000	0.0000	0.0000

PHI(3)	THETA(3)
14.4775	165.5225

LINEAR VELOCITIES

V(B)	V(C)	V(BC)
1.5000	-0.4146	1.3416

REAL	IMAG	ABS	PHASE	
-0.7500	-1.2990	1.5000	-120.0000	(B)
-0.4146	0.0000	0.4146	180.0000	(C)
-0.3354	-1.2990	1.3416	-104.4775	(BC)

NORMAL ACCELERATIONS

REAL	IMAG	ABS	PHASE	
1.2990	-0.7500	1.5000	-30.0000	(B)
0.0000	0.0000	0.0000	0.0000	(C)
0.5809	-0.1500	0.6000	-14.4775	(BC)

TANGENTIAL ACCELERATIONS

REAL	IMAG	ABS	PHASE	
0.0000	0.0000	0.0000	0.0000	(B)
0.8730	0.0000	0.8730	0.0000	(C)
-0.1549	-0.6000	0.6197	-104.4775	(BC)

ABSOLUTE ACCELERATIONS

REAL	IMAG	ABS	PHASE	
1.2990	-0.7500	1.5000	-30.0000	(B)
0.8730	0.0000	0.8730	0.0000	(C)
0.4260	-0.7500	0.8626	-60.4018	(BC)

A1	A2	B1	B2
1.0000	0.0000	-0.2500	-0.9682

C1	C2	ATC	ATBC
0.7181	-0.6000	0.8730	0.6197

Slider-Crank: Simplified Vector Method

PROBLEM DATA

AB	BC
1.5000	3.0000

THETA(A)	OMEGA(A)	ALPHA(A)	ECC(E)
150.0000	1.0000	0.0000	0.5000

PHI(3)	THETA(3)
4.7802	175.2198

LINEAR VELOCITIES

V(B)	V(C)	V(BC)
1.5000	-0.6414	1.3036

REAL	IMAG	ABS	PHASE	
-0.7500	-1.2990	1.5000	-120.0000	(B)
-0.6414	0.0000	0.6414	180.0000	(C)
-0.1086	-1.2990	1.3036	-94.7802	(BC)

NORMAL ACCELERATIONS

REAL	IMAG	ABS	PHASE	
1.2990	-0.7500	1.5000	-30.0000	(B)
0.0000	0.0000	0.0000	0.0000	(C)
0.5645	-0.0472	0.5664	-4.7802	(BC)

TANGENTIAL ACCELERATIONS

REAL	IMAG	ABS	PHASE	
0.0000	0.0000	0.0000	0.0000	(B)
0.7933	0.0000	0.7933	0.0000	(C)
-0.0588	-0.7028	0.7053	-94.7802	(BC)

ABSOLUTE ACCELERATIONS

REAL	IMAG	ABS	PHASE	
1.2990	-0.7500	1.5000	-30.0000	(B)
0.7933	0.0000	0.7933	0.0000	(C)
0.5057	-0.7500	0.9046	-56.0099	(BC)

A1	A2	B1	B2
1.0000	0.0000	-0.0833	-0.9965

C1	C2	ATC	ATBC
0.7346	-0.7028	0.7933	0.7053

Slider-Crank: Simplified Vector Method

PROBLEM DATA

AB	BC
1.5000	3.0000

THETA(A)	OMEGA(A)	ALPHA(A)	ECC(E)
150.0000	1.0000	0.0000	-0.5000

PHI(3)	THETA(3)
24.6243	155.3757

LINEAR VELOCITIES

V(B)	V(C)	V(BC)
1.5000	-0.1546	1.4290

REAL	IMAG	ABS	PHASE	
-0.7500	-1.2990	1.5000	-120.0000	(B)
-0.1546	0.0000	0.1546	180.0000	(C)
-0.5954	-1.2990	1.4290	-114.6243	(BC)

NORMAL ACCELERATIONS

REAL	IMAG	ABS	PHASE	
1.2990	-0.7500	1.5000	-30.0000	(B)
0.0000	0.0000	0.0000	0.0000	(C)
0.6188	-0.2836	0.6807	-24.6243	(BC)

TANGENTIAL ACCELERATIONS

REAL	IMAG	ABS	PHASE	
0.0000	0.0000	0.0000	0.0000	(B)
0.8940	0.0000	0.8940	0.0000	(C)
-0.2138	-0.4664	0.5130	-114.6243	(BC)

ABSOLUTE ACCELERATIONS

REAL	IMAG	ABS	PHASE	
1.2990	-0.7500	1.5000	-30.0000	(B)
0.8940	0.0000	0.8940	0.0000	(C)
0.4050	-0.7500	0.8524	-61.6307	(BC)

A1	A2	B1	B2
1.0000	0.0000	-0.4167	-0.9091

C1	C2	ATC	ATBC
0.6803	-0.4664	0.8940	0.5130

Quick-Return Linkage Analysis: Simplified Vector Method

```
C       AB,CD,AD = LINK LENGTHS
C       C       = DEGREES/RADIAN(CONSTANT)
C       THETAA = THETA(2), POSITION ANGLE OF LINK AB
C       PHID    = PHI(3)
C       THETAD = THETA(3), POSITION ANGLE OF LINK CD
C       OMEGAA = OMEGA(2), ANGULAR VEL. OF LINK AB
C       ALPHAA = ALPHA(2), ANGULAR VEL. OF AB
C       SNB     = 90(DEGS)
C       SNC     = 90.0+THETA(D)+THETA(A)
C       SNBC    = THETA(D)-THETA(A)
C       VB      = VELOCITY OF B
C       VC      = VELOCITY OF C
C       VBC     = VELOCITY OF B RELATIVE TO C
C       CVB     = VEL OF B (COMPLEX)
C       CVC     = VEL OF C (COMPLEX)
C       CVBC    = VEL OF B RELATIVE TO C (COMPLEX)
C       ABSVB   = ABSOLUTE VELOCITY OF B
C       ABSVC   = ABSOLUTE VELOCITY OF C
C       ABSVBC  = ABSOLUTE VELOCITY OF B REL. TO C
C       ACCNB   = NORMAL ACC OF B (COMPLEX)
C       ACCTB   = TANGENTIAL ACC OF B (COMPLEX)
C       ACCNC   = NORMAL ACC OF C (COMPLEX)
C       ACCTC   = TANGENTIAL ACC OF C (COMPLEX)
C       ACCNBC  = NORMAL ACC OF B REL.TO C (COMPLEX)
C       ACCTBC  = TANGENTIAL ACC OF B RELATIVE
C                 TO C(COMPLEX)
C       ACCCOR  = CORIOLIS ACCELERATION (COMPLEX)
C       ATC     = TANGENTIAL ACC OF C
C       ATBC    = TANGENTIAL ACC OF B RELATIVE TO C
C       ACCB    = ACC OF B (COMPLEX)
C       ACCC    = ACC OF C (COMPLEX)
C       ACCBC   = ACC OF B RELATIVE TO C (COMPLEX)
C       ABSNB   = ABSOLUTE NORMAL ACC OF B
C       ABSTB   = ABSOLUTE TANGENTIAL ACC OF B
C       ABSNC   = ABSOLUTE NORMAL ACC OF C
C       ABSTC   = ABSOLUTE TANGENTIAL ACC OF C
C       ABSNBC  = ABSOLUTE NORMAL ACC OF B REL. TO C
C       ABSTBC  = ABSOLUTE TANGENTIAL ACC OF B
C                 RELATIVE TO C
C       ABSCOR  = ABSOLUTE CORIOLIS ACCELERATION
C       ABSAB   = ABSOLUTE ACC OF B
C       ABSAC   = ABSOLUTE ACC OF C
C       ABSABC  = ABSOLUTE ACC OF B RELATIVE TO C
C       PHZVB   = PHASE ANGLE OF VELOCITY OF B
C       PHZVC   = PHASE ANGLE OF VELOCITY OF C
C       PHZVBC  = PHASE ANGLE OF VELOCITY OF B
C                 RELATIVE TO C
C       PHZNB   = PHASE ANGLE OF NORMAL ACC OF B
```

```
C       PHZTB  = PHASE ANGLE OF TANGENTIAL ACC OF B
C       PHZNC  = PHASE ANGLE OF NORMAL ACC OF C
C       PHZTC  = PHASE ANGLE OF TANGENTIAL ACC OF C
C       PHZNBC = PHASE ANGLE OF NORMAL ACC OF B
C                RELATIVE TO C
C       PHZTBC = PHASE ANGLE OF TANGENTIAL ACC OF B
C                RELATIVE TO C
C       PHZCOR = PHASE ANGLE OF CORIOLIS ACC
C       PHZAB  = PHASE ANGLE OF ABSOLUTE ACC OF B
C       PHZAC  = PHASE ANGLE OF ABSOLUTE ACC OF C
C       PHZABC = PHASE ANGLE OF ABSOLUTE ACC OF B
C                RELATIVE TO C
        COMPLEX CVB,CVC,CVBC,ACCNC,ACCTC,ACCC
        COMPLEX ACCNB,ACCTB,ACCB,ACCCOR,ACCTBC
        COMPLEX ACCBC,ACCNBC
        C=57.29578

        OPEN (UNIT=6, FILE='QRTRN.LIS', STATUS='NEW')
        OPEN (UNIT=10, FILE='QRTRN.DAT', STATUS='OLD')

      1 READ(10,*)AB,AD,THETAA,OMEGAA,ALPHAA
C    99 FORMAT(5F12.4)
        IF(AB.EQ.0.0)GO TO 999
        WRITE(6,899)
    899 FORMAT(1H1,9X,'QUICK-RETURN: SIMPLIFIED VEC
       +TOR METHOD',//10X,
       +'PROBLEM DATA')

        WRITE(6,900)
    900 FORMAT(/12X,'AB',10X,'AD')
        WRITE(6,101)AB,AD
        WRITE(6,1900)
   1900 FORMAT(/7X,'THETA(2)',4X,'OMEGA(2)',4X,'ALPH
       +A(2)')
        WRITE(6,101)THETAA,OMEGAA,ALPHAA

        THETAA=THETAA/C
C       CALCULATE CD USING COSINE RULE
        CD=SQRT(AB**2+AD**2-2.0*AB*AD*COS(THETAA))
C       CALCULATE PHID AND THETAD
        AA = (CD**2 + AD**2 - AB**2)/(2.0*CD*AD)
        IF(ABS(AA).LE.1.0) GO TO 11
        IF(AA.LT.0.0) AA=-1.0
        IF(AA.GT.1.0) AA= 1.0
     11 PHID = C*ACOS(AA)
        THETAA = THETAA*C
        IF(THETAA.LT.180.0)GO TO 111
        PHID = -PHID
    111 IF(ABS(PHID) .LT. 0.01)PHID=0.0
```

```
      THETAD = 180.0 - PHID
      THETAA = THETAA/C
      WRITE(6,901)
  901 FORMAT(/8X,'PHI(3)',4X,'THETA(3)')
      WRITE(6,101)PHID,THETAD
      THETAD=THETAD/C
      SNB=90.0
      SNC=90.0-THETAD*C+THETAA*C
      SNBC=THETAD*C-THETAA*C
      WRITE(6,902)
  902 FORMAT(/12X,'CD')
      WRITE(6,888)CD
  888 FORMAT(2X,4F12.4)
      SMB=90.0/C
      SMC=SNC/C
      SMBC=SNBC/C
      VB=AB*OMEGAA
      VC=VB*SIN(SMC)
      VBC=VB*SIN(SMBC)
      WRITE(6,903)
  903 FORMAT(/10X,'LINEAR VELOCITIES')
      WRITE(6,904)
  904 FORMAT(/10X,'V(B)',8X,'V(C)',7X,'V(BC)')
      WRITE(6,101)VB,VC,VBC
      THETAX=THETAA+90.0/C
      THETAY=THETAD+90.0/C
      X=COS(THETAX)
      Y=SIN(THETAX)
      CVB=VB*CMPLX(X,Y)
      ABSVB=CABS(CVB)
      PHZVB=C*ATAN2(AIMAG(CVB),REAL(CVB))
      WRITE(6,905)
  905 FORMAT(/10X,'REAL',8X,'IMAG',9X,'ABS',7X,
     +'PHASE')
      WRITE(6,301)CVB,ABSVB,PHZVB
      X=COS(THETAY)
      Y=SIN(THETAY)
      CVC=VC*CMPLX(X,Y)
      ABSVC=CABS(CVC)
C     TEST VEL OF C
      IF (ABSVC.GT.0.0001)GO TO 20
      PHZVC = 0.0
      GO TO 25
   20 PHZVC=C*ATAN2(AIMAG(CVC),REAL(CVC))
   25 CONTINUE
      WRITE(6,302)CVC,ABSVC,PHZVC
      X=COS(THETAD)
      Y=SIN(THETAD)
```

```
      CVBC=VBC*CMPLX(X,Y)
      ABSVBC=CABS(CVBC)
C     TEST ABS VEL OF B WRT C
      IF(ABSVBC.GT.0.0001)GO TO 30
      PHZVBC=0.0
      GO TO 35
   30 PHZVBC=C*ATAN2(AIMAG(CVBC),REAL(CVBC))
   35 CONTINUE
      WRITE(6,303)CVBC,ABSVBC,PHZVBC
C     CALCULATE ATC AND ATBC
      A1=COS(THETAY)
      A2=SIN(THETAY)
      B1=COS(THETAD)
      B2=SIN(THETAD)
      C1=-VB**2/AB*COS(THETAA)+AB*ALPHAA*COS
     +(THETAX)-2*VBC*VC/CD*COS(THE
     +TAY)+VC**2/CD*COS(THETAD)

      C2=-VB**2/AB*SIN(THETAA)+AB*ALPHAA*SIN
     +(THETAX)-2*VBC*VC/CD*SIN(THE
     +TAY)+VC**2/CD*SIN(THETAD)
      ATC=(C1*B2-C2*B1)/(A1*B2-A2*B1)
      ATBC=(A1*C2-A2*C1)/(A1*B2-A2*B1)
C     CALCULATE ACCELERATION COMPONENTS
      X=COS(THETAA)
      Y=SIN(THETAA)
      ACCNB=-VB**2/AB*CMPLX(X,Y)
      ABSNB=CABS(ACCNB)
      PHZNB=C*ATAN2(AIMAG(ACCNB),REAL(ACCNB))
      X=COS(THETAX)
      Y=SIN(THETAX)
      ACCTB=AB*ALPHAA *CMPLX(X,Y)
      ABSTB=CABS(ACCTB)
C     TEST ABS TAN ACC OF B
      IF(ABSTB.GT.0.0001)GO TO 40
      PHZTB = 0.0
      GO TO 45
   40 PHZTB=C*ATAN2(AIMAG(ACCTB),REAL(ACCTB))
   45 CONTINUE
      X=COS(THETAD)
      Y=SIN(THETAD)
      ACCNC=-VC**2/CD* CMPLX(X,Y)
      ABSNC=CABS(ACCNC)
C     TEST ABS NOR ACC OF C
      IF(ABSNC.GT.0.0001)GO TO 440
      PHZNC = 0.0
      GO TO 445
  440 PHZNC=C*ATAN2(AIMAG(ACCNC),REAL(ACCNC))
  445 CONTINUE
```

```
        X=COS (THETAY)
        Y=SIN (THETAY)
        ACCTC=ATC*CMPLX (X, Y)
        ABSTC=CABS (ACCTC)
C       TEST ABS TAN ACC OF C
        IF (ABSTC.GT.0.0001) GO TO 50
        PHZTC = 0.0
        GO TO 55
     50 PHZTC=C*ATAN2 (AIMAG (ACCTC) , REAL (ACCTC) )
     55 CONTINUE
        ACCNBC = 0.0
        ABSNBC = 0.0
        PHZNBC = 0.0
        X=COS (THETAD)
        Y=SIN (THETAD)
        ACCTBC=ATBC*CMPLX (X, Y)
        ABSTBC=CABS (ACCTBC)
C       TEST TAN   ACC B WRT C
        IF (ABSTBC.GT.0.0001) GO TO 70
        PHZTBC=0.0
        GO TO 75
     70 PHZTBC=C*ATAN2 (AIMAG (ACCTBC) , REAL (ACCTBC) )
     75 CONTINUE
        X=COS (THETAY)
        Y=SIN (THETAY)
        ACCCOR= (2*VBC*VC/CD) *CMPLX (X, Y)
        ABSCOR=CABS (ACCCOR)
C       TEST ABS CORIOLIS ACC
        IF (ABSCOR.GT.0.0001) GO TO 60
        PHZCOR=0.0
        GO TO 65
     60 PHZCOR=C*ATAN2 (AIMAG (ACCCOR) , REAL (ACCCOR) )
     65 CONTINUE

        WRITE (6,906)
    906 FORMAT (/10X,'NORMAL AND CORIOLIS ACCELERA
       +TIONS')
        WRITE (6,915)
    915 FORMAT (/10X,'REAL',8X,'IMAG',9X,'ABS',7X,
       +'PHASE')
        WRITE (6,301) ACCNB,ABSNB,PHZNB
        WRITE (6,302) ACCNC,ABSNC,PHZNC
        WRITE (6,303) ACCNBC,ABSNBC,PHZNBC
        WRITE (6,304) ACCCOR,ABSCOR,PHZCOR
        WRITE (6,1906)
   1906 FORMAT (/10X,'TANGENTIAL ACCELERATIONS')
        WRITE (6,1916)
```

```
 1916 FORMAT(/10X,'REAL',8X,'IMAG',9X,'ABS',7X,
     +'PHASE')
      WRITE(6,301)ACCTB,ABSTB,PHZTB
      WRITE(6,302)ACCTC,ABSTC,PHZTC
      WRITE(6,303)ACCTBC,ABSTBC,PHZTBC
      ACCB=ACCNB+ACCTB
      ABSAB=CABS(ACCB)
      PHZAB=C*ATAN2(AIMAG(ACCB),REAL(ACCB))
      WRITE(6,907)
  907 FORMAT(/10X,'ABSOLUTE ACCELERATIONS')
      WRITE(6,905)
      WRITE(6,301)ACCB,ABSAB,PHZAB
      ACCC=ACCNC+ACCTC
      ABSAC=CABS(ACCC)
C     TEST ABS ACC OF C
      IF(ABSAC.GT.0.0001)GO TO 770
      PHZAC = 0.0
      GO TO 775
  770 PHZAC=C*ATAN2(AIMAG(ACCC),REAL(ACCC))
  775 CONTINUE
      WRITE(6,302)ACCC,ABSAC,PHZAC
      ACCBC=ACCCOR+ACCTBC+ACCNBC
      ABSABC=CABS(ACCBC)
C     TEST ABS ACC B WRT C
      IF(ABSABC.GT.0.0001)GO TO 80
      PHZABC=0.0
      GO TO 85
   80 PHZABC=C*ATAN2(AIMAG(ACCBC),REAL(ACCBC))
   85 CONTINUE
      WRITE(6,303)ACCBC,ABSABC,PHZABC
      THETAA=THETAA*C
      THETAD=THETAD*C
      THETAX=THETAX*C
      THETAY=THETAY*C
      WRITE(6,909)
  909 FORMAT(/12X,'A1',10X,'A2',10X,'B1',10X,'B2')
      WRITE(6,101)A1,A2,B1,B2
      WRITE(6,1909)
 1909 FORMAT(/12X,'C1',10X,'C2',9X,'ATC',8X,'ATBC')
      WRITE(6,101)C1,C2,ATC,ATBC
  101 FORMAT(2X,8F12.4)
  201 FORMAT(2X,4F12.4,12X,4F12.4)
  301 FORMAT(2X,4F12.4,8X,'(B) ')
  302 FORMAT(2X,4F12.4,8X,'(C) ')
  303 FORMAT(2X,4F12.4,8X,'(BC)')
  304 FORMAT(2X,4F12.4,8X,'(COR)')
      GO TO 1
  999 STOP
      END
```

Quick-Return: Simplified Vector Method

```
PROBLEM DATA

    AB              AD
 0.1666          0.3333

 THETA(2)        OMEGA(2)        ALPHA(2)
30.0000         62.8300          0.0000

 PHI(3)          THETA(3)             CD
23.7828         156.2172         0.2066

    LINEAR VELOCITIES

   V(B)            V(C)            V(BC)
10.4675         -6.1847          8.4450

   REAL            IMAG             ABS          PHASE
 -5.2337          9.0651         10.4675       120.0000      (B)
  2.4941          5.6595          6.1847        66.2172      (C)
 -7.7278          3.4056          8.4450       156.2172      (BC)

    NORMAL AND CORIOLIS ACCELERATIONS

   REAL            IMAG             ABS          PHASE
-569.5604       -328.8359        657.6718      -150.0000      (B)
 169.4521        -74.6765        185.1772       -23.7828      (C)
   0.0000          0.0000          0.0000         0.0000      (BC)
 203.9361        462.7619        505.7060        66.2172      (COR)

    TANGENTIAL ACCELERATIONS

   REAL            IMAG             ABS          PHASE
   0.0000          0.0000          0.0000         0.0000      (B)
-417.9106       -948.3024       1036.3043      -113.7828      (C)
-525.0381        231.3810        573.7615       156.2172      (BC)

    ABSOLUTE ACCELERATIONS

   REAL            IMAG             ABS          PHASE
-569.5604       -328.8359        657.6718      -150.0000      (B)
-248.4585      -1022.9788       1052.7190      -103.6515      (C)
-321.1020        694.1429        764.8143       114.8247      (BC)

     A1              A2              B1              B2
 -0.4033         -0.9151         -0.9151          0.4033

     C1              C2             ATC            ATBC
-942.9487       -716.9213       1036.3043       573.7615
```

Quick-Return: Simplified Vector Method

```
PROBLEM DATA

      AB              AD
   0.2500          0.1666

 THETA(2)        OMEGA(2)       ALPHA(2)
  60.0000         30.0000        0.0000

  PHI(3)         THETA(3)            CD
  79.1236        100.8764        0.2205

   LINEAR VELOCITIES

    V(B)            V(C)           V(BC)
   7.5000          5.6709         4.9082

     REAL            IMAG            ABS         PHASE
   -6.4952          3.7500         7.5000      150.0000        (B)
   -5.5691         -1.0700         5.6709     -169.1236        (C)
   -0.9261          4.8200         4.9082      100.8764        (BC)

   NORMAL AND CORIOLIS ACCELERATIONS

      REAL            IMAG            ABS         PHASE
 -112.5000       -194.8557       225.0000     -120.0000        (B)
   27.5242       -143.2493       145.8696      -79.1236        (C)
    0.0000          0.0000         0.0000        0.0000        (BC)
 -247.9662        -47.6447       252.5020     -169.1236        (COR)

   TANGENTIAL ACCELERATIONS

      REAL            IMAG            ABS         PHASE
    0.0000          0.0000         0.0000        0.0000        (B)
  103.3647         19.8607       105.2554       10.8764        (C)
    4.5773        -23.8224        24.2581      -79.1236        (BC)

   ABSOLUTE ACCELERATIONS

      REAL            IMAG            ABS         PHASE
 -112.5000       -194.8557       225.0000     -120.0000        (B)
  130.8889       -123.3886       179.8796      -43.3105        (C)
 -243.3889        -71.4671       253.6645     -163.6360        (BC)

        A1              A2              B1              B2
   -0.9820         -0.1887         -0.1887         0.9820

        C1              C2             ATC            ATBC
  107.9419         -3.9617       -105.2554       -24.2581
```

Sliding Coupler Mechanism: Simplified Vector Method

```
C      AB,BC,AC = LINK LENGTHS
C      C       = DEGREES/RADIAN(CONSTANT)
C      PHIC    = PHI(C)
C      THETAA = THETA(A), POSITION ANGLE OF LINK AB
C      ALPHAA = ALPHA(A),ANGULAR ACC  OF AB
C      OMEGAA = OMEGA(A),ANGULAR VELOCITY OF AB
C      VB      = VELOCITY OF B
C      VC      = VELOCITY OF C
C      VCB     = VELOCITY OF C RELATIVE TO B
C      CVB     = VEL OF B (COMPLEX)
C      CVC     = VEL OF C (COMPLEX)
C      CVCB    = VEL OF C RELATIVE TO BE (COMPLEX)
C      ABSVB  = ABSOLUTE VELOCITY OF B
C      ABXVC  = ABSOLUTE VELOCITY OF C
C      ABSVCB = ABSOLUTE VELOCITY OF C REL. TO B
C      ACCNB  = NORMAL ACC OF B (COMPLEX)
C      ACCTB  = TANGENTIAL ACC OF B (COMPLEX)
C      ACCNC  = NORMAL ACC OF C (COMPLEX)
C      ACCTC  = TANGENTIAL ACC OF C (COMPLEX)
C      ACCNCB = NORMAL ACC OF C REL. TO B(COMPLEX)
C      ACCTCB = TANGENTIAL ACC OF C RELATIVE TO
C               B(COMPLEX)
C      ATC     = TANGENTIAL ACC OF C
C      ATCB    = TAGENTIAL ACC OF C RELATIVE TO B
C      ACCB    = ACC OF B (COMPLEX)
C      ACCC    = ACC OF C (COMPLEX)
C      ACCCB  = ACC OF C RELATIVE TO B (COMPLEX)
C      ABSNB  = ABSOLUTE NORMAL ACC OF B
C      ABSTB  = ABSOLUTE TANGENTIAL ACC OF B
C      ABSNC  = ABSOLUTE NORMAL ACC OF C
C      ABSTB  = ABSOLUTE TANGENTIAL ACC OF C
C      ABSNCB = ABSOLUTE NORMAL ACC OF C REL. TO B
C      ABSTCB = ABSOLUTE TANGENTIAL ACC OF C
C               RELATIVE TO B
C      ABSAB  = ABSOLUTE ACCELERATION OF B
C      ABSAC  = ABSOLUTE ACCELERATION OF C
C      ABSACB = ABSOLUTE ACC OF C RELATIVE TO B
C      PHZVB  = PHASE ANGLE OF VELOCITY OF B
C      PHZVC  = PHASE ANGLE OF VELOCITY OF C
C      PHZVCB = PHASE ANGLE OF VELOCITY OF C
C               RELATIVE TO B
C      PHZNB  = PHASE ANGLE OF NORMAL ACC OF B
C      PHZTB  = PHASE ANGLE OF TANGENTIAL ACC OF B
C      PHZNC  = PHASE ANGLE OF NORMAL ACC OF C
C      PHZTC  = PHASE ANGLE OF TANGENTIAL ACC OF C
C      PHZNCB = PHASE ANGLE OF NORMAL ACC OF C
C               RELATIVE TO B
C      PHZTCB = PHASE ANGLE OF TANGENTIAL ACC OF C
C               RELATIVE TO B
```

```
C      PHZAB  = PHASE ANGLE OF ABSOLUTE ACC OF B
C      PHZAC  = PHASE ANGLE OF ABSOLUTE ACC OF C
C      PHZACB = PHASE ANGLE OF ABSOLUTE ACC OF C
C               RELATIVE TO B

       COMPLEX CVB,CVC,CVCB
       COMPLEX ACCNB,ACCTB,ACCB,ACCNC,ACCTC,ACCCOR
       COMPLEX ACCC,ACCNCB,ACCTCB,ACCCB
       C = 57.29578

       OPEN (UNIT=6, FILE='SCPLR.LIS', STATUS='NEW')
       OPEN (UNIT=10, FILE='SCPLR.DAT', STATUS='OLD')

     1 READ(10,*)AB,AC,THETAA,OMEGAA,ALPHAA
C 100 FORMAT(7F10.4)
       IF(AB.EQ.0.0) GO TO 999
       WRITE(6,104)
   104 FORMAT(1H1,9X,'SLIDING COUPLER: SIMPLIFIED
      + VECTOR METHOD',
      +//10X,'PROBLEM DATA')

       WRITE(6,105)
   105 FORMAT(//12X,'AB',10X,'AC')
       WRITE(6,106)AB,AC
       WRITE(6,1105)
  1105 FORMAT(/6X,'THETA(A)',4X,'OMEGA(A)',4X,'ALPH
      +A(A)')
       WRITE(6,106)THETAA,OMEGAA,ALPHAA

       THETAA=THETAA/C
       BC=SQRT(AB**2+AC**2-2*AB*AC*COS(THETAA))
       AA = (BC**2 + AC**2 - AB**2)/(2.0*BC*AC)
       IF(ABS(AA).LE.1.0) GO TO 11
       IF(AA.LT.0.0) AA=-1.0
       IF(AA.GT.1.0) AA= 1.0

    11 PHI  = C*ACOS(AA)
       PHIC = -PHI
       THETAA=THETAA*C
       IF(THETAA.LT.180.0)GO TO 111
       PHIC = PHI
   111 IF(ABS(PHIC) .LT. 0.01)PHIC=0.0
       WRITE(6,901)
   901 FORMAT(/11X,'PHI',6X,'PHI(C)')
       WRITE(6,101)PHI,PHIC

       SNB=90.0
       SNC=(THETAA-PHIC)
       SNCB=(90.0-THETAA+PHIC)
```

```
      WRITE(6,902)
  902 FORMAT(/12X,'BC')
      WRITE(6,888)BC
  888 FORMAT(2X,4F12.4)
      VB=AB*OMEGAA
      VC = -VB*SIN(SNC/C)/SIN(SNB/C)
      VCB = -VB*SIN(SNCB/C)/SIN(SNB/C)

      X=COS((THETAA+90.0)/C)
      Y=SIN((THETAA+90.0)/C)
      CVB=VB*CMPLX(X,Y)
      ABSVB = CABS(CVB)
      PHZVB=C*ATAN2(AIMAG(CVB),REAL(CVB))
      X=COS(PHIC/C)
      Y=SIN(PHIC/C)
      CVC=VC*CMPLX(X,Y)
      ABSVC = CABS(CVC)

C     TEST ABSVC
      IF(ABSVC.GT. 0.0001)GO TO 80
      PHZVC=0.0
      GO TO 85
   80 PHZVC=C*ATAN2(AIMAG(CVC),REAL(CVC))
   85 CONTINUE
      X=COS((PHIC+90.0)/C)
      Y=SIN((PHIC+90.0)/C)
      CVCB=VCB*CMPLX(X,Y)
      ABSVCB = CABS(CVCB)

C     TEST ABSVCB
      IF(ABSVCB.GT. 0.0001)GO TO 90
      PHZVCB=0.0
      GO TO 95
   90 PHZVCB=C*ATAN2(AIMAG(CVCB),REAL(CVCB))
   95 CONTINUE
      WRITE(6,903)
  903 FORMAT(/10X,'LINEAR VELOCITIES')
      WRITE(6,904)
  904 FORMAT(/10X,'V(B)',8X,'V(C)',7X,'V(CB)')
      WRITE(6,107)VB,VC,VCB
      WRITE(6,905)
  905 FORMAT(/10X,'REAL',8X,'IMAG',9X,'ABS',7X,
     +'PHASE')
      WRITE(6,301)CVB,ABSVB,PHZVB
      WRITE(6,302)CVC,ABSVC,PHZVC
      WRITE(6,303)CVCB,ABSVCB,PHZVCB

      A1=COS(PHIC/C)
      A2=SIN(PHIC/C)
      B1=-COS((PHIC+90.0)/C)
      B2=-SIN((PHIC+90.0)/C)
```

```
      C1=-OMEGAA**2*AB*COS (THETAA/C) +ALPHAA*AB*
      +COS ( (THETAA+90.0) /C)
      +-VCB**2/BC*COS (PHIC/C) -2.0*VC*VCB/BC*COS
      + ( (PHIC+90.0) /C)

      C2=-OMEGAA**2*AB*SIN (THETAA/C) +ALPHAA*AB*
      +SIN ( (THETAA+90.0) /C)
      +-VCB**2/BC*SIN (PHIC/C) -2.0*VC*VCB/BC*SIN
      + ( (PHIC+90.0) /C)
      ATC= (C1*B2-C2*B1) / (A1*B2-A2*B1)
      ATCB= (A1*C2-A2*C1) / (A1*B2-A2*B1)

      X=COS (THETAA/C)
      Y=SIN (THETAA/C)
      ACCNB=-OMEGAA**2*AB*CMPLX (X, Y)
      ABSNB=CABS (ACCNB)
      PHZNB = C*ATAN2 (AIMAG (ACCNB) , REAL (ACCNB) )
      X=COS ( (THETAA+90.0) /C)
      Y=SIN ( (THETAA+90.0) /C)
      ACCTB=ALPHAA*AB*CMPLX (X, Y)
      ABSTB=CABS (ACCTB)

C     TEST ABSTB
      IF (ABSTB.GT.0.0001) GO TO 10
      PHZTB = 0.0
      GO TO 15
   10 PHZTB = C*ATAN2 (AIMAG (ACCTB) , REAL (ACCTB) )
   15 CONTINUE

      ACCNC = 0.0
      PHZNC=0.0
      ABSNC=0.0
      X=COS (PHIC/C)
      Y=SIN (PHIC/C)
      ACCTC=ATC*CMPLX (X, Y)
      ABSTC=CABS (ACCTC)

C     TEST ABSTC
      IF (ABSTC.GT. 0.0001) GO TO 30
      PHZTC=0.0
      GO TO 35
   30 PHZTC = C*ATAN2 (AIMAG (ACCTC) , REAL (ACCTC) )
   35 CONTINUE
      X=COS ( (PHIC+90.0) /C)
      Y=SIN ( (PHIC+90.0) /C)
      ACCCOR=2.0*VC*VCB/BC*CMPLX (X, Y)
      ABSCOR=CABS (ACCCOR)

C     TEST ABS CORIOLIS ACC
      IF (ABSCOR.GT.0.0001) GO TO 110
      PHZCOR=0.0
```

```
          GO TO 115
      110 PHZCOR=C*ATAN2(AIMAG(ACCCOR),REAL(ACCCOR))
      115 CONTINUE

          X=COS(PHIC/C)
          Y=SIN(PHIC/C)
          ACCNCB=-VCB**2/BC*CMPLX(X,Y)
          ABSNCB=CABS(ACCNCB)
C         TEST ABSNCB
          IF(ABSNCB.GT.0.0001)GO TO 40
          PHZNCB=0.0
          GO TO 45
       40 PHZNCB = C*ATAN2(AIMAG(ACCNCB),REAL(ACCNCB))
       45 CONTINUE
          X=COS((PHIC+90.0)/C)
          Y=SIN((PHIC+90.0)/C)
          ACCTCB=ATCB*CMPLX(X,Y)
          ABSTCB=CABS(ACCTCB)

C         TEST ABSTCB
          IF(ABSTCB.GT.0.0001)GO TO 50
          PHZTCB=0.0
          GO TO 55
       50 PHZTCB = C*ATAN2(AIMAG(ACCTCB),REAL(ACCTCB))
       55 CONTINUE

          WRITE(6,906)
      906 FORMAT(/10X,'NORMAL AND CORIOLIS ACCELERA
         +TIONS')
          WRITE(6,915)
      915 FORMAT(/10X,'REAL',8X,'IMAG',9X,'ABS',7X,
         +'PHASE')
          WRITE(6,301)ACCNB,ABSNB,PHZNB
          WRITE(6,302)ACCNC,ABSNC,PHZNC
          WRITE(6,303)ACCNCB,ABSNCB,PHZNCB
          WRITE(6,304)ACCCOR,ABSCOR,PHZCOR

          WRITE(6,1906)
     1906 FORMAT(/10X,'TANGENTIAL ACCELERATIONS')
          WRITE(6,1916)
     1916 FORMAT(/10X,'REAL',8X,'IMAG',9X,'ABS',7X,
         +'PHASE')
          WRITE(6,301)ACCTB,ABSTB,PHZTB
          WRITE(6,302)ACCTC,ABSTC,PHZTC
          WRITE(6,303)ACCTCB,ABSTCB,PHZTCB

          ACCB = ACCNB + ACCTB
          ABSAB = CABS(ACCB)
          PHZAB = C*ATAN2(AIMAG(ACCB),REAL(ACCB))
```

```
      ACCC= ACCNC + ACCTC + ACCCOR
      ABSAC = CABS(ACCC)
C     TEST ABSAC
      IF(ABSAC.GT.0.0001)GO TO 60
      PHZAC=0.0
      GO TO 65
   60 PHZAC = C*ATAN2(AIMAG(ACCC),REAL(ACCC))
   65 CONTINUE
      ACCCB = ACCNCB + ACCTCB
      ABSACB = CABS(ACCCB)
C     TEST ABSACB
      IF(ABSACB.GT.0.0001)GO TO 70
      PHZACB=0.0
      GO TO 75
   70 PHZACB = C*ATAN2(AIMAG(ACCCB),REAL(ACCCB))
   75 CONTINUE
      WRITE(6,907)
  907 FORMAT(/10X,'ABSOLUTE ACCELERATIONS')
      WRITE(6,905)
      WRITE(6,301)ACCB,ABSAB,PHZAB
      WRITE(6,302)ACCC,ABSAC,PHZAC
      WRITE(6,303)ACCCB,ABSACB,PHZACB

      WRITE(6,909)
  909 FORMAT(//12X,'A1',10X,'A2',10X,'B1',10X,'B2')
      WRITE(6,107)A1,A2,B1,B2
      WRITE(6,1909)
 1909 FORMAT(/12X,'C1',10X,'C2',9X,'ATC',8X,'ATCB')
      WRITE(6,107)C1,C2,ATC,ATCB

  101 FORMAT(2X,8F12.4)
  106 FORMAT(2X,8F12.4)
  107 FORMAT(2X,8F12.4)
  301 FORMAT(2X,4F12.4,8X,'(B)')
  302 FORMAT(2X,4F12.4,8X,'(C)')
  303 FORMAT(2X,4F12.4,8X,'(CB)')
  304 FORMAT(2X,4F12.4,8X,'(COR)')
      GO TO 1
  999 STOP
      END
```

Sliding Coupler Mechanism: Simplified Vector Method

PROBLEM DATA

AB	AC
0.6667	0.8333

THETA(A)	OMEGA(A)	ALPHA(A)
120.0000	18.0000	0.0000

PHI	PHI(C)	BC
26.3310	-26.3310	1.3017

LINEAR VELOCITIES

V(B)	V(C)	V(CB)
12.0006	-6.6531	9.9875

REAL	IMAG	ABS	PHASE	
-10.3928	-6.0003	12.0006	-150.0000	(B)
-5.9628	2.9510	6.6531	153.6691	(C)
4.4300	8.9513	9.9875	63.6690	(CB)

NORMAL AND CORIOLIS ACCELERATIONS

REAL	IMAG	ABS	PHASE	
108.0054	-187.0708	216.0108	-60.0000	(B)
0.0000	0.0000	0.0000	0.0000	(C)
-68.6803	33.9901	76.6310	153.6691	(CB)
-45.2841	-91.5009	102.0935	-116.3310	(COR)

TANGENTIAL ACCELERATIONS

REAL	IMAG	ABS	PHASE	
0.0000	0.0000	0.0000	0.0000	(B)
92.4432	-45.7505	103.1448	-26.3310	(C)
7.8340	15.8293	17.6618	63.6690	(CB)

ABSOLUTE ACCELERATIONS

REAL	IMAG	ABS	PHASE	
108.0054	-187.0708	216.0108	-60.0000	(B)
47.1591	-137.2514	145.1273	-71.0375	(C)
-60.8463	49.8195	78.6400	140.6902	(CB)

A1	A2	B1	B2
0.8962	-0.4436	-0.4436	-0.8962

C1	C2	ATC	ATCB
84.6092	-61.5798	103.1448	17.6618

Sliding Coupler Mechanism: Simplified Vector Method

PROBLEM DATA

```
    AB          AC
 3.0000      2.0000

THETA(A)    OMEGA(A)    ALPHA(A)
210.0000    30.0000      0.0000

    PHI       PHI(C)          BC
 18.0675    18.0675      4.8366
```

LINEAR VELOCITIES

```
   V(B)       V(C)        V(CB)
90.0000     18.6083     88.0553

   REAL        IMAG        ABS        PHASE
 45.0000    -77.9423     90.0000     -60.0000     (B)
 17.6907      5.7711     18.6083      18.0675     (C)
-27.3093     83.7134     88.0553     108.0676     (CB)
```

NORMAL AND CORIOLIS ACCELERATIONS

```
    REAL        IMAG         ABS         PHASE
 2338.2686   1349.9999    2700.0000     30.0000     (B)
    0.0000      0.0000       0.0000      0.0000     (C)
-1524.1018   -497.1978    1603.1506   -161.9325     (CB)
 -210.1406    644.1615     677.5715    108.0676     (COR)
```

TANGENTIAL ACCELERATIONS

```
   REAL        IMAG         ABS         PHASE
 0.0000      0.0000       0.0000       0.0000      (B)
987.3007    322.0806    1038.5078     18.0675      (C)
-37.0068    113.4400     119.3237    108.0676      (CB)
```

ABSOLUTE ACCELERATIONS

```
    REAL        IMAG         ABS         PHASE
 2338.2686   1349.9999    2700.0000     30.0000     (B)
  777.1601    966.2421    1240.0006     51.1899     (C)
-1561.1086   -383.7578    1607.5852   -166.1892     (CB)
```

```
      A1          A2          B1          B2
  0.9507      0.3101      0.3101     -0.9507

      C1          C2         ATC        ATCB
1024.3074    208.6406   1038.5078    119.3237
```

Slider-Crank Mechanism Analysis: Modified Vector Method

```
C    C       = DEGREES/RADIAN(CONSTANT)
C    AB      = CRANK
C    BC      = CONNECTING ROD
C    PHI     = PHI(C),ANGLE BETWEEN CONNECTING
C              ROD AND DEAD CENTER
C    THETAA  = POSITION ANGLE OF LINK AB
C    THETAC  = POSITION ANGLE OF LINK BC
C    OMEGAA  = ANGULAR VELOCITY OF LINK AB
C    ALPHAA  = ANGULAR ACC OF LINK AB
C    VB      = VELOCITY OF B
C    VC      = VELOCITY OF C
C    VBC     = VELOCITY OF B RELATIVE TO C
C    CVB     = VEL OF B (COMPLEX)
C    CVC     = VEL OF C (COMPLEX)
C    CVBC    = VEL OF B RELATIVE TO C (COMPLEX)
C    ABSVB   = ABSOLUTE VELOCITY OF B
C    ABSVC   = ABSOLUTE VELOCITY OF C
C    ABSVBC  = ABSOLUTE VELOCITY OF B REL. TO C
C    ACCNB   = NORMAL ACC OF B (COMPLEX)
C    ACCTB   = TANGENTIAL ACC OF B (COMPLEX)
C    ACCNC   = NORMAL ACC OF C (COMPLEX)
C    ACCTC   = TANGENTIAL ACC OF C (COMPLEX)
C    ACCB    = ACC OF B (COMPLEX)
C    ACCC    = ACC OF C (COMPLEX)
C    ACCBC   = ACC OF B RELATIVE TO C (COMPLEX)
C    ABSNB   = ABSOLUTE NORMAL ACC OF B
C    ABSTB   = ABSOLUTE TANGENTIAL ACC OF B
C    ABSNC   = ABSOLUTE NORMAL ACC OF C
C    ABSTC   = ABSOLUTE TANGENTIAL ACC OF C
C    ABSAB   = ABSOLUTE ACC OF B
C    ABSAC   = ABSOLUTE ACC OF C
C    ABSABC  = ABSOLUTE ACC OF B RELATIVE TO C
C    PHZVB   = PHASE ANGLE OF VELOCITY OF B
C    PHZVC   = PHASE ANGLE OF VELOCITY OF C
C    PHZVBC  = PHASE ANGLE OF VEL. OF B REL TO C
C    PHZNB   = PHASE ANGLE OF NORMAL ACC OF B
C    PHZTB   = PHASE ANGLE OF TANGENTIAL ACC OF B
C    PHZNC   = PHASE ANGLE OF NORMAL ACC OF C
C    PHZTC   = PHASE ANGLE OF TANGENTIAL ACC OF C
C    PHZAB   = PHASE ANGLE OF ABSOLUTE ACC OF B
C    PHZAC   = PHASE ANGLE OF ABSOLUTE ACC OF C
C    PHZABC  = PHASE ANGLE OF ABSOLUTE ACC OF B
C              RELATIVE TO C
     REAL NUMR1,NUMR2,NUMR3
     COMPLEX CVB,CVC,CVBC,ACCC,ACCNC,ACCTC,CXY
     COMPLEX ACCNB,ACCTB,ACCB,ACC1BC,ACC2BC,ACCBC
     C=57.29578
```

```
       OPEN (UNIT=6, FILE='SCRNK2.LIS', STATUS='NEW')
       OPEN (UNIT=10, FILE='SCRNK2.DAT', STATUS='OLD')

    1 READ(10,*)AB,BC,THETAA,OMEGAA,ALPHAA
C  99 FORMAT(5F12.4)
      IF(AB.EQ.0.0) GO TO 999
      WRITE(6,899)

  899 FORMAT(1H1,9X,'SLIDER-CRANK: MODIFIED VECTOR
     + METHOD',//10X,
     +'PROBLEM DATA')

      WRITE(6,900)
  900 FORMAT(//12X,'AB',10X,'BC')
      WRITE(6,100)AB,BC
      WRITE(6,1900)
 1900 FORMAT(/6X,'THETA(A)',4X,'OMEGA(A)',4X,'ALPH
     +A(A)')
      WRITE(6,100)THETAA,OMEGAA,ALPHAA

      THETAA=THETAA/C
      PHI=ASIN(AB/BC*SIN(THETAA))
      PHI=PHI*C
      WRITE(6,901)
  901 FORMAT(//8X,'PHI(C)')
      WRITE(6,101)PHI
      THETAA=THETAA*C
      SNB=90.0-PHI
      SNC=THETAA+PHI
      SNBC=90.0-THETAA
      VB=AB*OMEGAA
      VC=-VB*SIN(SNC/C)/SIN(SNB/C)
      VBC=-VB*SIN(SNBC/C)/SIN(SNB/C)
      THETAA=THETAA/C
      PHI=PHI/C
      WRITE(6,903)
  903 FORMAT(/10X,'LINEAR VELOCITIES')
      WRITE(6,904)
  904 FORMAT(/10X,'V(B)',8X,'V(C)',7X,'V(BC)')
      WRITE(6,101)VB,VC,VBC
      X=SIN(THETAA)
      Y=-COS(THETAA)
      CXY=CMPLX(X,Y)
      CVB=-AB*OMEGAA*(CXY)
      ABSVB=CABS(CVB)
      PHZVB=C*ATAN2(AIMAG(CVB),REAL(CVB))
      WRITE(6,905)
  905 FORMAT(/10X,'REAL',8X,'IMAG',9X,'ABS',7X,
     +'PHASE')
```

```
      WRITE(6,301)CVB,ABSVB,PHZVB
      CVC=VC
      ABSVC=CABS(CVC)
C     TEST FOR ABS VEL OF C
      IF(ABSVC.GT.0.0001)GO TO 20
      PHZVC=0.0
      GO TO 25
   20 PHZVC=C*ATAN2(AIMAG(CVC),REAL(CVC))
   25 CONTINUE
      WRITE(6,302)CVC,ABSVC,PHZVC
      X=SIN(THETAA)-SIN(SNC/C)/SIN(SNB/C)
      Y=-COS(THETAA)
      CXY=CMPLX(X,Y)
      CVBC=-AB*OMEGAA*(CXY)
      ABSVBC=CABS(CVBC)
C     TEST VEL OF B REL TO C
      IF(ABSVBC.GT.0.0001) GO TO 30
      PHZVBC=0.0
      GO TO 35
   30 PHZVBC=C*ATAN2(AIMAG(CVBC),REAL(CVBC))
   35 CONTINUE
      WRITE(6,303)CVBC,ABSVBC,PHZVBC

      X=COS(THETAA)
      Y=SIN(THETAA)
      CXY=CMPLX(X,Y)
      ACCNB=-AB*OMEGAA**2*(CXY)
      ABSNB=CABS(ACCNB)
      PHZNB=C*ATAN2(AIMAG(ACCNB),REAL(ACCNB))
      X=SIN(THETAA)
      Y=-COS(THETAA)
      CXY=CMPLX(X,Y)
      ACCTB=-AB*ALPHAA*(CXY)
      ABSTB=CABS(ACCTB)
C     TEST ABS TAN ACC OF B
      IF(ABSTB.GT.0.0001) GO TO 40
      PHZTB=0.0
      GO TO 45
   40 PHZTB=C*ATAN2(AIMAG(ACCTB),REAL(ACCTB))
   45 CONTINUE

      NUMR1 = AB*COS(THETAA)*SIN(THETAA)
      NUMR2=AB*(COS(THETAA)**2-SIN(THETAA)**2)
      NUMR3=AB**3* COS(THETAA)**2*SIN(THETAA)**2
      DNUMR=SQRT(BC**2-AB**2*(SIN(THETAA))**2)
      ACCC=-AB*OMEGAA**2*(COS(THETAA)+NUMR2/DNUM
     +R+NUMR3/DNUMR**3)-AB*
     +ALPHAA*(SIN(THETAA)+NUMR1/DNUMR)
```

```
         ACCNC=0.0
         ABSNC=0.0
         PHZNC=0.0
         ACCTC=ACCC
         ABSTC=CABS(ACCTC)
C        TEST FOR TAN ACC OF C
         IF(ABSTC.GT.0.0001)GO TO 90
         PHZTC= 0.0
         GO TO 95
      90 PHZTC=C*ATAN2(AIMAG(ACCTC),REAL(ACCTC))
      95 CONTINUE

         WRITE(6,906)
     906 FORMAT(/10X,'NORMAL ACCELERATIONS')
         WRITE(6,915)
     915 FORMAT(/10X,'REAL',8X,'IMAG',9X,'ABS',7X,
        +'PHASE')
         WRITE(6,301)ACCNB,ABSNB,PHZNB
         WRITE(6,302)ACCNC,ABSNC,PHZNC
         WRITE(6,1906)
    1906 FORMAT(/10X,'TANGENTIAL ACCELERATIONS')
         WRITE(6,916)
     916 FORMAT(/10X,'REAL',8X,'IMAG',9X,'ABS',7X,
        +'PHASE')
         WRITE(6,301)ACCTB,ABSTB,PHZTB
         WRITE(6,302)ACCTC,ABSTC,PHZTC

         X=NUMR2/DNUMR+NUMR3/DNUMR**3
         Y=-SIN(THETAA)
         CXY=CMPLX(X,Y)
         ACC1BC=AB*OMEGAA**2*(CXY)
         ABS1BC=CABS(ACC1BC)
C        TEST FOR TERM 1 OF ABS ACC OF B REL TO C
         IF(ABS1BC.GT.0.0001) GO TO 50
         PHZ1BC=0.0
         GO TO 55
      50 PHZ1BC=C*ATAN2(AIMAG(ACC1BC),REAL(ACC1BC))

      55 CONTINUE
         X=NUMR1/DNUMR
         Y=COS(THETAA)
         CXY=CMPLX(X,Y)
         ACC2BC=AB*ALPHAA*(CXY)
         ABS2BC=CABS(ACC2BC)

C        TEST TERM 2 OF ACC OF B REL TO C
         IF(ABS2BC.GT.0.0001) GO TO 60
         PHZ2BC=0.0
         GO TO 65
```

```
   60 PHZ2BC=C*ATAN2(AIMAG(ACC2BC),REAL(ACC2BC))

   65 CONTINUE
      WRITE(6,907)
  907 FORMAT(/10X,'ABSOLUTE ACCELERATIONS')

      WRITE(6,905)
      ACCB=ACCNB+ACCTB
      ABSAB=CABS(ACCB)
      PHZAB=C*ATAN2(AIMAG(ACCB),REAL(ACCB))

      WRITE(6,301)ACCB,ABSAB,PHZAB
      ACCC=ACCNC+ACCTC
      ABSAC=CABS(ACCC)

C     TEST FOR ABS ACC OF C
      IF(ABSAC.GT.0.0001)GO TO 10
      PHZAC=0.0
      GO TO 15
   10 PHZAC=C*ATAN2(AIMAG(ACCC),REAL(ACCC))

   15 CONTINUE
      WRITE(6,302)ACCC,ABSAC,PHZAC
      ACCBC=ACC1BC+ACC2BC
      ABSABC=CABS(ACCBC)

C     TEST ABS ACC OF B REL TO C
      IF(ABSABC.GT.0.0001) GO TO 70
      PHZABC=0.0
      GO TO 75
   70 PHZABC=C*ATAN2(AIMAG(ACCBC),REAL(ACCBC))

   75 CONTINUE
      WRITE(6,303)ACCBC,ABSABC,PHZABC
      THETAA=THETAA*C
      PHI=PHI*C

  100 FORMAT(2X,5F12.4)
  201 FORMAT(2X,4F12.4,12X,4F12.4)
  101 FORMAT(2X,8F12.4)
  301 FORMAT(2X,4F12.4,8X,'(B) ')
  302 FORMAT(2X,4F12.4,8X,'(C) ')
  303 FORMAT(2X,4F12.4,8X,'(BC)')
      GO TO 1
  999 STOP
      END
```

Slider-Crank Mechanism: Modified Vector Method

```
    PROBLEM DATA

        AB            BC
    1.5000        3.0000

THETA(A)      OMEGA(A)      ALPHA(A)
120.0000        1.0000        0.0000


    PHI(C)
   25.6589

    LINEAR VELOCITIES

     V(B)           V(C)          V(BC)
    1.5000        -0.9387         0.8321

     REAL           IMAG           ABS          PHASE
   -1.2990        -0.7500        1.5000      -150.0000       (B)
   -0.9387         0.0000        0.9387       180.0000       (C)
   -0.3603        -0.7500        0.8321      -115.6589       (BC)

    NORMAL ACCELERATIONS

     REAL           IMAG           ABS          PHASE
    0.7500        -1.2990        1.5000       -60.0000       (B)
    0.0000         0.0000        0.0000         0.0000       (C)

    TANGENTIAL ACCELERATIONS

     REAL           IMAG           ABS          PHASE
    0.0000         0.0000        0.0000         0.0000       (B)
    1.1180         0.0000        1.1180         0.0000       (C)

    ABSOLUTE ACCELERATIONS

     REAL           IMAG           ABS          PHASE
    0.7500        -1.2990        1.5000       -60.0000       (B)
    1.1180         0.0000        1.1180         0.0000       (C)
   -0.3680        -1.2990        1.3502      -105.8176       (BC)
```

Slider-Crank: Modified Vector Method

PROBLEM DATA

AB	BC
1.5000	3.0000

THETA(A)	OMEGA(A)	ALPHA(A)
30.0000	-1.0000	0.0000

PHI(C)
14.4775

LINEAR VELOCITIES

V(B)	V(C)	V(BC)
-1.5000	1.0854	1.3416

REAL	IMAG	ABS	PHASE	
0.7500	-1.2990	1.5000	-60.0000	(B)
1.0854	0.0000	1.0854	0.0000	(C)
-0.3354	-1.2990	1.3416	-104.4775	(BC)

NORMAL ACCELERATIONS

REAL	IMAG	ABS	PHASE	
-1.2990	-0.7500	1.5000	-150.0000	(B)
0.0000	0.0000	0.0000	0.0000	(C)

TANGENTIAL ACCELERATIONS

REAL	IMAG	ABS	PHASE	
0.0000	0.0000	0.0000	0.0000	(B)
-1.7251	0.0000	1.7251	180.0000	(C)

ABSOLUTE ACCELERATIONS

REAL	IMAG	ABS	PHASE	
-1.2990	-0.7500	1.5000	-150.0000	(B)
-1.7251	0.0000	1.7251	180.0000	(C)
0.4260	-0.7500	0.8626	-60.4018	(BC)

A.2 LINKAGES: CALCULATOR PROGRAMS

Presented as an alternative to the computer programs listed above, the calculator programs in this section are based on the same mathematical methods covered in Part III.

CALCULATOR OPERATING PROCEDURE

This procedure is based on use of the Hewlett-Packard HP-41C programmable calculator in conjunction with the Math Pack Module for computation involving complex numbers and a HP peripheral printer for printing out the results.

PROCEDURE

1. Install the Math Pack Module and connect the printer.
2. Turn on the calculator.
3. Allocate the number of storage registers required for program.

   Press: XEQ
   alpha
   SIZE
   alpha
   080

4. Set a flag to facilitate use of the Math Pack Module that computes complex numbers.

   Press: □ SF04

5. Prepare to load the program. It may be necessary to remove a previously stored program from program memory to create a room for the new program.

   Press: □ GTO. .

6. Load the program.

   Press: PRGM

Write the program steps into the program memory following the step-by-step instructions given in the printed program. Upon entering the last instruction, "END," exit the program mode.

Press: PRGM

7. Execute and run the program.

Press: XEQ
alpha
"PROGRAM NAME"
alpha

If the program name is spelled incorrectly, the display panel will flash "NONEXISTENT." To correct this, repeat the procedure more carefully. Also, if the printer is not on-line, the signal "PRINTER OFF" will be flashed on the display panel. Be sure that the printer is on-line, switched on, and set in the manual position or mode. Check to be sure that the correct program is being executed.

8. Input the data for variables as requested in the display panel.

Press: (numerical keys for the appropriate value of each variable displayed), then
Press: R/S (after each variable input)

Note that, depending on the nature of the problem, some variable values are given, whereas others are calculated by the program.

9. Program execution and printing will begin after the data for the last requested variable are entered.

10. At the completion of program execution and printing, turn the calculator and printer to the "OFF" position.

Four-Bar: Simplified Vector Method

```
01◆LBL "4-BAR L"              51 RCL 12
02 " 4-BAR LINKAGE"           52 X↑2
03 XEQ 09                     53 -
04 " ---- -------"            54 RCL 11
05 XEQ 04                     55 /
06 FIX 4                      56 2
07 SF 04                      57 /
08 "AB?"                      58 RCL 18
09 PROMPT                     59 /
10 STO 10                     60 ACOS
11 "BC?"                      61 STO 19
12 PROMPT                     62 RCL 18
13 STO 11                     63 X↑2
14 "CD?"                      64 RCL 12
15 PROMPT                     65 X↑2
16 STO 12                     66 +
17 "AD?"                      67 RCL 11
18 PROMPT                     68 X↑2
19 STO 13                     69 -
20 "THETA A?"                 70 RCL 18
21 PROMPT                     71 /
22 STO 14                     72 2
23 "OMEGA A?"                 73 /
24 PROMPT                     74 RCL 12
25 STO 15                     75 /
26 "ALPHA A?"                 76 ACOS
27 PROMPT                     77 STO 23
28 STO 16                     78 RCL 14
29 RCL 14                     79 SIN
30 COS                        80 STO 20
31 STO 17                     81 RCL 10
32 RCL 10                     82 *
33 *                          83 RCL 18
34 RCL 13                     84 /
35 *                          85 ASIN
36 2                          86 STO 21
37 *                          87 CHS
38 CHS                        88 RCL 23
39 RCL 10                     89 -
40 X↑2                        90 180
41 +                          91 +
42 RCL 13                     92 STO 24
43 X↑2                        93 RCL 19
44 +                          94 RCL 21
45 SQRT                       95 -
46 STO 18                     96 STO 25
47 X↑2                        97 RCL 15
48 RCL 11                     98 RCL 10
49 X↑2                        99 *
50 +                          100 STO 26
```

Four-Bar: Simplified Vector Method

```
101 RCL 25          151 "--- --- ---"
102 RCL 14          152 XEQ 04
103 -               153 RCL 24
104 STO 69          154 90
105 SIN             155 +
106 STO 27          156 STO 34
107 RCL 25          157 XEQ 05
108 RCL 24          158 RCL 30
109 -               159 XEQ 07
110 STO 70          160 XEQ 08
111 SIN             161 ":C="
112 ST/ 27          162 ASTO 33
113 STO 28          163 XEQ 03
114 RCL 24          164 RCL 14
115 RCL 14          165 90
116 -               166 +
117 STO 68          167 STO 32
118 SIN             168 XEQ 05 ⊁
119 RCL 28          169 RCL 26
120 /               170 XEQ 07
121 STO 28          171 XEQ 08
122 RCL 26          172 ":B="
123 *               173 ASTO 33
124 STO 29          174 XEQ 03
125 RCL 27          175 RCL 25
126 RCL 26          176 90
127 *               177 +
128 STO 30          178 STO 35
129 "IN/SEC"        179 XEQ 05
130 ASTO 31         180 RCL 29
131 "LINEAR VELOCITY"  181 XEQ 07
132 XEQ 09          182 XEQ 08
133 "------ -------"   183 ":C/B="
134 XEQ 04          184 ASTO 33
135 "V:C="          185 XEQ 03
136 ARCL 30         186 RCL 34
137 ARCL 31         187 COS
138 XEQ 04          188 STO 36
139 "V:B="          189 LASTX
140 ARCL 26         190 SIN
141 ARCL 31         191 STO 37
142 XEQ 04          192 RCL 35
143 "V:C/B="        193 COS
144 ARCL 29         194 STO 38
145 ARCL 31         195 CHS
146 XEQ 04          196 STO 39
147 "IPS∠"          197 LASTX
148 ASTO 71         198 SIN
149 "VEC LIN VEL"   199 STO 40
150 XEQ 09          200 CHS
```

Four-Bar: Simplified Vector Method

201 STO 41
202 RCL 24
203 COS
204 STO 42
205 STO 43
206 RCL 30
207 X↑2
208 RCL 12
209 /
210 STO 44
211 ST* 43
212 RCL 15
213 X↑2
214 RCL 10
215 *
216 STO 45
217 RCL 17
218 *
219 ST- 43
220 RCL 16
221 RCL 10
222 *
223 STO 46
224 RCL 32
225 COS
226 STO 47
227 *
228 ST+ 43
229 RCL 29
230 X↑2
231 RCL 11
232 /
233 STO 48
234 RCL 25
235 COS
236 STO 49
237 *
238 ST- 43
239 RCL 24
240 SIN
241 STO 50
242 RCL 44
243 *
244 STO 51
245 RCL 20
246 RCL 45
247 *
248 ST- 51
249 RCL 14
250 90

251 +
252 SIN
253 STO 52
254 RCL 46
255 *
256 ST+ 51
257 RCL 25
258 SIN
259 STO 63
260 RCL 48
261 *
262 ST- 51
263 RCL 43
264 RCL 41
265 *
266 STO 54
267 RCL 51
268 RCL 39
269 *
270 ST- 54
271 RCL 36
272 RCL 41
273 *
274 STO 55
275 RCL 37
276 RCL 39
277 *
278 ST- 55
279 RCL 36
280 RCL 51
281 *
282 STO 56
283 RCL 37
284 RCL 43
285 *
286 ST- 56
287 RCL 54
288 RCL 55
289 /
290 STO 57
291 RCL 56
292 RCL 55
293 /
294 STO 58
295 "NORM ACC"
296 XEQ 09
297 "---- ---"
298 XEQ 04
299 "IPS↑2∠"
300 ASTO 63

Four-Bar: Simplified Vector Method

```
301 RCL 24          351 XEQ 07
302 XEQ 05          352 XEQ 08
303 RCL 44          353 ":C/B="
304 CHS             354 ASTO 33
305 XEQ 07          355 XEQ 16
306 XEQ 08          356 "ABS ACC"
307 ":C="           357 XEQ 09
308 ASTO 33         358 "--- ---"
309 XEQ 16          359 XEQ 04
310 RCL 14          360 RCL 24
311 XEQ 05          361 XEQ 05
312 RCL 45          362 RCL 59
313 CHS             363 STO 64
314 XEQ 07          364 RCL 60
315 XEQ 08          365 STO 65
316 ":B/C="         366 RCL 27
317 ASTO 33         367 RCL 26
318 XEQ 16          368 *
319 RCL 25          369 X↑2
320 XEQ 05          370 RCL 12
321 RCL 48          371 /
322 CHS             372 CHS
323 XEQ 07          373 ST* 64
324 XEQ 08          374 ST* 65
325 ":C/B="         375 RCL 34
326 ASTO 33         376 XEQ 05
327 XEQ 16          377 RCL 59
328 "TAN ACC"       378 STO 66
329 XEQ 09          379 RCL 60
330 "--- ---"       380 STO 67
331 XEQ 04          381 RCL 57
332 RCL 34          382 ST* 66
333 XEQ 05          383 ST* 67
334 RCL 57          384 XEQ 06
335 XEQ 07          385 1
336 XEQ 08          386 XEQ 07
337 ":C="           387 XEQ 08
338 ASTO 33         388 ":C="
339 XEQ 16          389 ASTO 33
340 RCL 32          390 XEQ 16
341 XEQ 05          391 RCL 14
342 RCL 46          392 XEQ 05
343 XEQ 07          393 RCL 59
344 XEQ 08          394 STO 64
345 ":B="           395 RCL 60
346 ASTO 33         396 STO 65
347 XEQ 16          397 RCL 26
348 RCL 35          398 RCL 15
349 XEQ 05          399 *
350 RCL 58          400 CHS
```

Four-Bar: Simplified Vector Method

401 ST* 64	451 XEQ 09
402 ST* 65	452 "------- ----"
403 RCL 32	453 XEQ 04
404 XEQ 05	454 "AB="
405 RCL 59	455 ARCL 10
406 STO 66	456 XEQ 04
407 RCL 60	457 "BC="
408 STO 67	458 ARCL 11
409 RCL 46	459 XEQ 04
410 ST* 66	460 "CD="
411 ST* 67	461 ARCL 12
412 XEQ 06	462 XEQ 04
413 1	463 "AD="
414 XEQ 07	464 ARCL 13
415 XEQ 08	465 XEQ 04
416 ":B="	466 "THETA A="
417 ASTO 33	467 ARCL 14
418 XEQ 16	468 XEQ 04
419 RCL 25	469 "OMEGA A="
420 XEQ 05	470 ARCL 15
421 RCL 59	471 XEQ 04
422 STO 64	472 "ALPHA A="
423 RCL 60	473 ARCL 16
424 STO 65	474 XEQ 04
425 RCL 28	475 "BD="
426 RCL 26	476 ARCL 18
427 *	477 XEQ 04
428 X↑2	478 "PHI D="
429 RCL 11	479 ARCL 21
430 /	480 XEQ 04
431 CHS	481 "GAMMA B="
432 ST* 64	482 ARCL 19
433 ST* 65	483 XEQ 04
434 RCL 35	484 "GAMMA D="
435 XEQ 05	485 ARCL 23
436 RCL 59	486 XEQ 04
437 STO 66	487 "THETA B="
438 RCL 60	488 ARCL 25
439 STO 67	489 XEQ 04
440 RCL 58	490 "THETA C="
441 ST* 66	491 ARCL 24
442 ST* 67	492 XEQ 04
443 XEQ 06	493 "SMB="
444 1	494 ARCL 70
445 XEQ 07	495 XEQ 04
446 XEQ 08	496 "SMC="
447 ":C/B="	497 ARCL 69
448 ASTO 33	498 XEQ 04
449 XEQ 16	499 "SMCB="
450 "PROBLEM DATA"	500 ARCL 68

Four-Bar: *Simplified Vector Method*

```
501 XEQ 04
502 "A 1="
503 ARCL 36
504 XEQ 04
505 "A 2="
506 ARCL 37
507 XEQ 04
508 "B 1="
509 ARCL 39
510 XEQ 04
511 "B 2="
512 ARCL 41
513 XEQ 04
514 "C 1="
515 ARCL 43
516 XEQ 04
517 "C 2="
518 ARCL 51
519 XEQ 04
520 "ATC="
521 ARCL 57
522 XEQ 04
523 "ATCB="
524 ARCL 58
525 XEQ 04
526 XEQ 13
527◆LBL 05
528 57.296
529 /
530 0
531 XROM "e↑Z"
532 STO 59
533 X<>Y
534 STO 60
535 RTN
536◆LBL 08
537 RCL 59
538 X<0?
539 GTO 14
540 RTN
541◆LBL 14
542 RCL 60
543 X<0?
544 GTO 15
545 180
546 ST+ 62
547 RTN
548◆LBL 15
549 180
550 ST- 62
```

```
551 RTN
552◆LBL 06
553 RCL 65
554 RCL 64
555 RCL 67
556 RCL 66
557 XROM "C+"
558 STO 59
559 X<>Y
560 STO 60
561 RTN
562◆LBL 07
563 ST* 59
564 ST* 60
565 RCL 60
566 RCL 59
567 XROM "MAGZ"
568 STO 61
569 RCL 59
570 X=0?
571 GTO 12
572 RCL 60
573 RCL 59
574 /
575 ATAN
576 STO 62
577 RTN
578◆LBL 12
579 0
580 ATAN
581 STO 62
582 RTN
583◆LBL 04
584 AVIEW
585 ADV
586 CLA
587 RTN
588◆LBL 09
589 AVIEW
590 RTN
591◆LBL 03
592 FIX 2
593 26
594 ACCHR
595 CLA
596 ARCL 33
597 ARCL 61
598 ARCL 71
599 ARCL 62
600 ACA
```

Four-Bar: Simplified Vector Method

```
601 PRBUF
602 ADV
603 FIX 4
604 RTN
605◆LBL 16
606 FIX 2
607 22
608 ACCHR
609 CLA
610 ARCL 33
611 ARCL 61
612 ARCL 63
613 ARCL 62
614 ACA
615 PRBUF
616 ADV
617 FIX 4
618 RTN
619◆LBL 13
620 AOFF
621 STOP
622 .END.
```

Four-Bar: Simplified Vector Method

PROBLEM DATA   4-BAR LINKAGE
‾‾‾‾‾‾ ‾‾‾‾   ‾‾‾‾‾ ‾‾‾‾‾‾‾

AB=1.5000   LINEAR VELOCITY
                                                    ‾‾‾‾‾‾ ‾‾‾‾‾‾‾‾
BC=3.0000
                                                    V:C=-1.0392IN/SEC
CD=3.0000
                                                    V:B=3.0000IN/SEC
AD=4.0000
                                                    V:C/B=-3.4601IN/SEC
THETA A=30.0000
                                                    VEC LIN VEL
OMEGA A=2.0000   ‾‾‾ ‾‾‾ ‾‾‾

ALPHA A=1.0000   ū:C=1.04IPS∠12.33

BD=2.8032   ū:B=3.00IPS∠120.00

PHI D=15.5188   ū:C/B=3.46IPS∠-43.37

GAMMA B=62.1478   NORM ACC
                                                    ‾‾‾‾ ‾‾‾
GAMMA D=62.1478
                                                    ā:C=0.36IPS↑2∠-77.67
THETA B=46.6289
                                                    ā:B/C=6.00IPS↑2∠-150.00
THETA C=102.3334
                                                    ā:C/B=3.99IPS↑2∠-133.37
SMB=-55.7045
                                                    TAN ACC
SMC=16.6289   ‾‾‾ ‾‾‾

SMCB=72.3334   ā:C=11.02IPS↑2∠-167.67

A 1=-0.9769   ā:B=1.50IPS↑2∠120.00

A 2=-0.2136   ā:C/B=2.76IPS↑2∠136.63

B 1=0.7269   ABS ACC
                                                    ‾‾‾ ‾‾‾
B 2=-0.6867
                                                    ā:C=11.03IPS↑2∠-165.80
C 1=-8.7635
                                                    ā:B=6.18IPS↑2∠-164.04
C 2=-4.2502
                                                    ā:C/B=4.85IPS↑2∠-168.04
ATC=11.0244

ATCB=2.7601

Slider-Crank: Simplified Vector Method

```
01◆LBL "SL-CR V"              51 RCL 17
02 " SLIDER CR ANL"          52 -
03 XEQ 09                     53 SIN
04 " ------ -- ---"           54 STO 20
05 XEQ 04                     55 RCL 19
06 FIX 4                      56 RCL 20
07 SF 04                      57 /
08 "AB?"                      58 STO 21
09 PROMPT                     59 RCL 18
10 STO 10                     60 *
11 "BC?"                      61 STO 22
12 PROMPT                     62 90
13 STO 11                     63 RCL 12
14 "THETA 2?"                 64 -
15 PROMPT                     65 SIN
16 STO 12                     66 STO 23
17 "OMEGA 2?"                 67 RCL 20
18 PROMPT                     68 /
19 STO 13                     69 STO 24
20 "ALPHA 2?"                 70 RCL 18
21 PROMPT                     71 *
22 STO 09                     72 STO 25
23 "ECC?"                     73 "LINEAR VELOCITY"
24 PROMPT                     74 XEQ 09
25 STO 14                     75 "------ --------"
26 RCL 12                     76 XEQ 04
27 SIN                        77 "IN/SEC"
28 STO 15                     78 ASTO 26
29 RCL 10                     79 "V:B="
30 *                         80 ARCL 18
31 RCL 14                     81 ARCL 26
32 -                          82 XEQ 04
33 RCL 11                     83 "V:C="
34 /                          84 ARCL 22
35 ASIN                       85 ARCL 26
36 STO 16                     86 XEQ 04
37 180                        87 "V:B/C="
38 RCL 16                     88 ARCL 25
39 -                          89 ARCL 26
40 STO 17                     90 XEQ 04
41 RCL 10                     91 RCL 12
42 RCL 13                     92 90
43 *                         93 +
44 STO 18                     94 STO 27
45 RCL 17                     95 RCL 17
46 RCL 12                     96 90
47 -                          97 +
48 SIN                        98 STO 28
49 STO 19                     99 "VEC LIN VEL"
50 90                        100 XEQ 09
```

Slider-Crank: Simplified Vector Method

```
101 "—— ——— ——"          151 RCL 10
102 XEQ 04                152 *
103 RCL 27                153 STO 37
104 XEQ 05                154 RCL 27
105 RCL 18                155 COS
106 XEQ 07                156 STO 38
107 XEQ 08                157 RCL 37
108 "IPS∠"                158 *
109 ASTO 29               159 ST+ 36
110 ":B="                 160 RCL 17
111 ASTO 30               161 COS
112 XEQ 03                162 STO 39
113 0                     163 RCL 25
114 XEQ 05                164 X↑2
115 RCL 22                165 RCL 11
116 XEQ 07                166 /
117 XEQ 08                167 STO 43
118 ":C="                 168 RCL 39
119 ASTO 30               169 *
120 XEQ 03                170 ST+ 36
121 RCL 28                171 RCL 15
122 XEQ 05                172 RCL 40
123 RCL 25                173 *
124 XEQ 07                174 CHS
125 XEQ 08                175 STO 41
126 ":B/C="              176 RCL 27
127 ASTO 30               177 SIN
128 XEQ 03                178 STO 42
129 1                     179 RCL 37
130 STO 31                180 *
131 0                     181 ST+ 41
132 STO 32                182 RCL 17
133 RCL 28                183 SIN
134 COS                   184 STO 44
135 STO 33                185 RCL 43
136 LASTX                 186 *
137 SIN                   187 ST+ 41
138 STO 34                188 RCL 36
139 RCL 12                189 RCL 34
140 COS                   190 *
141 STO 35                191 STO 45
142 RCL 18                192 RCL 41
143 RCL 13                193 RCL 33
144 *                     194 *
145 STO 40                195 ST- 45
146 RCL 35                196 RCL 31
147 *                     197 RCL 34
148 CHS                   198 *
149 STO 36                199 STO 46
150 RCL 09                200 RCL 32
```

Slider-Crank: Simplified Vector Method

```
201 RCL 33            251 ASTO 30
202 *                 252 XEQ 16
203 ST- 46            253 "TAN ACC"
204 RCL 31            254 XEQ 09
205 RCL 41            255 "--- ---"
206 *                 256 XEQ 04
207 STO 47            257 RCL 27
208 RCL 32            258 XEQ 05
209 RCL 36            259 RCL 37
210 *                 260 XEQ 07
211 ST- 47            261 XEQ 08
212 RCL 45            262 ":B="
213 RCL 46            263 ASTO 30
214 /                 264 XEQ 16
215 STO 48            265 0
216 RCL 47            266 XEQ 05
217 RCL 46            267 RCL 48
218 /                 268 XEQ 07
219 STO 49            269 XEQ 08
220 "NORM ACC"        270 ":C="
221 XEQ 09            271 ASTO 30
222 "--- ---"         272 XEQ 16
223 XEQ 04            273 RCL 28
224 RCL 12            274 XEQ 05
225 XEQ 05            275 RCL 49
226 RCL 40            276 XEQ 07
227 CHS               277 XEQ 08
228 XEQ 07            278 ":B/C="
229 XEQ 08            279 ASTO 30
230 "IPS↑2∠"          280 XEQ 16
231 ASTO 50           281 "ABS ACC"
232 ":B="             282 XEQ 09
233 ASTO 30           283 "--- ---"
234 XEQ 16            284 XEQ 04
235 0                 285 RCL 12
236 STO 59            286 XEQ 05
237 STO 60            287 RCL 59
238 1                 288 STO 51
239 XEQ 07            289 RCL 60
240 XEQ 08            290 STO 52
241 ":C="             291 RCL 40
242 ASTO 30           292 CHS
243 XEQ 16            293 ST* 51
244 RCL 17            294 ST* 52
245 XEQ 05            295 RCL 27
246 RCL 43            296 XEQ 05
247 CHS               297 RCL 59
248 XEQ 07            298 STO 53
249 XEQ 08            299 RCL 60
250 ":B/C="           300 STO 54
```

Slider-Crank: Simplified Vector Method

```
301 RCL 37
302 ST* 53
303 ST* 54
304 XEQ 06
305 1
306 XEQ 07
307 XEQ 08
308 ":B="
309 ASTO 30
310 XEQ 16
311 0
312 XEQ 05
313 RCL 48
314 XEQ 07
315 XEQ 08
316 ":C="
317 ASTO 30
318 XEQ 16
319 RCL 17
320 XEQ 05
321 RCL 59
322 STO 51
323 RCL 60
324 STO 52
325 RCL 25
326 X↑2
327 RCL 11
328 /
329 CHS
330 ST* 51
331 ST* 52
332 RCL 28
333 XEQ 05
334 RCL 59
335 STO 53
336 RCL 60
337 STO 54
338 RCL 49
339 ST* 53
340 ST* 54
341 XEQ 06
342 1
343 XEQ 07
344 XEQ 08
345 ":B/C="
346 ASTO 30
347 XEQ 16
348 "PROBLEM DATA"
349 XEQ 09
350 "------- ----"
```

```
351 XEQ 04
352 "AB="
353 ARCL 10
354 XEQ 04
355 "BC="
356 ARCL 11
357 XEQ 04
358 "THETA 2="
359 ARCL 12
360 XEQ 04
361 "OMEGA 2="
362 ARCL 13
363 XEQ 04
364 "ALPHA 2="
365 ARCL 09
366 XEQ 04
367 "ECC="
368 ARCL 14
369 XEQ 04
370 "PHI 3="
371 ARCL 16
372 XEQ 04
373 "THETA 3="
374 ARCL 17
375 XEQ 04
376 90
377 RCL 17
378 -
379 STO 55
380 RCL 17
381 RCL 12
382 -
383 STO 56
384 90
385 RCL 12
386 -
387 STO 57
388 "SNB="
389 ARCL 55
390 XEQ 04
391 "SNC="
392 ARCL 56
393 XEQ 04
394 "SNBC="
395 ARCL 57
396 XEQ 04
397 "A 1="
398 ARCL 31
399 XEQ 04
400 "A 2="
```

Slider-Crank: Simplified Vector Method

```
401 ARCL 32
402 XEQ 04
403 "B 1="
404 ARCL 33
405 XEQ 04
406 "B 2="
407 ARCL 34
408 XEQ 04
409 "C 1="
410 ARCL 36
411 XEQ 04
412 "C 2="
413 ARCL 41
414 XEQ 04
415 "ATC="
416 ARCL 48
417 XEQ 04
418 "ATBC="
419 ARCL 49
420 XEQ 04
421 XEQ 13
422◆LBL 05
423 57.296
424 /
425 0
426 XROM "etZ"
427 STO 59
428 X<>Y
429 STO 60
430 RTN
431◆LBL 08
432 RCL 59
433 X<0?
434 GTO 14
435 RTN
436◆LBL 14
437 RCL 60
438 X<0?
439 GTO 15
440 180
441 ST+ 62
442 RTN
443◆LBL 15
444 180
445 ST- 62
446 RTN
447◆LBL 06
448 RCL 52
449 RCL 51
450 RCL 54
```

```
451 RCL 53
452 XROM "C+"
453 STO 59
454 X<>Y
455 STO 60
456 RTN
457◆LBL 07
458 ST* 59
459 ST* 60
460 RCL 60
461 RCL 59
462 XROM "MAGZ"
463 STO 61
464 RCL 59
465 X=0?
466 GTO 12
467 RCL 60
468 RCL 59
469 /
470 ATAN
471 STO 62
472 RTN
473◆LBL 12
474 0
475 ATAN
476 STO 62
477 RTN
478◆LBL 04
479 AVIEW
480 ADV
481 CLA
482 RTN
483◆LBL 09
484 AVIEW
485 RTN
486◆LBL 03
487 FIX 2
488 26
489 ACCHR
490 CLA
491 ARCL 30
492 ARCL 61
493 ARCL 29
494 ARCL 62
495 ACA
496 PRBUF
497 ADV
498 FIX 4
499 RTN
500◆LBL 16
```

Slider-Crank: Simplified Vector Method

```
501 FIX 2
502 22
503 ACCHR
504 CLA
505 ARCL 30
506 ARCL 61
507 ARCL 50
508 ARCL 62
509 ACA
510 PRBUF
511 ADV
512 FIX 4
513 RTN
514◆LBL 13
515 AOFF
516 STOP
517 .END.
```

Slider-Crank: Simplified Vector Method

PROBLEM DATA
------- ---

AB=1.5000

BC=3.0000

THETA 2=150.0000

OMEGA 2=1.0000

ALPHA 2=0.0000

ECC=0.0000

PHI 3=14.4775

THETA 3=165.5225

SHB=-75.5225

SHC=15.5225

SHBC=-60.0000

A 1=1.0000

A 2=0.0000

B 1=-0.2500

B 2=-0.9682

C 1=0.7181

C 2=-0.6000

ATC=0.8730

ATBC=0.6197

SLIDER CR ANL
------ -- ---

LINEAR VELOCITY
------ --------

V:B=1.5000IN/SEC

V:C=-0.4146IN/SEC

V:B/C=1.3416IN/SEC

VEC LIN VEL
--- --- ---

$\bar{u}$:B=1.50IPS∠-120.00

$\bar{u}$:C=0.41IPS∠180.00

$\bar{u}$:B/C=1.34IPS∠-104.48

NORM ACC
---- ---

$\bar{a}$:B=1.50IPS↑2∠-30.00

$\bar{a}$:C=0.00IPS↑2∠0.00

$\bar{a}$:B/C=0.60IPS↑2∠-14.48

TAN ACC
--- ---

$\bar{a}$:B=0.00IPS↑2∠0.00

$\bar{a}$:C=0.87IPS↑2∠0.00

$\bar{a}$:B/C=0.62IPS↑2∠-104.48

ABS ACC
--- ---

$\bar{a}$:B=1.50IPS↑2∠-30.00

$\bar{a}$:C=0.87IPS↑2∠0.00

$\bar{a}$:B/C=0.86IPS↑2∠-60.40

Slider-Crank: Simplified Vector Method

PROBLEM DATA
——————— ————

AB=1.5000

BC=3.0000

THETA 2=150.0000

OMEGA 2=1.0000

ALPHA 2=0.0000

ECC=0.5000

PHI 3=4.7802

THETA 3=175.2198

SNB=-85.2198

SNC=25.2198

SNBC=-60.0000

A 1=1.0000

A 2=0.0000

B 1=-0.0833

B 2=-0.9965

C 1=0.7346

C 2=-0.7028

ATC=0.7933

ATBC=0.7053

SLIDER CR ANL
——————— —— ———

LINEAR VELOCITY
——————— ————————

V:B=1.5000IN/SEC

V:C=-0.6414IN/SEC

V:B/C=1.3036IN/SEC

VEC LIN VEL
——— ——— ———

ū:B=1.50IPS∠-120.00

ū:C=0.64IPS∠180.00

ū:B/C=1.30IPS∠-94.78

NORM ACC
———— ———

ā:B=1.50IPS↑2∠-30.00

ā:C=0.00IPS↑2∠0.00

ā:B/C=0.57IPS↑2∠-4.78

TAN ACC
——— ——

ā:B=0.00IPS↑2∠0.00

ā:C=0.79IPS↑2∠0.00

ā:B/C=0.71IPS↑2∠-94.78

ABS ACC
——— ———

ā:B=1.50IPS↑2∠-30.00

ā:C=0.79IPS↑2∠0.00

ā:B/C=0.90IPS↑2∠-56.01

Slider-Crank: Simplified Vector Method

PROBLEM DATA
------- ----

AB=1.5000

BC=3.0000

THETA 2=150.0000

OMEGA 2=1.0000

ALPHA 2=0.0000

ECC=-0.5000

PHI 3=24.6243

THETA 3=155.3757

SNB=-65.3757

SNC=5.3757

SNBC=-60.0000

A 1=1.0000

A 2=0.0000

B 1=-0.4167

B 2=-0.9091

C 1=0.6803

C 2=-0.4664

ATC=0.8940

ATBC=0.5130

SLIDER CR ANL
------ -- ---

LINEAR VELOCITY
------ --------

V:B=1.5000IN/SEC

V:C=-0.1546IN/SEC

V:B/C=1.4290IN/SEC

VEC LIN VEL
--- --- ---

ū:B=1.50IPS∠-120.00

ū:C=0.15IPS∠180.00

ū:B/C=1.43IPS∠-114.63

NORM ACC
---- ---

ā:B=1.50IPS↑2∠-30.00

ā:C=0.00IPS↑2∠0.00

ā:B/C=0.68IPS↑2∠-24.62

TAN ACC
--- ---

ā:B=0.00IPS↑2∠0.00

ā:C=0.89IPS↑2∠0.00

ā:B/C=0.51IPS↑2∠-114.63

ABS ACC
--- ---

ā:B=1.50IPS↑2∠-30.00

ā:C=0.89IPS↑2∠0.00

ā:B/C=0.85IPS↑2∠-61.63

Quick-Return: Simplified Vector Method

```
01◆LBL "Q-R MEC"              51 -
02 " Q-R MECHANISM"           52 2
03 XEQ 09                     53 /
04 " --- ---------"           54 RCL 15
05 XEQ 04                     55 /
06 FIX 4                      56 RCL 11
07 SF 04                      57 /
08 "AB?"                      58 ACOS
09 PROMPT                     59 STO 65
10 12                         60 180
11 /                          61 RCL 12
12 STO 10                     62 X(=Y?
13 "AD?"                      63 GTO 10
14 PROMPT                     64 -1
15 12                         65 ST* 65
16 /                          66◆LBL 10
17 STO 11                     67 180
18 "THETA 2?"                 68 RCL 65
19 PROMPT                     69 -
20 STO 12                     70 STO 17
21 "OMEGA 2?"                 71 RCL 12
22 PROMPT                     72 SIN
23 STO 13                     73 STO 16
24 "ALPHA 2?"                 74◆LBL 02
25 PROMPT                     75 RCL 10
26 STO 14                     76 RCL 13
27 RCL 12                     77 *
28 COS                        78 STO 09
29 STO 08                     79 RCL 17
30 RCL 11                     80 RCL 12
31 *                          81 -
32 RCL 10                     82 STO 18
33 *                          83 COS
34 2                          84 STO 19
35 *                          85 RCL 09
36 CHS                        86 *
37 RCL 11                     87 STO 20
38 X↑2                        88 RCL 18
39 +                          89 SIN
40 RCL 10                     90 STO 21
41 X↑2                        91 RCL 09
42 +                          92 *
43 SQRT                       93 STO 22
44 STO 15                     94 "LIN VELOCITY"
45 X↑2                        95 XEQ 09
46 RCL 11                     96 "--- --------"
47 X↑2                        97 XEQ 04
48 +                          98 FIX 2
49 RCL 10                     99 "FPS"
50 Y↑2                        100 ASTO 23
```

Quick-Return: Simplified Vector Method

```
101 "V:B="                          151 COS
102 ARCL 09                         152 STO 28
103 ARCL 23                         153 LASTX
104 XEQ 04                          154 SIN
105 "V:C="                          155 STO 29
106 ARCL 20                         156 RCL 17
107 ARCL 23                         157 SIN
108 XEQ 04                          158 STO 30
109 "V:B/C="                        159 LASTX
110 ARCL 22                         160 COS
111 ARCL 23                         161 STO 31
112 XEQ 04                          162 RCL 20
113 FIX 4                           163 X↑2
114 "VEC LIN VEL"                   164 *
115 XEQ 09                          165 RCL 15
116 "--- --- ---"                   166 /
117 XEQ 04                          167 STO 32
118 RCL 12                          168 RCL 28
119 90                              169 RCL 20
120 +                              170 *
121 STO 24                          171 RCL 15
122 XEQ 05                          172 /
123 RCL 09                          173 RCL 22
124 XEQ 07                          174 *
125 XEQ 08                          175 2
126 ":B="                           176 *
127 ASTO 25                         177 ST- 32
128 "FPS∠"                          178 RCL 24
129 ASTO 26                         179 COS
130 XEQ 03                          180 STO 33
131 RCL 17                          181 RCL 10
132 90                              182 *
133 +                              183 RCL 14
134 STO 27                          184 *
135 XEQ 05                          185 ST+ 32
136 RCL 20                          186 RCL 08
137 XEQ 07                          187 RCL 10
138 XEQ 08                          188 *
139 ":C="                           189 RCL 13
140 ASTO 25                         190 X↑2
141 XEQ 03                          191 *
142 RCL 17                          192 ST- 32
143 XEQ 05                          193 RCL 30
144 RCL 22                          194 RCL 20
145 XEQ 07                          195 X↑2
146 XEQ 08                          196 *
147 ":B/C="                         197 RCL 15
148 ASTO 25                         198 /
149 XEQ 03                          199 STO 34
150 RCL 27                          200 RCL 29
```

Quick-Return: Simplified Vector Method

```
201 RCL 20
202 *
203 RCL 15
204 /
205 RCL 22
206 *
207 2
208 *
209 ST- 34
210 RCL 24
211 SIN
212 STO 35
213 RCL 10
214 *
215 RCL 14
216 *
217 ST+ 34
218 RCL 16
219 RCL 10
220 *
221 RCL 13
222 X↑2
223 *
224 ST- 34
225 RCL 32
226 RCL 30
227 *
228 STO 36
229 RCL 34
230 RCL 31
231 *
232 ST- 36
233 RCL 28
234 RCL 30
235 *
236 STO 37
237 RCL 29
238 RCL 31
239 *
240 ST- 37
241 RCL 36
242 RCL 37
243 /
244 STO 38
245 RCL 28
246 RCL 34
247 *
248 STO 39
249 RCL 29
250 RCL 32
```

```
251 *
252 ST- 39
253 RCL 39
254 RCL 37
255 /
256 STO 40
257 "NORM ACC"
258 XEQ 09
259 "---- ---"
260 XEQ 04
261 RCL 12
262 XEQ 05
263 RCL 09
264 RCL 13
265 *
266 STO 41
267 CHS
268 XEQ 07
269 XEQ 08
270 ":B="
271 ASTO 42
272 "FPS↑2∡"
273 ASTO 43
274 XEQ 16
275 RCL 17
276 XEQ 05
277 RCL 20
278 X↑2
279 RCL 15
280 /
281 STO 44
282 CHS
283 XEQ 07
284 XEQ 08
285 ":C="
286 ASTO 42
287 XEQ 16
288 RCL 27
289 XEQ 05
290 0
291 XEQ 07
292 XEQ 08
293 ":B/C="
294 ASTO 42
295 XEQ 16
296 "TANG ACC"
297 XEQ 09
298 "---- ---"
299 XEQ 04
300 RCL 24
```

Quick-Return: Simplified Vector Method

```
301 XEQ 05          351 STO 48
302 RCL 10          352 RCL 60
303 RCL 14          353 STO 49
304 *               354 RCL 45
305 STO 45          355 ST* 48
306 XEQ 07          356 ST* 49
307 XEQ 08          357 RCL 12
308 ":B="           358 XEQ 05
309 ASTO 42         359 RCL 59
310 XEQ 16          360 STO 50
311 RCL 27          361 RCL 60
312 XEQ 05          362 STO 51
313 RCL 38          363 RCL 41
314 XEQ 07          364 CHS
315 XEQ 08          365 ST* 50
316 ":C="           366 ST* 51
317 ASTO 42         367 XEQ 06
318 XEQ 16          368 1
319 RCL 17          369 XEQ 07
320 XEQ 05          370 XEQ 08
321 RCL 40          371 ":B="
322 XEQ 07          372 ASTO 42
323 XEQ 08          373 XEQ 16
324 ":B/C="         374 RCL 27
325 ASTO 42         375 XEQ 05
326 XEQ 16          376 RCL 59
327 "COR ACC"       377 STO 48
328 XEQ 09          378 RCL 60
329 "--- ---"       379 STO 49
330 XEQ 04          380 RCL 38
331 RCL 27          381 ST* 48
332 XEQ 05          382 ST* 49
333 RCL 20          383 RCL 17
334 RCL 15          384 XEQ 05
335 /               385 RCL 59
336 RCL 22          386 STO 50
337 *               387 RCL 60
338 2               388 STO 51
339 *               389 RCL 19
340 STO 46          390 RCL 10
341 XEQ 07          391 *
342 XEQ 08          392 RCL 13
343 XEQ 16          393 *
344 "ABS ACC"       394 X↑2
345 XEQ 09          395 RCL 15
346 "--- ---"       396 /
347 XEQ 04          397 CHS
348 RCL 24          398 ST* 50
349 XEQ 05          399 ST* 51
350 RCL 59          400 XEQ 06
```

Quick-Return: Simplified Vector Method

```
401 1
402 XEQ 07
403 XEQ 08
404 ":C="
405 ASTO 42
406 XEQ 16
407 RCL 17
408 XEQ 05
409 RCL 59
410 STO 48
411 RCL 60
412 STO 49
413 RCL 40
414 ST* 48
415 ST* 49
416 RCL 27
417 XEQ 05
418 RCL 59
419 STO 50
420 RCL 60
421 STO 51
422 RCL 19
423 RCL 21
424 *
425 RCL 41
426 *
427 RCL 10
428 *
429 2
430 *
431 RCL 15
432 /
433 ST* 50
434 ST* 51
435 XEQ 06
436 1
437 XEQ 07
438 XEQ 08
439 ":B/C="
440 ASTO 42
441 XEQ 16
442 "PROBLEM DATA"
443 XEQ 09
444 "------- ----"
445 XEQ 04
446 "AB="
447 ARCL 10
448 XEQ 04
449 "AD="
450 ARCL 11
```

```
451 XEQ 04
452 "THETA 2="
453 ARCL 12
454 XEQ 04
455 "OMEGA 2="
456 ARCL 13
457 XEQ 04
458 "ALPHA 2="
459 ARCL 14
460 XEQ 04
461 "PHI 3="
462 ARCL 65
463 XEQ 04
464 "THETA 3="
465 ARCL 17
466 XEQ 04
467 "CD="
468 ARCL 15
469 XEQ 04
470 "A 1="
471 ARCL 28
472 XEQ 04
473 "A 2="
474 ARCL 29
475 XEQ 04
476 "B 1="
477 ARCL 31
478 XEQ 04
479 "B 2="
480 ARCL 30
481 XEQ 04
482 "C 1="
483 ARCL 32
484 XEQ 04
485 "C 2="
486 ARCL 34
487 XEQ 04
488 "ATC="
489 ARCL 38
490 XEQ 04
491 "ATBC="
492 ARCL 40
493 XEQ 04
494 XEQ 13
495*LBL 05
496 57.296
497 /
498 0
499 XROM "e↑Z"
500 STO 59
```

Quick-Return: Simplified Vector Method

```
501 X<>Y
502 STO 60
503 RTN
504*LBL 08
505 RCL 59
506 X<0?
507 GTO 14
508 RTN
509*LBL 14
510 RCL 60
511 X<0?
512 GTO 15
513 180
514 ST+ 62
515 RTN
516*LBL 15
517 180
518 ST- 62
519 RTN
520*LBL 06
521 RCL 49
522 RCL 48
523 RCL 51
524 RCL 50
525 XROM "C+"
526 STO 59
527 X<>Y
528 STO 60
529 RTN
530*LBL 07
531 ST* 59
532 ST* 60
533 RCL 60
534 RCL 59
535 XROM "MAGZ"
536 STO 61
537 RCL 59
538 X=0?
539 GTO 12
540 RCL 60
541 RCL 59
542 /
543 ATAN
544 STO 62
545 RTN
546*LBL 12
547 0
548 ATAN
549 STO 62
550 RTN
```

```
551*LBL 04
552 AVIEW
553 ADV
554 CLA
555 RTN
556*LBL 09
557 AVIEW
558 RTN
559*LBL 03
560 FIX 1
561 26
562 ACCHR
563 CLA
564 ARCL 25
565 ARCL 61
566 ARCL 26
567 ARCL 62
568 ACA
569 PRBUF
570 ADV
571 FIX 4
572 RTN
573*LBL 16
574 FIX 1
575 22
576 ACCHR
577 CLA
578 ARCL 42
579 ARCL 61
580 ARCL 43
581 ARCL 62
582 ACA
583 PRBUF
584 ADV
585 FIX 4
586 RTN
587*LBL 13
588 AOFF
589 .END.
```

Quick-Return: Simplified Vector Method

PROBLEM DATA
------- ----

AB=0.1667

AD=0.3333

THETA 2=30.0000

OMEGA 2=62.8300

ALPHA 2=0.0000

PHI 3=23.7940

THETA 3=156.2060

CD=0.2066

A 1=-0.4034

A 2=-0.9150

B 1=-0.9150

B 2=0.4034

C 1=-943.4515

C 2=-717.2896

ATC=1,036.9562

ATBC=573.8702

Q-R MECHANISM
--- ---------

LIN VELOCITY
--- --------

V:B=10.47FPS

V:C=-6.19FPS

V:B/C=8.45FPS

VEC LIN VEL
--- --- ---

ū:B=10.5FPS∠120.0

ū:C=6.2FPS∠66.2

ū:B/C=8.4FPS∠156.2

NORM ACC
---- ---

ā:B=657.9FPS↑2∠-150.0

ā:C=185.2FPS↑2∠-23.8

ā:B/C=0.0FPS↑2∠0.0

TANG ACC
---- ---

ā:B=0.0FPS↑2∠0.0

ā:C=1,037.0FPS↑2∠-113.8

ā:B/C=573.9FPS↑2∠156.2

COR ACC
--- ---

ā:B/C=506.1FPS↑2∠66.2

ABS ACC
--- ---

ā:B=657.9FPS↑2∠-150.0

ā:C=1,053.4FPS↑2∠-103.7

ā:B/C=765.1FPS↑2∠114.8

Quick-Return: Simplified Vector Method

PROBLEM DATA
------- ----

AB=0.2500

AD=0.1667

THETA 2=60.0000

OMEGA 2=30.0000

ALPHA 2=0.0000

PHI 3=79.1066

THETA 3=100.8934

CD=0.2205

A 1=-0.9820

A 2=-0.1890

B 1=-0.1890

B 2=0.9820

C 1=107.9082

C 2=-3.9766

ATC=-105.2122

ATBC=-24.2977

Q-R MECHANISM
--- ---------

LIN VELOCITY
--- --------

V:B=7.50FPS

V:C=5.67FPS

V:B/C=4.91FPS

VEC LIN VEL
--- --- ---

ū:B=7.5FPS∠150.0

ū:C=5.7FPS∠-169.1

ū:B/C=4.9FPS∠100.9

NORM ACC
---- ---

ā:B=225.0FPS↑2∠-120.0

ā:C=145.8FPS↑2∠-79.1

ā:B/C=0.0FPS↑2∠0.0

TANG ACC
---- ---

ā:B=0.0FPS↑2∠0.0

ā:C=105.2FPS↑2∠10.9

ā:B/C=24.3FPS↑2∠-79.1

COR ACC
--- ---

ā:B/C=252.5FPS↑2∠-169.1

ABS ACC
--- ---

ā:B=225.0FPS↑2∠-120.0

ā:C=179.8FPS↑2∠-43.3

ā:B/C=253.7FPS↑2∠-163.6

Sliding Coupler: Simplified Vector Method

```
01+LBL "SL-CPLR"          51 X↑2
02 "SLIDING COUPLER"      52 RCL 11
03 XEQ 09                 53 X↑2
04 "------- -------"      54 +
05 XEQ 04                 55 RCL 10
06 SF 04                  56 X↑2
07 FIX 4                  57 -
08 "AB?"                  58 2
09 PROMPT                 59 /
10 12                     60 RCL 17
11 /                      61 /
12 STO 10                 62 RCL 11
13 "AC?"                  63 /
14 PROMPT                 64 ACOS
15 12                     65 CHS
16 /                      66 STO 18
17 STO 11                 67 180
18 "THETA A?"             68 ENTER↑
19 PROMPT                 69 RCL 12
20 STO 12                 70 X<=Y?
21 "OMEGA A?"             71 GTO 02
22 PROMPT                 72 -1
23 STO 13                 73 ST* 18
24 "ALPHA A?"             74 GTO 02
25 PROMPT                 75+LBL 01
26 STO 14                 76 180
27 RCL 12                 77 ENTER↑
28 SIN                    78 RCL 12
29 STO 15                 79 X<=Y?
30 LASTX                  80 GTO 18
31 COS                    81 0
32 STO 16                 82 ACOS
33 RCL 11                 83 CHS
34 *                      84 STO 18
35 RCL 10                 85+LBL 18
36 *                      86 0
37 2                      87 ACOS
38 *                      88 STO 18
39 CHS                    89+LBL 02
40 RCL 11                 90 RCL 13
41 X↑2                    91 RCL 10
42 +                      92 *
43 RCL 10                 93 STO 19
44 X↑2                    94 RCL 12
45 +                      95 RCL 18
46 SQRT                   96 -
47 STO 17                 97 STO 20
48 X=0?                   98 SIN
49 GTO 01                 99 STO 21
50 RCL 17                 100 RCL 19
```

Sliding Coupler: Simplified Vector Method

```
101 *              151 ":C="
102 CHS            152 ASTO 28
103 STO 22         153 XEQ 03
104 RCL 20         154 RCL 18
105 COS            155 90
106 STO 23         156 +
107 RCL 19         157 STO 29
108 *              158 XEQ 05
109 CHS            159 RCL 24
110 STO 24         160 XEQ 07
111 "LIN VELOCITY" 161 XEQ 08
112 XEQ 09         162 ":C/B="
113 "--- --------" 163 ASTO 28
114 XEQ 04         164 XEQ 03
115 "FPS"          165 RCL 18
116 ASTO 25        166 COS
117 "V:B="         167 STO 30
118 ARCL 19        168 LASTX
119 ARCL 25        169 SIN
120 XEQ 04         170 STO 31
121 "V:C="         171 RCL 29
122 ARCL 22        172 COS
123 ARCL 25        173 CHS
124 XEQ 04         174 STO 32
125 "V:C/B="       175 LASTX
126 ARCL 24        176 SIN
127 ARCL 25        177 CHS
128 XEQ 04         178 STO 33
129 "VEC LIN VEL"  179 RCL 13
130 XEQ 09         180 X↑2
131 "--- --- ---"  181 RCL 10
132 XEQ 04         182 *
133 RCL 12         183 STO 34
134 90             184 RCL 24
135 +              185 X↑2
136 STO 26         186 RCL 17
137 XEQ 05         187 /
138 RCL 19         188 STO 35
139 XEQ 07         189 RCL 14
140 XEQ 08         190 RCL 10
141 "FPS∠"         191 *
142 ASTO 27        192 STO 36
143 ":B="          193 RCL 24
144 ASTO 28        194 RCL 17
145 XEQ 03         195 /
146 RCL 18         196 STO 37
147 XEQ 05         197 RCL 22
148 RCL 22         198 *
149 XEQ 07         199 2
150 XEQ 08         200 *
```

Sliding Coupler: Simplified Vector Method

```
201 STO 38          251 RCL 33
202 RCL 16          252 *
203 RCL 34          253 STO 44
204 *               254 RCL 31
205 CHS             255 RCL 32
206 STO 39          256 *
207 RCL 26          257 ST- 44
208 COS             258 RCL 30
209 STO 40          259 RCL 41
210 RCL 36          260 *
211 *               261 STO 45
212 ST+ 39          262 RCL 31
213 RCL 30          263 RCL 39
214 RCL 35          264 *
215 *               265 ST- 45
216 ST- 39          266 RCL 43
217 RCL 32          267 RCL 44
218 CHS             268 /
219 RCL 38          269 STO 46
220 *               270 RCL 45
221 ST- 39          271 RCL 44
222 RCL 15          272 /
223 RCL 34          273 STO 47
224 *               274 "NORM ACC"
225 CHS             275 XEQ 09
226 STO 41          276 "---- ---"
227 RCL 26          277 XEQ 04
228 SIN             278 RCL 12
229 STO 42          279 XEQ 05
230 RCL 36          280 RCL 34
231 *               281 CHS
232 ST+ 41          282 XEQ 07
233 RCL 31          283 XEQ 08
234 RCL 35          284 "FPS↑2<"
235 *               285 ASTO 48
236 ST- 41          286 ":B="
237 RCL 33          287 ASTO 49
238 CHS             288 XEQ 16
239 RCL 38          289 RCL 29
240 *               290 XEQ 05
241 ST- 41          291 0
242 RCL 39          292 XEQ 07
243 RCL 33          293 XEQ 08
244 *               294 ":C="
245 STO 43          295 ASTO 49
246 RCL 41          296 XEQ 16
247 RCL 32          297 RCL 18
248 *               298 XEQ 05
249 ST- 43          299 RCL 35
250 RCL 30          300 CHS
```

Sliding Coupler: Simplified Vector Method

```
301 XEQ 07
302 XEQ 08
303 ":C/B="
304 ASTO 49
305 XEQ 16
306 "TANG ACC"
307 XEQ 09
308 "---- ---"
309 XEQ 04
310 RCL 26
311 XEQ 05
312 RCL 36
313 XEQ 07
314 XEQ 08
315 ":B="
316 ASTO 49
317 XEQ 16
318 RCL 18
319 XEQ 05
320 RCL 46
321 XEQ 07
322 XEQ 08
323 ":C="
324 ASTO 49
325 XEQ 16
326 RCL 29
327 XEQ 05
328 RCL 47
329 XEQ 07
330 XEQ 08
331 ":C/B="
332 ASTO 49
333 XEQ 16
334 "CORR ACC"
335 XEQ 09
336 "---- ---"
337 XEQ 04
338 RCL 29
339 XEQ 05
340 RCL 38
341 XEQ 07
342 XEQ 08
343 ":C="
344 ASTO 49
345 XEQ 16
346 "ABS ACC"
347 XEQ 09
348 "--- ---"
349 XEQ 04
350 RCL 12
```

```
351 XEQ 05
352 RCL 59
353 STO 50
354 RCL 60
355 STO 51
356 RCL 34
357 CHS
358 ST* 50
359 ST* 51
360 RCL 26
361 XEQ 05
362 RCL 59
363 STO 52
364 RCL 60
365 STO 53
366 RCL 36
367 ST* 52
368 ST* 53
369 XEQ 06
370 1
371 XEQ 07
372 XEQ 08
373 ":B="
374 ASTO 49
375 XEQ 16
376 RCL 29
377 XEQ 05
378 RCL 59
379 STO 50
380 RCL 60
381 STO 51
382 RCL 21
383 RCL 23
384 *
385 RCL 34
386 *
387 RCL 10
388 *
389 2
390 *
391 RCL 17
392 /
393 ST* 50
394 ST* 51
395 RCL 18
396 XEQ 05
397 RCL 59
398 STO 52
399 RCL 60
400 STO 53
```

Sliding Coupler: Simplified Vector Method

```
401 RCL 46                          451 "AC="
402 ST* 52                          452 ARCL 11
403 ST* 53                          453 XEQ 04
404 XEQ 06                          454 "THETA A="
405 1                               455 ARCL 12
406 XEQ 07                          456 XEQ 04
407 XEQ 08                          457 "OMEGA A="
408 ":C="                           458 ARCL 13
409 ASTO 49                         459 XEQ 04
410 XEQ 16                          460 "ALPHA A="
411 RCL 18                          461 ARCL 14
412 XEQ 05                          462 XEQ 04
413 RCL 59                          463 "PHI C="
414 STO 50                          464 ARCL 18
415 RCL 60                          465 XEQ 04
416 STO 51                          466 "A 1="
417 RCL 23                          467 ARCL 30
418 X↑2                             468 XEQ 04
419 RCL 34                          469 "A 2="
420 *                               470 ARCL 31
421 RCL 10                          471 XEQ 04
422 *                               472 "B 1="
423 CHS                             473 ARCL 32
424 RCL 17                          474 XEQ 04
425 /                               475 "B 2="
426 ST* 50                          476 ARCL 33
427 ST* 51                          477 XEQ 04
428 RCL 29                          478 "C 1="
429 XEQ 05                          479 ARCL 39
430 RCL 59                          480 XEQ 04
431 STO 52                          481 "C 2="
432 RCL 60                          482 ARCL 41
433 STO 53                          483 XEQ 04
434 RCL 47                          484 "ATC="
435 ST* 52                          485 ARCL 46
436 ST* 53                          486 XEQ 04
437 XEQ 06                          487 "ATBC="
438 1                               488 ARCL 47
439 XEQ 07                          489 XEQ 04
440 XEQ 08                          490 XEQ 13
441 ":C/B="                         491+LBL 05
442 ASTO 49                         492 57.296
443 XEQ 16                          493 /
444 "PROBLEM DATA"                  494 0
445 XEQ 09                          495 XROM "e↑Z"
446 "------- ----"                  496 STO 59
447 XEQ 04                          497 X<>Y
448 "AB="                           498 STO 60
449 ARCL 10                         499 RTN
450 XEQ 04                          500+LBL 08
```

Sliding Coupler: Simplified Vector Method

```
501 RCL 59              551 RTN
502 X<0?                552◆LBL 09
503 GTO 14              553 AVIEW
504 RTN                 554 RTN
505◆LBL 14              555◆LBL 03
506 RCL 60              556 FIX 2
507 X<0?                557 26
508 GTO 15              558 ACCHR
509 180                 559 CLA
510 ST+ 62              560 ARCL 28
511 RTN                 561 ARCL 61
512◆LBL 15              562 ARCL 27
513 180                 563 ARCL 62
514 ST- 62              564 ACA
515 RTN                 565 PRBUF
516◆LBL 06              566 ADV
517 RCL 51              567 FIX 4
518 RCL 50              568 RTN
519 RCL 53              569◆LBL 16
520 RCL 52              570 FIX 1
521 XROM "C+"           571 22
522 STO 59              572 ACCHR
523 X<>Y                573 CLA
524 STO 60              574 ARCL 49
525 RTN                 575 ARCL 61
526◆LBL 07              576 ARCL 48
527 ST* 59              577 ARCL 62
528 ST* 60              578 ACA
529 RCL 60              579 PRBUF
530 RCL 59              580 ADV
531 XROM "MAGZ"         581 FIX 4
532 STO 61              582 RTN
533 RCL 59              583◆LBL 13
534 X=0?                584 AOFF
535 GTO 12              585 .END.
536 RCL 60
537 RCL 59
538 /
539 ATAN
540 STO 62
541 RTN
542◆LBL 12
543 0
544 ATAN
545 STO 62
546 RTN
547◆LBL 04
548 AVIEW
549 ADV
550 CLA
```

Sliding Coupler: Simplified Vector Method

SLIDING COUPLER   PROBLEM DATA
------- -------   ------- ----

LIN VELOCITY   AB=0.6667
--- --------
                                         AC=0.8333
V:B=12.0000FPS
                                         THETA A=120.0000
V:C=-6.6530FPS
                                         OMEGA A=18.0000
V:C/B=9.9869FPS
                                         ALPHA A=0.0000
VEC LIN VEL
--- --- ---   PHI C=-26.3295

ū:B=12.00FPS∠-150.00   A 1=0.8963

ū:C=6.65FPS∠153.67   A 2=-0.4435

ū:C/B=9.99FPS∠63.67   B 1=-0.4435

NORM ACC   B 2=-0.8963
---- ---
                                         C 1=84.6063
ā:B=216.0FPS↑2∠-60.0
                                         C 2=-61.5830
ā:C=0.0FPS↑2∠0.0
                                         ATC=103.1432
ā:C/B=76.6FPS↑2∠153.7
                                         ATBC=17.6686
TANG ACC
---- ---

ā:B=0.0FPS↑2∠0.0

ā:C=103.1FPS↑2∠-26.3

ā:C/B=17.7FPS↑2∠63.7

CORR ACC
---- ---

ā:C=102.1FPS↑2∠-116.3

ABS ACC
--- ---

ā:B=216.0FPS↑2∠-60.0

ā:C=145.1FPS↑2∠-71.0

ā:C/B=78.6FPS↑2∠140.7

Sliding Coupler: Simplified Vector Method

SLIDING COUPLER
------- -------

LIN VELOCITY
--- --------

V:B=7.5000FPS

V:C=1.5507FPS

V:C/B=7.3379FPS

VEC LIN VEL
--- --- ---

ū:B=7.50FPS∠-60.00

ū:C=1.55FPS∠18.07

ū:C/B=7.34FPS∠108.07

NORM ACC
---- ---

ā:B=225.0FPS↑2∠30.0

ā:C=0.0FPS↑2∠0.0

ā:C/B=133.6FPS↑2∠-161.9

TANG ACC
---- ---

ā:B=0.0FPS↑2∠0.0

ā:C=86.5FPS↑2∠18.1

ā:C/B=9.9FPS↑2∠108.1

CORR ACC
---- ---

ā:C=56.5FPS↑2∠108.1

ABS ACC
--- ---

ā:B=225.0FPS↑2∠30.0

ā:C=103.3FPS↑2∠51.2

ā:C/B=134.0FPS↑2∠-166.2

PROBLEM DATA
------- ----

AB=0.2500

AC=0.1667

THETA A=210.0000

OMEGA A=30.0000

ALPHA A=0.0000

PHI C=18.0675

A 1=0.9507

A 2=0.3101

B 1=0.3101

B 2=-0.9507

C 1=85.3589

C 2=17.3868

ATC=86.5423

ATBC=9.9436

Slider-Crank: Modified Vector Method

```
01◆LBL "SL-CR C"          51 RCL 18
02 " SL-CR ANAL CL"       52 *
03 XEQ 09                 53 RCL 17
04 " ——— ——— ——"          54 *
05 XEQ 04                 55 CHS
06 FIX 4                  56 STO 20
07 SF 04                  57 90
08 "AB?"                  58 RCL 13
09 PROMPT                 59 -
10 STO 10                 60 SIN
11 "BC?"                  61 STO 21
12 PROMPT                 62 90
13 STO 11                 63 RCL 15
14 "OMEGA?"               64 -
15 PROMPT                 65 SIN
16 STO 12                 66 STO 22
17 "THETA A?"             67 RCL 21
18 PROMPT                 68 RCL 22
19 STO 13                 69 /
20 SIN                    70 RCL 17
21 STO 14                 71 *
22 "ALPHA A?"             72 CHS
23 PROMPT                 73 STO 23
24 STO 09                 74 "IN/SEC"
25 RCL 14                 75 ASTO 24
26 RCL 10                 76 "LINEAR VELOCITY"
27 *                      77 XEQ 09
28 RCL 11                 78 "——— ———"
29 /                      79 XEQ 04
30 ASIN                   80 "V:B="
31 STO 15                 81 ARCL 17
32 CHS                    82 ARCL 24
33 180                    83 XEQ 04
34 +                      84 "V:C="
35 STO 16                 85 ARCL 20
36 RCL 10                 86 ARCL 24
37 RCL 12                 87 XEQ 04
38 *                      88 "V:B/C="
39 STO 17                 89 ARCL 23
40 RCL 15                 90 ARCL 24
41 RCL 13                 91 XEQ 04
42 +                      92 "VEC LIN VEL"
43 SIN                    93 XEQ 09
44 STO 18                 94 "——— ——— ——"
45 90                     95 XEQ 04
46 RCL 15                 96 RCL 13
47 -                      97 XEQ 05
48 SIN                    98 RCL 60
49 STO 19                 99 STO 28
50 1/X                    100 RCL 59
```

Slider-Crank: Modified Vector Method

```
101 STO 27
102 RCL 17
103 CHS
104 XEQ 07
105 XEQ 08
106 "IPS∠"
107 ASTO 25
108 ":B="
109 ASTO 26
110 XEQ 03
111 0
112 XEQ 16
113 RCL 59
114 STO 29
115 RCL 60
116 STO 30
117 RCL 20
118 XEQ 07
119 XEQ 08
120 ":C="
121 ASTO 26
122 XEQ 03
123 RCL 17
124 CHS
125 ST* 27
126 ST* 28
127 RCL 20
128 ST* 29
129 ST* 30
130 XEQ 06
131 1
132 XEQ 07
133 XEQ 08
134 ":B/C="
135 ASTO 26
136 XEQ 03
137 -5
138 STO 49
139 "NORM ACC"
140 XEQ 09
141 "--- ---"
142 XEQ 04
143◆LBL 18
144 RCL 13
145 XEQ 16
146 RCL 59
147 STO 27
148 RCL 60
149 STO 28
150 RCL 10
```

```
151 RCL 12
152 X↑2
153 *
154 STO 31
155 CHS
156 ST* 27
157 ST* 28
158 RCL 13
159 XEQ 05
160 RCL 59
161 STO 29
162 RCL 60
163 STO 30
164 RCL 10
165 RCL 09
166 *
167 STO 32
168 CHS
169 ST* 30
170 ST* 29
171 XEQ 06
172 1
173 XEQ 07
174 XEQ 08
175 "IPS↑2∠"
176 ASTO 33
177 ":B="
178 ASTO 34
179 XEQ 17
180 RCL 49
181 X>0?
182 GTO 20
183 0
184 STO 61
185 STO 62
186 ":C="
187 ASTO 34
188 XEQ 17
189 "TAN ACC"
190 XEQ 09
191 "--- ---"
192 XEQ 04
193 0
194 STO 61
195 STO 62
196 ":B="
197 ASTO 34
198 XEQ 17
199◆LBL 19
200 RCL 13
```

Slider-Crank: Modified Vector Method

```
201 COS
202 STO 35
203 LASTX
204 SIN
205 STO 36
206 X↑2
207 CHS
208 RCL 35
209 X↑2
210 +
211 STO 37
212 RCL 10
213 *
214 STO 38
215 RCL 36
216 X↑2
217 RCL 10
218 X↑2
219 *
220 CHS
221 RCL 11
222 X↑2
223 +
224 STO 39
225 SQRT
226 STO 47
227 1/X
228 RCL 38
229 *
230 STO 40
231 RCL 35
232 +
233 STO 41
234 3
235 ENTER↑
236 2
237 /
238 STO 42
239 RCL 39
240 RCL 42
241 Y↑X
242 STO 43
243 RCL 10
244 ENTER↑
245 3
246 Y↑X
247 RCL 35
248 X↑2
249 *
250 RCL 36
```

```
251 X↑2
252 *
253 STO 44
254 RCL 43
255 /
256 STO 45
257 ST+ 41
258 RCL 41
259 RCL 31
260 CHS
261 *
262 STO 27
263 0
264 STO 28
265 RCL 10
266 RCL 35
267 *
268 RCL 36
269 *
270 STO 46
271 RCL 47
272 /
273 STO 48
274 RCL 36
275 +
276 RCL 32
277 CHS
278 *
279 STO 29
280 0
281 STO 30
282 XEQ 06
283 1
284 XEQ 07
285 XEQ 08
286 ":C="
287 ASTO 34
288 XEQ 17
289 RCL 49
290 X>0?
291 GTO 21
292 "ABS ACC"
293 XEQ 09
294 "—— ——"
295 XEQ 04
296 10
297 STO 49
298 XEQ 18
299◆LBL 20
300 XEQ 19
```

Slider-Crank: Modified Vector Method

```
301◆LBL 21              351 ARCL 09
302 RCL 38             352 XEQ 04
303 RCL 47             353 "PHI C="
304 /                  354 ARCL 15
305 STO 50             355 XEQ 04
306 RCL 44             356 90
307 RCL 43             357 RCL 15
308 /                  358 -
309 ST+ 50             359 STO 51
310 RCL 31             360 RCL 13
311 ST* 50             361 RCL 15
312 RCL 50             362 +
313 STO 27             363 STO 52
314 RCL 31             364 90
315 RCL 36             365 RCL 13
316 *                  366 -
317 CHS                367 STO 53
318 STO 28             368 "SWB="
319 RCL 48             369 ARCL 51
320 RCL 32             370 XEQ 04
321 *                  371 "SWC="
322 STO 29             372 ARCL 52
323 RCL 35             373 XEQ 04
324 RCL 32             374 "SWBC="
325 *                  375 ARCL 53
326 STO 30             376 XEQ 04
327 XEQ 01             377 GTO 13
328 1                  378◆LBL 05
329 XEQ 07             379 57.296
330 XEQ 08             380 /
331 "-:B/C="           381 0
332 ASTO 34            382 XROM "etZ"
333 XEQ 17             383 CHS
334 "PROBLEM DATA"     384 STO 60
335 XEQ 09             385 X<>Y
336 "------- ---"      386 STO 59
337 XEQ 04             387 RTN
338 "AB="             388◆LBL 08
339 ARCL 10            389 RCL 59
340 XEQ 04             390 X<0?
341 "BC="             391 GTO 14
342 ARCL 11            392 RTN
343 XEQ 04             393◆LBL 14
344 "THETA A="         394 RCL 60
345 ARCL 13            395 X<0?
346 XEQ 04             396 GTO 15
347 "OMEGA A="         397 180
348 ARCL 12            398 ST+ 62
349 XEQ 04             399 RTN
350 "ALPHA A=" '       400◆LBL 15
```

Slider-Crank: Modified Vector Method

```
401 100            451 RTN
402 ST- 62         452•LBL 01
403 RTN            453 RCL 28
404•LBL 06         454 RCL 27
405 RCL 28         455 RCL 30
406 RCL 27         456 RCL 29
407 RCL 30         457 XROM "C+"
408 RCL 29         458 STO 59
409 XROM "C-"      459 X<>Y
410 STO 59         460 STO 60
411 X<>Y           461 RTN
412 STO 60         462•LBL 03
413 RTN            463 FIX 2
414•LBL 07         464 26
415 ST* 59         465 ACCHR
416 ST* 60         466 CLA
417 RCL 60         467 ARCL 26
418 RCL 59         468 ARCL 61
419 XROM "MAGZ"    469 ARCL 25
420 STO 61         470 ARCL 62
421 RCL 59         471 ACA
422 X=0?           472 PRBUF
423 GTO 12         473 ADV
424 RCL 60         474 FIX 4
425 RCL 59         475 RTN
426 /              476•LBL 17
427 ATAN           477 FIX 2
428 STO 62         478 22
429 RTN            479 ACCHR
430•LBL 12         480 CLA
431 0              481 ARCL 34
432 ATAN           482 ARCL 61
433 STO 62         483 ARCL 33
434 RTN            484 ARCL 62
435•LBL 04         485 ACA
436 AVIEW          486 PRBUF
437 ADV            487 ADV
438 CLA            488 FIX 4
439 RTN            489 RTN
440•LBL 09         490•LBL 13
441 AVIEW          491 AOFF
442 RTN            492 STOP
443•LBL 16         493 END
444 57.296
445 /
446 0
447 XROM "etZ"
448 STO 59
449 X<>Y
450 STO 60
```

Slider-Crank: Modified Vector Method

SL-CR ANAL CL

—— —— —

LINEAR VELOCITY

—— ———

V:B=1.5000IN/SEC

V:C=-0.9387IN/SEC

V:B/C=0.8321IN/SEC

VEC LIN VEL

— —— ——

ū:B=1.50IPS∠-150.00

ū:C=0.94IPS∠180.00

ū:B/C=0.83IPS∠-115.66

NORM ACC

— ——

ā:B=1.50IPS↑2∠-60.00

ā:C=0.00IPS↑2∠0.00

TAN ACC

— ——

ā:B=0.00IPS↑2∠0.00

ā:C=1.12IPS↑2∠0.00

ABS ACC

— ——

ā:B=1.50IPS↑2∠-60.00

ā:C=1.12IPS↑2∠0.00

ā:B/C=1.35IPS↑2∠-105.82

PROBLEM DATA

—— ——

AB=1.5000

BC=3.0000

THETA A=120.0000

OMEGA A=1.0000

ALPHA A=0.0000

PHI C=25.6589

SNB=64.3411

SNC=145.6589

SNBC=-30.0000

Slider-Crank: Modified Vector Method

PROBLEM DATA
‾‾‾‾‾‾‾ ‾‾‾‾

AB=1.5000

BC=3.0000

THETA A=30.0000

OMEGA A=-1.0000

ALPHA A=0.0000

PHI C=14.4775

SWB=75.5225

SWC=44.4775

SWBC=60.0000

SL-CR ANAL CL
‾‾‾‾ ‾‾‾ ‾‾

LINEAR VELOCITY
‾‾‾‾‾‾ ‾‾‾‾‾‾‾‾

V:B=-1.5000IN/SEC

V:C=1.0854IN/SEC

V:B/C=1.3416IN/SEC

VEC LIN VEL
‾‾‾ ‾‾‾ ‾‾‾

ū:B=1.50IPS∠-60.00

ū:C=1.09IPS∠0.00

ū:B/C=1.34IPS∠-104.48

NORM ACC
‾‾‾ ‾‾‾

ā:B=1.50IPS↑2∠-150.00

ā:C=0.00IPS↑2∠0.00

TAN ACC
‾‾ ‾‾‾

ā:B=0.00IPS↑2∠0.00

ā:C=1.73IPS↑2∠180.00

ABS ACC
‾‾ ‾‾

ā:B=1.50IPS↑2∠-150.00

ā:C=1.73IPS↑2∠180.00

ā:B/C=0.86IPS↑2∠-60.40

A.3   GEARS: BASIC PROGRAM

The program presented in this section will calculate design layout and general analytical data for an external gear pair (or gear and pinion) assembly, for three standard involute tooth systems, namely:

- 14.5 deg. full depth system
- 20 deg. full depth system
- 20 deg. stub system

The program, for example, will calculate layout data such as tooth numbers and tooth dimensions, plus values for parameters such as angles of approach and recess, path of contact, and contact ratios.

Written in BASIC computer language, the program is developed to run on an IBM or IBM-compatible personal computer, and is interactive, requiring the following as input values:

- diametral pitch
- pressure angle
- any two of the following
  - gear 1 tooth number
  - gear 2 tooth number
  - speed ratio

By varying the above input values, one can achieve any desired design using this program. Output data for the program is stored in an output file, "GEAROUT", and are retrievable after each run, by employing the command "PRINT GEAROUT" at the DOS prompt. The output data for three typical design cases are listed as examples.

Gear and Pinion Design/Analysis Program

```
4000    OPEN "GEAROUT" FOR OUTPUT AS #1
4010    PRINT #1,"GEAR AND PINION DESIGN/ANALYSIS"
4200    INPUT "WHAT IS PD";PD
4400    INPUT "PRESSURE ANGLE-DEGREES(14.5/20)";E$
4500    IF E$="14.5" THEN 4700
4600    IF E$="20" THEN 5400
4700    THETAD=14.5
5000    A=1/PD
5010    B=1.57/PD
5020    HK=2/PD
5030    HT=2.157/PD
5040    C=.157/PD
5050    T=1.57/PD
5060    RF=.209/PD
5170    GOTO 6000
5400    THETAD=20
5410    INPUT "TYPE-STANDARD OR STUB(STD/STB)";D$
5700    IF D$="STD" THEN A=1/PD
5710    IF D$="STD" THEN B=1.157/PD
5720    IF D$="STD" THEN HK=2/PD
5730    IF D$="STD" THEN HT=2.157/PD
5740    IF D$="STD" THEN C=.157/PD
5750    IF D$="STD" THEN T=1.5708/PD
5760    IF D$="STD" THEN RF=.235/PD
5800    IF D$="STB" THEN A=.8/PD
5810    IF D$="STB" THEN B=1/PD
5820    IF D$="STB" THEN HK=1.6/PD
5830    IF D$="STB" THEN HT=1.8/PD
5840    IF D$="STB" THEN C=.2/PD
5850    IF D$="STB" THEN T=1.5708/PD
5860    IF D$="STB" THEN RF=.2/PD
6000    PRINT #1,
8007    PRINT #1,"   PROBLEM DATA"
8009    PRINT #1,"PRESSURE ANGLE=",THETAD,D$
8010    PRINT #1,"DIAMETRAL PITCH(PD)=",PD
8011    INPUT "ARE T1 AND T2 GIVEN (Y/N)";A$
8012    IF A$="Y" THEN 8019
8014    INPUT "ARE T1 AND SR GIVEN (Y/N)";B$
8016    IF B$="Y" THEN 8027
8017    INPUT "ARE T2 AND SR GIVEN (Y/N)";C$
8018    IF C$="Y" THEN 8034
8019    INPUT " WHAT IS T1 ?";T1
```

```
8020    INPUT " WHAT IS T2 ?";T2
8021    SR=T1/T2
8022    PRINT #1,"GEAR1 TOOTH NUMBER(T1)=",T1
8023    PRINT #1,"GEAR2 TOOTH NUMBER(T2)=",T2
8024    PRINT #1,"SPEED2/SPEED1(SR)=       ",SR
8025    GOTO 8040
8026    PRINT #1,
8027    INPUT " WHAT IS T1 ?";T1
8028    INPUT " WHAT IS SR ?";SR
8029    T2=T1/SR
8030    PRINT #1,"GEAR1 TOOTH NUMBER(T1)=",T1
8031    PRINT #1,"SPEED2/SPEED1(SR)          ",SR
8032    PRINT #1,"GEAR2 TOOTH NUMBER(T2)=",T2
8033    GOTO 8040
8034    INPUT " WHAT IS T2 ?";T2
8035    INPUT " WHAT IS SR ?";SR
8036    T1=T2*SR
8037    PRINT #1,"GEAR2 TOOTH NUMBER(T2)=",T2
8038    PRINT #1,"SPEED2/SPEED1(SR)=       ",SR
8039    PRINT #1,"GEAR1 TOOTH NUMBER(T1)=",T1
8040    PRINT #1,
8041    PRINT #1,"  GEAR TOOTH DATA"
8044    PRINT #1,"ADDENDUM (A)=       ",A
8045    PRINT #1,"DEDENDUM (B)=       ",B
8046    PRINT #1,"WORKING DEPTH (HK)=",HK
8047    PRINT #1,"WHOLE DEPTH (HT)=",HT
8048    PRINT #1,"CLEARANCE (C)=",C
8049    PRINT #1,"TOOTH THICKNESS (T)=",T
8050    PRINT #1,"FILLET RADIUS(RF)=",RF
8051    PRINT #1,
8070    LET PI=3.141
8075    LET THETAR=THETAD*PI/180
8077    PRINT #1,"  GENERATED DATA"
8080    LET PC=PI/PD
8081    PRINT #1,"CIRCULAR PITCH=",PC
8085    LET PB=PC*COS(THETAR)
8086    PRINT #1,"BASE PITCH=       ",PB
8090    LET A1=A
8100    LET A2=A1
8120    LET B1=B
8122    LET B2=B1
8130    LET DP1=T1/PD
```

```
8131    PRINT #1,"GEAR1 PITCH DIAMETER=",DP1
8140    LET DP2=T2/PD
8141    PRINT #1,"GEAR2 PITCH DIAMETER=",DP2
8150    LET RP1=DP1/2
8151    PRINT #1,"GEAR1 PITCH RADIUS=",RP1
8160    LET RP2=DP2/2
8161    PRINT #1,"GEAR2 PITCH RADIUS=",RP2
8170    LET RA1=RP1+A1
8171    PRINT #1,"GEAR1 ADDENDUM RADIUS=",RA1
8172    LET RA2=RP2+A2
8173    PRINT #1,"GEAR2 ADDENDUM RADIUS=",RA2
8190    LET RD1=RP1-B1
8191    PRINT #1,"GEAR1 DEDENDUM RADIUS=",RD1
8200    LET RD2=RP2-B2
8201    PRINT #1,"GEAR2 DEDENDUM RADIUS=",RD2
8210    LET RB1=RP1*COS(THETAR)
8231    PRINT #1,"GEAR1 BASE RADIUS=",RB1
8240    LET RB2=RP2*COS(THETAR)
8241    PRINT #1,"GEAR2 BASE RADIUS=",RB2
8242    LET CD=(T1+T2)/(2*PD)
8243    PRINT #1,"CENTER DISTANCE= ",CD
8250    PA=SQR(RA1^2-RB1^2)-RB1*TAN(THETAR)
8251    PRINT #1,"PATH OF APPROACH=",PA
8260    LET BP=SQR(RA2^2-RB2^2)-RB2*TAN(THETAR)
8261    PRINT #1,"PATH OF RECESS=",BP
8270    LET BA=PA+BP
8271    PRINT #1,"PATH OF CONTACT=",BA
8272    LET CR=(PA+BP)/PB
8273    PRINT #1,"CONTACT RATIO=",CR
8280    LET ALPHA1=(PA/RB1)*180/PI
8281    PRINT #1,"ANGLE OF APPROACH, GEAR1=",ALPHA1
8290    LET BETA1=(BP/RB1)*180/PI
8291    PRINT #1,"ANGLE OF RECESS, GEAR1=", BETA1
8300    LET GAMMA1=ALPHA1+BETA1
8301    PRINT #1,"ANGLE OF ACTION, GEAR1=",GAMMA1
8380    LET ALPHA2=(PA/RB2)*180/PI
8381    PRINT #1,"ANGLE OF APPROACH, GEAR2=",ALPHA2
8390    LET BETA2=(BP/RB2)*180/PI
8391    PRINT #1,"ANGLE OF RECESS, GEAR2=", BETA2
8400    LET GAMMA2=ALPHA2+BETA2
8401    PRINT #1,"ANGLE OF ACTION, GEAR2=",GAMMA2
```

Gear and Pinion Design/Analysis

```
    PROBLEM DATA
PRESSURE ANGLE=                    14.5
DIAMETRAL PITCH(PD)=               2
GEAR1 TOOTH NUMBER(T1)=            20
GEAR2 TOOTH NUMBER(T2)=            25
SPEED2/SPEED1(SR)=                .8

    GEAR TOOTH DATA
ADDENDUM (A)=                     .5
DEDENDUM (B)=                     .785
WORKING DEPTH (HK)=               1
WHOLE DEPTH (HT)=                 1.0785
CLEARANCE (C)=                    .0785
TOOTH THICKNESS (T)=              .785
FILLET RADIUS(RF)=                .1045

    GENERATED DATA
CIRCULAR PITCH=                   1.5705
BASE PITCH=                       1.520495
GEAR1 PITCH DIAMETER=             10
GEAR2 PITCH DIAMETER=             12.5
GEAR1 PITCH RADIUS=               5
GEAR2 PITCH RADIUS=               6.25
GEAR1 ADDENDUM RADIUS=            5.5
GEAR2 ADDENDUM RADIUS=            6.75
GEAR1 DEDENDUM RADIUS=            4.215
GEAR2 DEDENDUM RADIUS=            5.465
GEAR1 BASE RADIUS=                4.840799
GEAR2 BASE RADIUS=                6.050998
CENTER DISTANCE=                  11.25
PATH OF APPROACH=                 1.359207
PATH OF RECESS=                   1.426722
PATH OF CONTACT=                  2.785929
CONTACT RATIO=                    1.832252
ANGLE OF APPROACH, GEAR1=         16.09063
ANGLE OF RECESS, GEAR1=           16.8899
ANGLE OF ACTION, GEAR1=           32.98053
ANGLE OF APPROACH, GEAR2=         12.87251
ANGLE OF RECESS, GEAR2=           13.51192
ANGLE OF ACTION, GEAR2=           26.38442
```

Gear and Pinion Design/Analysis

```
    PROBLEM DATA
PRESSURE ANGLE=                 20              STD
DIAMETRAL PITCH(PD)=            2
GEAR1 TOOTH NUMBER(T1)=         20
SPEED2/SPEED1(SR)               .8
GEAR2 TOOTH NUMBER(T2)=         25

    GEAR TOOTH DATA
ADDENDUM (A)=                   .5
DEDENDUM (B)=                   .5785
WORKING DEPTH (HK)=             1
WHOLE DEPTH (HT)=               1.0785
CLEARANCE (C)=                  .0785
TOOTH THICKNESS (T)=            .7854
FILLET RADIUS(RF)=              .1175

    GENERATED DATA
CIRCULAR PITCH=                 1.5705
BASE PITCH=                     1.475823
GEAR1 PITCH DIAMETER=           10
GEAR2 PITCH DIAMETER=           12.5
GEAR1 PITCH RADIUS=             5
GEAR2 PITCH RADIUS=             6.25
GEAR1 ADDENDUM RADIUS=          5.5
GEAR2 ADDENDUM RADIUS=          6.75
GEAR1 DEDENDUM RADIUS=          4.4215
GEAR2 DEDENDUM RADIUS=          5.6715
GEAR1 BASE RADIUS=              4.698576
GEAR2 BASE RADIUS=              5.87322
CENTER DISTANCE=                11.25
PATH OF APPROACH=               1.149122
PATH OF RECESS=                 1.189589
PATH OF CONTACT=                2.338711
CONTACT RATIO=                  1.584683
ANGLE OF APPROACH, GEAR1=       14.01536
ANGLE OF RECESS, GEAR1=         14.50893
ANGLE OF ACTION, GEAR1=         28.52429
ANGLE OF APPROACH, GEAR2=       11.21229
ANGLE OF RECESS, GEAR2=         11.60714
ANGLE OF ACTION, GEAR2=         22.81943
```

Gear and Pinion Design/Analysis

```
    PROBLEM DATA
PRESSURE ANGLE=                    20              STB
DIAMETRAL PITCH(PD)=               2
GEAR2 TOOTH NUMBER(T2)=            25
SPEED2/SPEED1(SR)=                 .8
GEAR1 TOOTH NUMBER(T1)=            20

    GEAR TOOTH DATA
ADDENDUM  (A)=                     .4
DEDENDUM  (B)=                     .5
WORKING DEPTH  (HK)=               .8
WHOLE DEPTH  (HT)=                 .9
CLEARANCE  (C)=                    .1
TOOTH THICKNESS  (T)=              .7854
FILLET RADIUS(RF)=                 .1

    GENERATED DATA
CIRCULAR PITCH=                    1.5705
BASE PITCH=                        1.475823
GEAR1 PITCH DIAMETER=              10
GEAR2 PITCH DIAMETER=              12.5
GEAR1 PITCH RADIUS=                5
GEAR2 PITCH RADIUS=                6.25
GEAR1 ADDENDUM RADIUS=             5.4
GEAR2 ADDENDUM RADIUS=             6.65
GEAR1 DEDENDUM RADIUS=             4.5
GEAR2 DEDENDUM RADIUS=             5.75
GEAR1 BASE RADIUS=                 4.698576
GEAR2 BASE RADIUS=                 5.87322
CENTER DISTANCE=                   11.25
PATH OF APPROACH=                  .9516714
PATH OF RECESS=                    .9817007
PATH OF CONTACT=                   1.933372
CONTACT RATIO=                     1.31003
ANGLE OF APPROACH, GEAR1=          11.60714
ANGLE OF RECESS, GEAR1=            11.9734
ANGLE OF ACTION, GEAR1=            23.58054
ANGLE OF APPROACH, GEAR2=          9.285714
ANGLE OF RECESS, GEAR2=            9.578718
ANGLE OF ACTION, GEAR2=            18.86443
```

Table A.1   Spur Gear Relationships*

Gear 1	G_1 = Input Gear
Gear 2	G_2 = Output Gear
Pressure Angle	$\theta = \cos^{-1} \dfrac{R_{B1}}{R_{P1}}$ or $\cos^{-1} \dfrac{R_{B2}}{R_{P2}}$
Speed Ratio or $\dfrac{\text{Speed of } G_2}{\text{Speed of } G_2}$	$SR = \dfrac{T_1}{T_2}$
Gear 1 Tooth Number	$T_1 = T_2 \times SR$
Gear 2 Tooth Number	$T_2 = \dfrac{T_1}{SR}$
Diametral Pitch	$P_D = \dfrac{T_1}{D_{P1}}$ or $\dfrac{T_2}{D_{P2}}$
Circular Pitch	$P_C = \dfrac{\pi}{P_D}$
Base Pitch	$P_B = P_C \cos \theta$
Gear 1 Addendum	A_1 = Defined by θ. See Table, 20.2.
Gear 2 Addendum	A_2 = Defined by θ. See Table, 20.2.
Gear 1 Dedendum	B_1 = Defined by θ. See Table, 20.2.
Gear 2 Dedendum	B_2 = Defined by θ. See Table, 20.2.
Gear 1 Pitch Diameter	$D_{P1} = \dfrac{T_1}{P_{D1}}$
Gear 2 Pitch Diameter	$D_{P2} = \dfrac{T_2}{P_{D2}}$
Gear 1 Pitch Radius	$R_{P1} = \dfrac{T_1}{2P_D} = \dfrac{D_{P1}}{2}$
Gear 2 Pitch Radius	$R_{P2} = \dfrac{T_2}{2P_D} = \dfrac{D_{P2}}{2}$
Center Distance	$C = \dfrac{T_1 + T_2}{2P_D}$
Gear 1 Addendum Radius	$R_{A1} = R_{P1} + A_1$
Gear 2 Addendum Radius	$R_{A2} = R_{P2} + A_2$
Gear 1 Dedendum Radius	$R_{D1} = R_{P1} - B_1$
Gear 2 Dedendum Radius	$R_{D2} = R_{P2} - B_2$
Gear 1 Base Radius	$R_{B1} = R_{P1} \cos \theta$
Gear 2 Base Radius	$R_{B2} = R_{P2} \cos \theta$
Path of Approach	$PA = \sqrt{R_{A1}^2 - R_{B1}^2} - R_{B1} \tan \theta$
Path of Recess	$BP = \sqrt{R_{A2}^2 - R_{B2}^2} - R_{B2} \tan \theta$
Path of Contact	$BA = PA + BP$
Contact Ratio	$CR = \dfrac{PA + BP}{P_B}$

Gear 1 Angle of Approach	$\alpha_1 = \dfrac{PA}{R_{B1}}$	
Gear 1 Angle of Recess	$\beta_1 = \dfrac{BP}{R_{B1}}$	
Gear 1 Angle of Action	$\gamma_1 = \alpha_1 + \beta_1$	
Gear 2 Angle of Approach	$\alpha_2 = \dfrac{PA}{R_{B2}}$	
Gear 2 Angle of Recess	$\beta_2 = \dfrac{BP}{R_{B2}}$	
Gear 2 Angle of Action	$\gamma_2 = \alpha_2 + \beta_2$	

*Table A.1 is organized in a programmed format so that a given variable is defined in terms of other variables previously defined.

Appendix B

B.1 NOMENCLATURE

Linkage Symbols

a	linear acceleration
$\bar{a}$	average linear acceleration
a_B	linear acceleration of point B
a, b, c, etc.	termini of velocity vectors $\overline{V}_A$, $\overline{V}_B$, $\overline{V}_C$, etc. on velocity polygon
a', b', c', etc.	termini of acceleration vectors $\overline{A}_A$, $\overline{A}_B$, $\overline{A}_C$, etc. on acceleration polygon
A	linear acceleration (magnitude)
$\overline{A}$	linear acceleration vector (magnitude and direction)
A_B	linear acceleration of point B
$A_{B/C}$	linear acceleration of point B relative to point C
A^{CD}	effective component of acceleration along CD
A^{Cor}	Coriolis acceleration
A^N	normal acceleration

A^r	rotational component of acceleration
A^t	translational component of acceleration
A^T	tangential acceleration
A, B, C, etc.	pivot points on a linkage
CCW	counterclockwise direction
CW	clockwise direction
e	eccentricity of a slider-crank mechanism
i	$\sqrt{-1}$
I	instant center
k_a	acceleration scale (actual acceleration represented by unit length of acceleration vector or acceleration axis of motion curve)
k_s	space scale (actual length of machine member or displacement represented by unit length of vector or displacement axis of motion curve)
k_t	time scale (actual time represented by unit length on time axis of motion curve)
k_v	velocity scale (actual velocity represented by unit length of velocity vector or velocity axis of motion curve)
n	number of links of a mechanism
N	number of instant centers; number of revolutions per minute
o	pole or origin of velocity polygon
o'	pole or origin of acceleration polygon
P	point of contact between two sliding bodies
$P(C)$	contact point P on C
$P(F)$	contact point P on F
$P(C)/P(F)$	point P on C relative to point P on F
R, r	radius
s	linear displacement
t	time

v	linear velocity
$\bar{v}$	average linear velocity, unit vector
v_1	initial linear velocity
v_2	final linear velocity
v_B	linear velocity of point B
V	linear velocity (magnitude)
$\overline{V}$	linear velocity vector (magnitude and direction)
V_B	linear velocity of point B
$V_{B/C}$	linear velocity of point B relative to point C
V^{CD}	effective component of velocity along CD
V^r	rotational component of velocity
V^t	translational component of velocity
V_A^i	initial velocity of point A
V_A^f	final velocity of point A initial velocity of point B
V_{B_i}	initial velocity of point B
V_{B_f}	final velocity of point B
α (alpha)	angular acceleration, or other angle
α_2	angular acceleration of link 2
α_{AB}	angular acceleration of link AB
β (beta)	angle
γ (gamma)	angle
ϕ (phi)	angle
θ (theta)	angular displacement
ω (omega)	angular velocity
ω_1	initial angular velocity
ω_2	angular velocity of link 2, final angular velocity
ω_{AB}	angular velocity of link AB
1, 2, 3, etc.	links 1, 2, 3, etc.

| 23 | instant center of links 2 and 3 |
| $\perp$ | perpendicular to |

Gear and Cam Symbols

A	addendum, planet arm, or carrier, linear acceleration
B	backlash, dedendum
C	center distance, clearance, planet carrier
CR	contact ratio
D_P	pitch diameter
D_O	outside diameter
D_R	root diameter
D_B	base circle diameter
F	follower, fillet
H_K	working depth
H_T	whole depth
L	lead, lift
M	module
N	tooth number, number of threads per lead
P	planet, pitch point
PA	pressure angle
P_D	diametral pitch
P_{DN}	diametral pitch in normal plane
P_C	circular pitch
P_{CN}	circular pitch in normal plane
P_B	base circle pitch
R	ring gear, radius
R_D	dedendum radius
R_P	pitch circle radius

R_B base circle radius

R_O outside radius

R_A addendum radius

R_F fillet radius

S   sun gear, linear displacement

SR   speed ratio

T   tooth number, tooth thickness

x   variable linear displacement

$\bar{x}$ maximum linear displacement

x_1 linear displacement to point 1

x_2 linear displacement to point 2

V   linear velocity

α angle of approach

β angle of recess, angular displacement

β_1 angular displacement to point 1

β_2 angular displacement to point 2

γ angle of action

λ lead angle

θ pressure angle, cam rotation angle, angular displacement

π 3.14159

ψ helix angle

ω angular velocity

ϕ any angle

B.2 TRIGONOMETRY REVIEW

Functions of a Right Triangle (Figure B.1)

$$\sin \alpha = \frac{\text{opposite}}{\text{hypotenuse}} = \frac{a}{c}$$

$$\cos \alpha = \frac{\text{adjacent}}{\text{hypotenuse}} = \frac{b}{c}$$

$$\tan \alpha = \frac{\text{opposite}}{\text{adjacent}} = \frac{a}{b} = \frac{\sin \alpha}{\cos \alpha}$$

$$\operatorname{cosec} \alpha = \frac{1}{\sin \alpha} = \frac{c}{a}$$

$$\sec \alpha = \frac{1}{\cos \alpha} = \frac{c}{b}$$

$$\cot \alpha = \frac{1}{\tan \alpha} = \frac{b}{a} = \frac{\cos \alpha}{\sin \alpha}$$

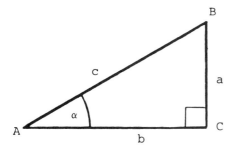

Figure B.1   Right triangle.

Functions of an Angle in the Interval $0° < \theta < 360°$

If a vector AB is rotated through the four quadrants as shown in Figure B.2, any function of the angle θ is numerically equal to the same function of the acute angle α between the terminal side of the vector and the x axis. That is,

Function of $\theta = \pm$ same function α

where the positive or negative sign depends on the quadrant in which the angle α falls. Signs are determined as shown in Figure B.2.

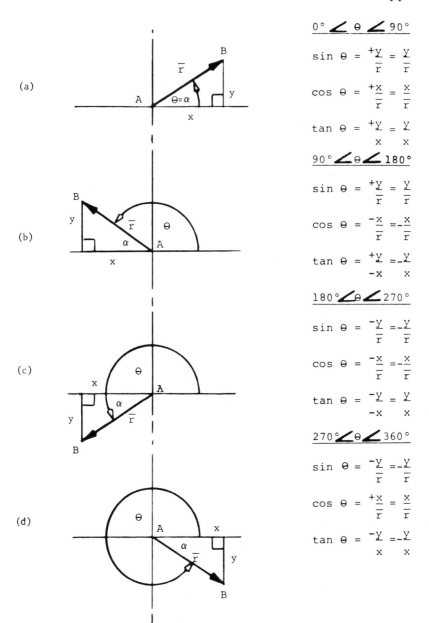

$$0° \angle \theta \angle 90°$$

$$\sin \theta = \frac{+y}{r} = \frac{y}{r}$$

$$\cos \theta = \frac{+x}{r} = \frac{x}{r}$$

$$\tan \theta = \frac{+y}{x} = \frac{y}{x}$$

$$90° \angle \theta \angle 180°$$

$$\sin \theta = \frac{+y}{r} = \frac{y}{r}$$

$$\cos \theta = \frac{-x}{r} = -\frac{x}{r}$$

$$\tan \theta = \frac{+y}{-x} = -\frac{y}{x}$$

$$180° \angle \theta \angle 270°$$

$$\sin \theta = \frac{-y}{r} = -\frac{y}{r}$$

$$\cos \theta = \frac{-x}{r} = -\frac{x}{r}$$

$$\tan \theta = \frac{-y}{-x} = \frac{y}{x}$$

$$270° \angle \theta \angle 360°$$

$$\sin \theta = \frac{-y}{r} = -\frac{y}{r}$$

$$\cos \theta = \frac{+x}{r} = \frac{x}{r}$$

$$\tan \theta = \frac{-y}{x} = -\frac{y}{x}$$

Figure B.2   Vector in rotation.

These signs of the functions of θ in the four quadrants may be summarized as follows:

$0° < θ < \ \ 90°$: $α = θ$ sin θ is positive
                                                cos θ is positive
                                                tan θ is positive

$90° < θ < 180°$: $α = 180° - θ$ sin θ is positive
                                                cos θ is negative
                                              tan θ is negative

$180° < θ < 270°$: $α = θ - 180°$ sin θ is negative
                                                cos θ is negative
                                              tan θ is positive

$270° < θ < 360°$: $α = 360° - θ$ sin θ is negative
                                                cos θ is positive
                                              tan θ is negative

Signs of the functions of θ are most conveniently remembered using Figure B.3.

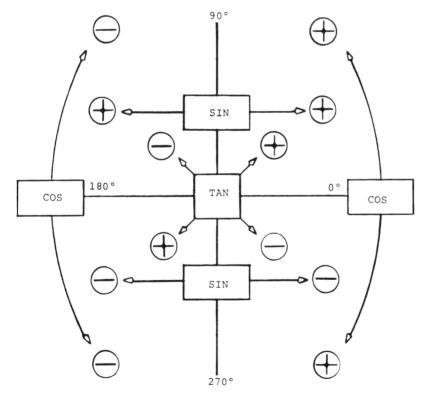

Figure B.3   Sign diagram.

Laws for Oblique Triangles

Law of Cosines

In any triangle, the square of any side is equal to the sum of the squares of the other sides minus twice their product times the cosine of their included angle. For example, in triangle ABC in Figure B.4,

$$a^2 = b^2 + c^2 - 2bc \cos \alpha$$

$$b^2 = a^2 + c^2 - 2ac \cos \beta$$

$$c^2 = a^2 + b^2 - 2ab \cos \gamma$$

where

$$\alpha + \beta + \gamma = 180°$$

Law of Sines

In any triangle, any two sides are proportional to the sides of the opposite angles. In triangle ABC in Figure B.4,

$$\frac{a}{\sin \alpha} = \frac{b}{\sin \beta} = \frac{c}{\sin \gamma}$$

Law of Tangents

In any triangle, the difference of the opposite angles divided by their sum equals the tangent of one-half the difference of the opposite angles

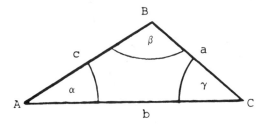

Figure B.4   Oblique triangle.

divided by the tangent of one-half their sum. In triangle ABC in Figure B.4,

$$\frac{a - b}{a + b} = \frac{\tan(1/2)(\alpha - \beta)}{\tan(1/2)(\alpha + \beta)}$$

$$\frac{a - c}{a + c} = \frac{\tan(1/2)(\alpha - \gamma)}{\tan(1/2)(\alpha + \gamma)}$$

$$\frac{b - c}{b + c} = \frac{\tan(1/2)(\beta - \gamma)}{\tan(1/2)(\beta + \gamma)}$$

Other Useful Relationships

$$\sin^2 \alpha + \cos^2 \alpha = 1$$

$$\sin(\alpha + \beta) = \sin \alpha \cos \beta + \cos \alpha \sin \beta$$

$$\sin(\alpha - \beta) = \sin \alpha \cos \beta - \cos \alpha \sin \beta$$

$$\cos(\alpha + \beta) = \cos \alpha \cos \beta - \sin \alpha \sin \beta$$

$$\cos(\alpha - \beta) = \cos \alpha \cos \beta + \sin \alpha \sin \beta$$

$$\tan(\alpha + \beta) = \frac{\tan \alpha + \tan \beta}{1 - \tan \alpha \tan \beta}$$

$$\tan(\alpha - \beta) = \frac{\tan \alpha - \tan \beta}{1 + \tan \alpha \tan \beta}$$

$$\sin 2\alpha = 2 \sin \alpha \cos \alpha$$

$$\cos 2\alpha = \cos^2 \alpha - \sin^2 \alpha$$

$$\tan 2\alpha = \frac{2 \tan \alpha}{1 - \tan^2 \alpha}$$

$$\sin \frac{\alpha}{2} = \sqrt{\frac{1 - \cos \alpha}{2}}$$

$$\cos \frac{\alpha}{2} = \sqrt{\frac{1 + \cos \alpha}{2}}$$

$$\tan \frac{\alpha}{2} = \sqrt{\frac{\sin \alpha}{1 + \cos \alpha}}$$

$$\sin \alpha + \sin \beta = 2 \sin \frac{\alpha + \beta}{2} \cos \frac{\alpha - \beta}{2}$$

$$\sin \alpha - \sin \beta = 2 \cos \frac{\alpha + \beta}{2} \sin \frac{\alpha - \beta}{2}$$

$$\cos \alpha + \cos \beta = 2 \cos \frac{\alpha + \beta}{2} \cos \frac{\alpha - \beta}{2}$$

$$\cos \alpha - \cos \beta = -2 \sin \frac{\alpha + \beta}{2} \sin \frac{\alpha - \beta}{2}$$

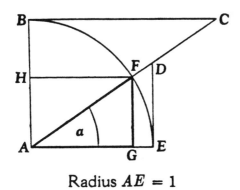

Radius $AE = 1$

$FG = \sin a$ $BC = \cot a$
$AG = \cos a$ $AD = \sec a$
$DE = \tan a$ $AC = \csc a$

$\sin a$		$\sqrt{1 - \cos^2 a}$	$\dfrac{\tan a}{\sqrt{1 + \tan^2 a}}$
$\cos a$	$\sqrt{1 - \sin^2 a}$		$\dfrac{1}{\sqrt{1 + \tan^2 a}}$
$\tan a$	$\dfrac{\sin a}{\sqrt{1 - \sin^2 a}}$	$\dfrac{\sqrt{1 - \cos^2 a}}{\cos a}$	

Figure B.5   Trigonometric formulas in terms of the unit circle.

$$\sin(90° - \theta) = +\cos \theta$$

$$\sin(90° + \theta) = +\cos \theta$$

$$\sin(180° - \theta) = +\sin \theta$$

$$\sin(180° + \theta) = -\sin \theta$$

$$\sin(270° - \theta) = -\cos \theta$$

$$\sin(270° + \theta) = -\cos \theta$$

$$\sin(360° - \theta) = -\sin \theta$$

$$\sin(360° + \theta) = +\sin \theta$$

$$\sin(-\theta) = -\sin \theta$$

$$\cos(90° - \theta) = +\sin \theta$$

$$\cos(90° + \theta) = -\sin \theta$$

$$\cos(180° - \theta) = -\cos \theta$$

$$\cos(180° + \theta) = -\cos \theta$$

$$\cos(270° - \theta) = -\sin \theta$$

$$\cos(270° + \theta) = +\sin \theta$$

$$\cos(360° - \theta) = +\cos \theta$$

$$\cos(360° + \theta) = +\cos \theta$$

$$\cos(-\theta) = +\cos \theta$$

$$\tan(90° - \theta) = +\cot \theta$$

$$\tan(90° + \theta) = -\cot \theta$$

$$\tan(180° - \theta) = -\tan \theta$$

$$\tan(180° + \theta) = +\tan \theta$$

$$\tan(270° - \theta) = +\cot \theta$$

$$\tan(270° + \theta) = -\cot \theta$$

$$\tan(360° - \theta) = -\tan \theta$$

$$\tan(360° + \theta) = +\tan \theta$$

$$\tan(-\theta) = -\tan \theta$$

Table B.3 Table of Trigonometric Functions

deg	rad	sin	cos	tan	deg	rad	sin	cos	tan
0	.000	.000	1.000	.000					
1	.017	.017	1.000	.017	46	.803	.719	.695	1.036
2	.035	.035	.999	.035	47	.820	.731	.682	1.072
3	.052	.052	.999	.052	48	.838	.743	.669	1.111
4	.070	.070	.998	.070	49	.855	.755	.656	1.150
5	.087	.087	.996	.087	50	.873	.766	.643	1.192
6	.105	.105	.995	.105	51	.890	.777	.629	1.235
7	.122	.122	.993	.123	52	.908	.788	.616	1.280
8	.140	.139	.990	.141	53	.925	.799	.602	1.327
9	.157	.156	.988	.158	54	.942	.809	.588	1.376
10	.175	.174	.985	.176	55	.960	.819	.574	1.428
11	.192	.191	.982	.194	56	.977	.829	.559	1.483
12	.209	.208	.978	.213	57	.995	.839	.545	1.540
13	.227	.225	.974	.231	58	1.012	.848	.530	1.600
14	.244	.242	.970	.249	59	1.030	.857	.515	1.664
15	.262	.259	.966	.268	60	1.047	.866	.500	1.732
16	.279	.276	.961	.287	61	1.065	.875	.485	1.804
17	.297	.292	.956	.306	62	1.082	.883	.470	1.881
18	.314	.309	.951	.325	63	1.100	.891	.454	1.963
19	.332	.326	.946	.344	64	1.117	.899	.438	2.050
20	.349	.342	.940	.364	65	1.134	.906	.423	2.145
21	.367	.358	.934	.384	66	1.152	.914	.407	2.246
22	.384	.375	.927	.404	67	1.169	.921	.391	2.356
23	.401	.391	.921	.424	68	1.187	.927	.375	2.475
24	.419	.407	.914	.445	69	1.204	.934	.358	2.605
25	.436	.423	.906	.466	70	1.222	.940	.342	2.747
26	.454	.438	.899	.488	71	1.239	.946	.326	2.904
27	.471	.454	.891	.510	72	1.257	.951	.309	3.078
28	.489	.470	.883	.532	73	1.274	.956	.292	3.271
29	.506	.485	.875	.554	74	1.292	.961	.276	3.487
30	.524	.500	.866	.577	75	1.309	.966	.259	3.732
31	.541	.515	.857	.601	76	1.326	.970	.242	4.011
32	.559	.530	.848	.625	77	1.344	.974	.225	4.331
33	.576	.545	.839	.649	78	1.361	.978	.208	4.705
34	.593	.559	.829	.675	79	1.379	.982	.191	5.145
35	.611	.574	.819	.700	80	1.396	.985	.174	5.671
36	.628	.588	.809	.727	81	1.414	.988	.156	6.314
37	.646	.602	.799	.754	82	1.431	.990	.139	7.115
38	.663	.616	.788	.781	83	1.449	.993	.122	8.144
39	.681	.629	.777	.810	84	1.466	.995	.105	9.514
40	.698	.643	.766	.839	85	1.484	.996	.087	11.430
41	.716	.656	.755	.869	86	1.501	.998	.070	14.301
42	.733	.669	.743	.900	87	1.518	.999	.052	19.081
43	.751	.682	.731	.933	88	1.536	.999	.035	28.636
44	.768	.695	.719	.966	89	1.553	1.000	.017	57.290
45	.785	.707	.707	1.000	90	1.571	1.000	.000	—

Selected References

AGMA Standard 112.05, ANSI B16.14, Gear Nomenclature (Geometry), American Gear Manufacturers Association, Alexandria, VA, 1978.

Annand, W. J. D., *Mechanics of Machines*, Chemical Publishing, New York, 1966.

Barton, L. O., "Finding Slider-Crank Acceleration Graphically," *Machine Design*, Vol. 50, No. 28, Dec. 7, 1978.

Barton, L. O., "Simplified Slider-Crank Equations," *Machine Design*, Vol. 51, No. 8, April 1979.

Barton, L. O., "The Acceleration Polygon—A Generalized Procedure," *Engineering Design Graphics Journal*, Vol. 43, No. 2, Spring 1979.

Barton, L. O., "Painless Analysis of Four-Bar Linkages," *Machine Design*, Vol. 51, No. 17, July 26, 1979.

Barton, L. O., "Simplifying Velocity Analysis for Mechanisms," *Machine Design*, Vol. 53, No. 13, June 11, 1981.

Barton, L. O., "Simplified Analysis of Quick Return Mechanisms," *Machine Design*, Vol. 52, No. 18, Aug. 7, 1980.

Barton, L. O., "A Diagrammatic Representation of the Basic Motion Equations," *Engineering Design Graphics Journal*, Vol. 44, No. 3, Fall 1980.

Barton, L. O., "A New Way to Analyze Slider-Cranks," *Machine Design*, Vol. 17, July 22, 1982.

Barton, L. O., "Simplifying the Analysis of Sliding Coupler Mechanisms," *Machine Design*, Vol. 55, No. 19, Aug. 25, 1983.

Barton, L. O., "Finding Four Bar Linkage Acceleration Graphically," *Machine Design*, Vol. 57, No. 22, September 26, 1985.

Barton, L. O., "Applying the Mean Proportional Principle to Graphical Solutions," *Engineering Design Graphics Journal*, Vol. 55, No. 2, Spring, 1991.

Barton, L. O., "A New Graphical Method for Solving the Euler-Savary Equation," *Journal of Engineering Technology*, Vol. 9, No. 1, Spring, 1992.

Beggs, J. S., *Mechanism*, McGraw-Hill, New York, 1955.

Bickford, J. H., *Mechanisms for Product Design*, Industrial Press, New York, 1972.

Billings, J. H., *Applied Kinematics*, 2nd ed., Van Nostrand Reinhold, New York, 1943.

Buchsbaum, F., et al., *Handbook of Gears*, Stock Drive Products, New York, 1983.

Buckingham Associates, *Analytical Mechanics of Gears*, Dover Publications, New York, 1988.

Chen, F. Y., *Mechanics and Design of Cam Mechanisms*, Pergamon Press, New York, 1982.

Chironis, N. P., *Mechanisms, Linkages and Mechanical Controls*, McGraw-Hill, New York, 1965.

Chironis, N. P., *Machine Devices and Instrumentation*, McGraw-Hill, New York, 1966.

Chironis, N. P., *Gear Design and Application*, McGraw-Hill, New York, 1967.

Colbourne, J. R., *The Geometry of Involute Gears*, Springer-Verlag, New York, 1987.

Deutschman, A., et al., *Machine Design, Theory & Practice*, Macmillan, 1975.

Dudley, D. W., *Practical Gear Design*, McGraw-Hill, New York, 1984.

Erdman, A. G., and G. N. Sandor, *Mechanism Design: Analysis and Synthesis*, Vol. 1, Prentice Hall, New Jersey, 1984.

Esposito, A., *Kinematics for Technology*, Charles E. Merrill, Columbus, Ohio, 1973.

Hain, K., *Applied Kinematics*, ed. by D. P. Adams and T. P. Goodman, McGraw-Hill, New York, 1967.

Hall Jr., A. S., *Kinematics and Linkage Design*, Prentice Hall, Englewood Cliffs, N.J., 1961.

Ham, C. W., E. J. Crane, and W. L. Rogers, *Mechanics of Machinery*, 4th ed., McGraw-Hill, New York, 1958.

Hartenberg, R. S., and J. Denavit, *Kinematic Synthesis of Linkages*, McGraw-Hill, New York, 1964.

Hinkle, R. T., *Kinematics of Machines*, 2nd ed., Prentice Hall, Englewood Cliffs, N.J., 1960.

Hirschhorn, J., *Kinematics and Dynamics of Plane Mechanisms*, McGraw-Hill, New York, 1962.

Holowenko, A. R., *Dynamics of Machinery*, Wiley, New York, 1955.

Howes, M. A. H., *Source Book on Gear Design Technology and Performance*, American Society for Metals, Metals Park, OH, 1980.

Hunt, K. H., *Mechanisms and Motion*, Wiley, New York, 1959.

Jensen, P. W., *Cam Design and Manufacture*, 4th Ed., Marcel Dekker, New York, 1987.

Jensen, P. W., *Classical and Modern Mechanisms for Engineers and Inventors*, Marcel Dekker, New York, 1991.

Keown, R. McA. and V. M. Faires, *Mechanism*, McGraw-Hill, New York, 1939.

Kepler, H. B., *Basic Graphical Kinematics*, 2nd ed., McGraw-Hill, New York, 1973.

Kolstee, H. M., *Motion and Power*, Prentice Hall, Englewood Cliffs, N.J., 1982.

Lent, D., *Analysis and Design of Mechanisms*, 2nd ed., Prentice Hall, Englewood Cliffs, N.J., 1970.

Mabie, H. H. and C. F. Reinholtz, *Mechanisms and Dynamics of Machinery*, 4th Ed., Wiley, New York, 1987.

Martin, G. H., *Kinematics and Dynamics of Machines*, 2nd ed., McGraw-Hill, New York, 1982.

Maxwell, R. L., *Kinematics and Dynamics of Machinery*, Prentice Hall, Englewood Cliffs, N.J., 1960.

Merritt, H. E., *Gear Engineering*, Wiley, New York, 1971.

Nickolaisen, R. H., *Machine Drafting and Design*, Reston, VA, 1986.

Nielsen, K. L., *Modern Trigonometry*, Barnes and Noble, New York, 1966.

Patton, W. J., *Kinematics*, Reston, Reston, Va., 1979.

Paul, B., *Kinematics and Dynamics of Planar Machinery*, Prentice Hall, Englewood Cliffs, N.J., 1979.

Pearce, C. E., *Principles of Mechanism*, Wiley, New York, 1934.

Prageman, I. H., *Mechanism*, International Textbook Press, Scranton, Pa., 1943.

Ramous, A. J., *Applied Kinematics*, Prentice Hall, Englewood Cliffs, N.J., 1972.

Rao, J. S., and R. V. Dukkipati, *Mechanism and Machine Theory*, Wiley, New York, 1989.

Reuleaux, F., *Kinematics of Machinery*, trans. and ed. by A. B. W. Kennedy, reprinted by Dover, New York, 1963.

Rosenauer, N. and A. H. Willis, *Kinematics of Mechanisms*, Associated General Publications, Sydney, Australia, 1953.

Rothbart, H. A., *Cams*, Wiley, New York, 1956.

Sahag, L. M., *Kinematics of Machines*, Ronald Press, New York, 1952.

Schwamb, P., A. L. Merrill, et al., *Elements of Mechanism*, 6th ed., Wiley, New York, 1947.

Shigley, J. E., *Kinematic Analysis of Mechanisms*, 2nd ed., McGraw-Hill, New York, 1969.

Shigley, J. E. and J. J. Uicker, Jr., *Theory of Machines and Mechanisms*, McGraw-Hill, New York, 1980.

Soni, A. H., *Mechanism Synthesis and Analysis*, McGraw-Hill, New York, 1974.

Tao, D. C., *Fundamentals of Applied Kinematics*, Addison-Wesley, Reading, Mass., 1967.

Tesar, D., and G. K. Matthew, *The Dynamic Synthesis, Analysis, and Design of Modeled Cam Systems*, Lexington Books, Lexington, MA, 1976.

Tuttle, S. B., *Mechanisms for Engineering Design*, Wiley, New York, 1967.

Walker, J. D., *Applied Mechanics*, English Universities Press, London, 1959.

Wilson, C. E., J. P. Sadler, W. J. Michels, *Kinematics and Dynamics of Machinery*, Harper & Row, New York, 1967.

Index